IV National Scientific Conference with Foreign Participants: Geodynamical Processes and Natural Hazards (4th GeoProNH 2021)

IOP Conference Series: Earth and Environmental Science
Volume 946

Yuzhno-Sakhalinsk, Russia
6 – 10 September 2021

ISBN: 978-1-7138-4928-5
ISSN: 1755-1307

Printed from e-media with permission by:

Curran Associates, Inc.
57 Morehouse Lane
Red Hook, NY 12571

CURRAN ASSOCIATES INC.
proceedings.com

Some format issues inherent in the e-media version may also appear in this print version.

Institute of Physics
Dirac House, Temple Back
Bristol BS1 6BE UK

Phone: 44 1 17 929 7481
Fax: 44 1 17 920 0979

techtracking@iop.org

Additional copies of this publication are available from:

Curran Associates, Inc.
57 Morehouse Lane
Red Hook, NY 12571 USA
Phone: 845-758-0400
Fax: 845-758-2633
Email: curran@proceedings.com
Web: www.proceedings.com

TABLE OF CONTENTS

IOP Conf. Series: Earth and Environmental Science **946** (2021) 011001 doi:10.1088/1755-1315/946/1/011001

Preface

IV National Scientific Conference with Foreign Participants "Geodynamical Processes and Natural Hazards"

4th GeoProNH 2021

This issue presents the proceedings of the National scientific conference with foreign participants "Geodynamical Processes and Natural Hazards" (4th GeoProNH). The 4th GeoProNH was held from September 6 to 10, 2021, at the Institute of Marine Geology and Geophysics. This Institute is located in Yuzhno-Sakhalinsk (Russia) and is rightfully considered a regional center for the study of earth sciences.

This conference provides a forum for international researchers, academics and practitioners to discuss, share and exchange their latest research and experience progress associated with the geosciences. The Sakhalin and Kuril Islands are parts of the Pacific Ring of Fire, so this region undergoes almost all possible geohazards: earthquakes, tsunamis, volcano eruptions, etc. We have called for the research papers on natural hazards and geodynamics, hydrophysics (dangerous marine phenomena), geophysics, geochemistry, volcanology, geoecology, geomorphology and other earth sciences. At the same time, we did not limit ourselves exclusively to research papers, many recent papers and reviews can be found in the conference collection. More than a hundred scientists, experts and postgraduates from France, Japan, Kyrgyzstan, and Russia took part in the 4th GeoProNH Conference and discussed the current state of the art and advances in geosciences related to the analysis of natural hazards and/or their preventions.

Due to the COVID-19 pandemic situation during the 2021 year, it was not easy to hold a regular conference. There were travel restrictions and regulations from the World Health Organization and the local Government to be complied with to stop the spread of the COVID-19 in terms of physical distancing. In this uncertain condition we didn't postponed Conference since the most part of participants have confirmed their intention to arrive to Sakhalin and address their research findings at this conference. Thus, the committee of 4th GeoProNH has decided to organize the real meeting with high safety degree. Also, we provided the virtual meeting for our participants from France and Japan, who couldn't be with us, using ZOOM.

The Conference Program included five days of plenary and topical sessions, poster presentations, and two days of the field geological excursions. The ultimate goal of the conference was to highlight and share the best ways to reduce negative impacts of the above hazards using modern monitoring systems. The published proceedings, hopefully, will serve the same purpose. To create these proceedings, we have selected more than 50 scientific reports and with the help of our experts the best of them. The issue is addressed to researchers, experts, and students who deal with geohazard assessments and relevant topics of geosciences.

We would like to acknowledge all of those who supported the 4th GeoProNH and express our sincere gratitude to all scientific committees, who made in-depth reviews of full papers, to sponsors, who give their support, and to each individual and institutional help that is very important for the success of this conference.

We are hopeful that the proceedings will serve as important references, which will lead not only to scientific research and academic purposes but also in practical, industry and policy aspects associated with the geosciences.

Best regards, Dr. Bogomolov Leonid, Chairman of the 4th GeoProNH
Institute of Marine Geology and Geophysics, Russia, Yuzhno-Sakhalinsk

The list of notable speakers in 2021 includes such outstanding scientists like:

- **Alexandre Chemenda**, Dr. Sci. in Phys. and Math, Professor, The University of Nice Sophia Antipolis, France.
- **Andrey A. Kurkin**, Dr. Sci. in Phys. and Math, Professor, vice-rector of the NNSTU, Russia.
- **Hiroaki Takahashi**, Ph. D, Professor, Institute of Seismology and Volcanology, Japan.
- **Jean-Robert Grasso**, Dr. Sci. in Phys. and Math, Professor, Physicien du Globe, ISTerre, University Grenoble Alpes, France.
- **Leonid M. Bogomolov**, Dr. Sci. in Phys. and Math., director of the IMGG FEB RAS, Russia.
- **Pavel V. Krestov**, Dr. Sci. in Biology, corresponding member of RAS, director of Botanical Garden-Institute FEB RAS, Russia.
- **Sergey A. Bulanov**, Dr. Sci. in Geography, leading researcher of geomorphology laboratory, Institute of Geography RAS, Russia.
- **Yuri L. Rebetsky**, Dr. Sci. in Phys. and Math, chief researcher, Institute of physics of the Earth RAS, Russia.

The list of members of scientific and organizing committees of the conference include next scientists:

- **Alexandre Chemenda**, Dr. Sci. in Phys. and Math, professor, The University of Nice Sophia Antipolis, France.
- **Andrey A. Kurkin**, Dr. Sci. in Phys. and Math, Professor, vice-rector of the NNSTU, Russia.
- **Boris W. Levin**, Dr. Sci. in Phys. and Math, Professor, corresponding member of the RAS, scientific supervisor of the IMGG FEB RAS, Russia.
- **Dmitry P. Kovalev**, Dr. Sci. in Phys. and Math, leading researcher, IMGG FEB RAS.
- **Evgeny I. Gordeev**, Dr. Sci. in Phys. and Math, Professor, academician of the RAS, scientific supervisor of the Institute of Volcanology and Seismology FEB RAS, Russia.
- **Grigory I. Dolgikh**, Dr. Sci. in Phys. and Math, Professor, academician of the RAS, vice-chairman of the FEB RAS, Russia.
- **Hiroaki Takahashi**, Ph. D, professor, Institute of Seismology and Volcanology, Japan.
- **Jean-Robert Grasso**, Dr. Sci. in Phys. and Math, Professor, Physicien du Globe, ISTerre, University Grenoble Alpes, France.
- **Leonid M. Bogomolov**, Dr. Sci. in Phys. and Math., director of the IMGG FEB RAS, Russia.
- **Matthew A. Franchek**, Professor of mech. engineering, director of the Subsea Eng. Program, Houston University, USA.
- **Mikhail A. Nosov**, Dr. Sci. Phys. and Math., Professor, head of the department of physics of sea and land water, Moscow State University, Russia.
- **Pavel Izbekov**, Professor of Alaska Volcano Observatory, Geophysical Institute, University of Alaska Fairbanks, USA.
- **Pavel V. Krestov**, Dr. Sci. in Biology, corresponding member of RAS, director of Botanical Garden-Institute FEB RAS, Russia.
- **Sergey A. Bulanov**, Dr. Sci. in Geography, leading researcher of geomorphology laboratory, Institute of Geography RAS, Russia.
- **Sergey Z. Smirnov**, Dr. Sci. in Geology and Mineralogy, associate director of Institute of Geology and Mineralogy SB RAS, Russia.
- **Valentin I. Sergienko**, Academician of the RAS, chairman of the FEB RAS, Russia.
- **Vitaly V. Bratkov**, Dr. Sci. in Geography, professor, head of the department of geography, Moscow State University of Geodesy and Cartography, Russia.
- **Yury N. Levin**, director of SB FRC UGS RAS, Russia.

IOP Conf. Series: Earth and Environmental Science **946** (2021) 011001 doi:10.1088/1755-1315/946/1/011001

Conference Photos

Plenary session, Grand Hall of the Governnment of Sakhalin Region

Plenary session, Grand Hall of the Governnment of Sakhalin Region

Plenary session, Grand Hall of the Governnment of Sakhalin Region

Plenary session, Grand Hall of the Governnment of Sakhalin Region

Rodkin Mikhail Vladimirovich

Rebetsky Yuriy Leonidovich

Section "Modern geodynamics and problems of seismicity in the Far East and Eastern Siberia"

Valeriy Viktorovich Snakin

IOP Conf. Series: Earth and Environmental Science **946** (2021) 011001 doi:10.1088/1755-1315/946/1/011001

Muzychenko Leonid Evgen'evich

Section "Environmental problems and geo-ecological risks"

Polets Anastasiya Yur'evna and Dolgaya Anna Andreevna

Prytkov Alexander Sergeevich and Shestakov Nikolay Vladimirovich

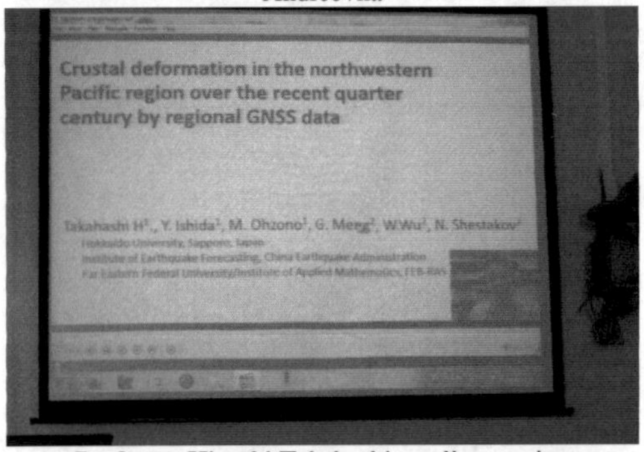

Professor Hiroaki Takahashi, on-line session

Kishkina Anastasiya Konstantinovna and Shestakov Nikolay Vladimirovich

Merkulova Tatiana Vladimirovna, Cherneva Nina Volodarovna, Andreeva Marina Yur'evna

Batugin Andrey Sergeevich and Fedotova Yuliya Viktorovna

All papers published in this volume (IOP Conference Series: Earth and Environmental Science) have been peer reviewed through processes administered by the Editors. Reviews were conducted by expert referees to the professional and scientific standards expected of a proceedings journal published by IOP Publishing.

• **Type of peer review:** Single-anonymous: authors' identities are known to the reviewers, reviewers' identities are hidden from authors.

Main criteria used by Reviewers when accepting/declining papers:

1. Relevance to the scope of the Conference
2. Suitability & length of the title
3. Scientific originality
4. Adequacy of the abstract
5. Scientific quality
6. Text quality
7. English level

• **Conference submission management system:**
Full paper submission was fully managed by editorial board of Conference

• **Number of submissions received:** 53 full papers

• **Number of submissions sent for review:** 51 full papers

• **Number of submissions accepted:** 48 full papers

• **Acceptance Rate (Number of Submissions Accepted/Number of Submissions Received X 100):** 90.56 %

• **Average number of reviews per paper:** 2 (117 reviews)

• **Total number of reviewers involved:** 51 reviewers (15 institutes) from 3 countries

• **Any additional info on review process:**
The review process and revision were conducted 3 rounds on average for each paper since the first submission after presentation at the conference. The first step was to check the format of paper and base quality of English level. The second-round of review was organized for papers with minor remarks from reviewers, for example the graphics or style. The third round required only for papers that need major modifications (scientific discussion on topic). All reviewers were asked to complete the review within enough time ranged no more than three weeks. We have asked authors to revise according to suggestions by reviewers within time ranged between 7-10 days. After being accepted, the final English proofread and format checking our specialists were carried out to ensure the quality prior to submission to IOP EES.

• **Contact person for queries:**
Name: Aleksandr Zakupin, Ph.D.
Affiliation: Institute for Marine Geology and Geophysics, Department of Seismology
Email: a.zakupin@imgg.ru

Study and modelling of rapid shoreline displacement due to coseismic subsidence along the Kamchatka subduction zone

A Khomchanovsky[1*], F Batanov[1] and T Pinegina[1,2]

[1]Institute of Volcanology and Seismology, Far East Branch of the Russian Academy of Sciences, Petropavlovsk-Kamchatsky, Russia
[2]Institute of Physics of the Earth, Russian Academy of Sciences, Moscow, Russia

E-mail: *khomscience@mail.ru

Abstract. The first studies of geological evidences of coastal coseismic subsidence (associated with subduction-zone earthquakes) were carried out in Russia at the Institute of Volcanology and Seismology, in Kamchatka. We developed a special method based on tephrostratigraphy and tephrochronology, descriptions and dating of the soil-pyroclastic sequence (SPS) overlying the coastal wave-build beach ridges. Three seismic events accompanied by coastal coseismic subsidence were detected in the northern part of Avachinsky Bay during the past ~5 thousand years. We found subsidence from one of the greatest historical tsunamigenic earthquake (1952 AD) south of Petropavlovsk-Kamchatsky. We identified, that 5 events of coastal coseismic subsidence had occurred during the past ~6 thousand years at the coast of Kronotsky Bay and Shipunsky Peninsula. Amplitudes of subsidence were estimated by geological data using three different methods. Erosion of the active beach and marine accumulative terrace becomes active after coastal subsidence. We calculated the shoreline retreat process and the amount of horizontal erosion by numerical simulation using Bruun rule. In some areas, shoreline retreat was about 300 m according to the model results.

1. Introduction

Field studies of geological evidences of coastal coseismic subsidence were carried out on the Pacific coast of the Kamchatka Peninsula, the southern part of which is located along the Kuril-Kamchatka subduction zone, where earthquakes with ~ Mw 8 and higher occur quite often (figure 1). During our studies we have identified that sometimes mega-earthquakes with ~ Mw 9 and coseismic coastal subsidence associated with unusually wide source occurred at Kamchatka subduction zone. The occurrence of such events is about once in a thousand of years [1–6].

The geological evidences of coseismic subsidence were identified and studied on wave-built terraces consisting of a beach ridges sequence, between which there are trough, sometimes filled with marshes or lagoons. Such coastal relief is associated with a sediment supply, as a result of which the shoreline moves out toward the sea (prograding). On the coast of Avachinsky and Kronotsky bays the height of beach ridges is decreased landwards, or in other words, young beach ridges are higher than ancient ones. As a rule, this indicates a relative lowering of the land level, in the case of Kamchatka it is associated with the coseismic subsidence during great earthquakes in the subduction zone (the coast of Avachinsky Bay, for example (figure 2)). The study area, shown in Figure 2, extends for about 75 km from Mayachny Cape to the Shipunsky Peninsula. The width of the Holocene accumulative terrace is from 500 to 2000 m. The age of the oldest beach ridges is about 5 thousand years [7]. Similar studies of vertical coseismic displacement were conducted on the coasts of Kronotsky Bay and Shipunsky Peninsula (in Morzhovaya Bay) (figure 2).

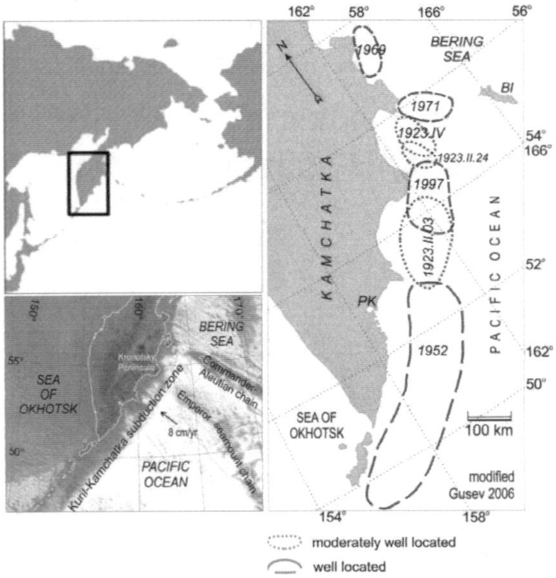

Figure 1. Location of the Kuril-Kamchatka subduction zone, with outlines of 20th century large and great Kamchatka earthquakes (modified after Gusev [8]). Heavy dashed lines indicate better located earthquake sources. PK – Petropavlovsk-Kamchatsky, BI – Bering Island. 1923 earthquakes are identified by month (Roman numerals) and day; e.g., 1923.II.24 is February 24, 1923 [5].

2. Methods

We studied the structure of beach ridges, which contain geological evidences of large seismic events, accompanied by vertical coseismic displacement (uplift and subsidence). The age of catastrophic earthquakes was estimated based on tephrochronology and radiocarbon dating of beach ridges deposits.

The development of the wave-built terrace under the influence of subduction earthquakes goes according to the following scenario [9, 5]: if there is enough sediment supply, the coastal shoreline moves seawards with the formation of new beach ridges (prograding). The shoreline begins to retreat after earthquake and subsidence; after a while (months or years), the equilibrium beach profile develops, and the shoreline stops to retreat, accumulative storm deposits overlap the active erosional scarp, this is how a buried scarp is forming (figure 3).

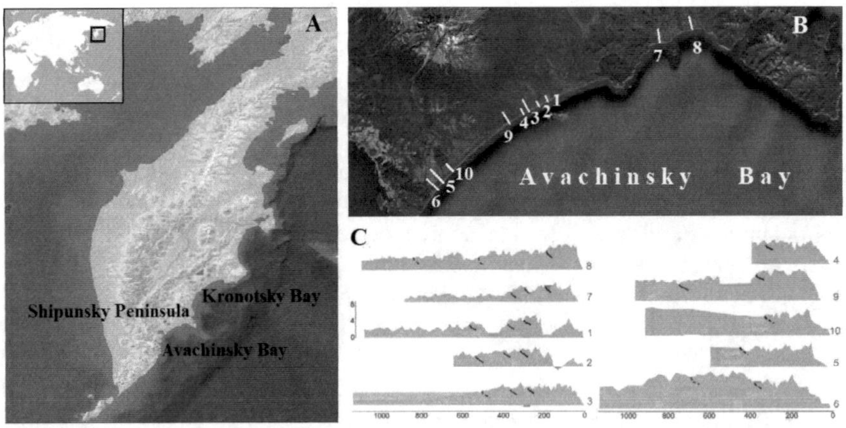

Figure 2. Study areas of geological evidences of coseismic subsidence (a); location of topographic profiles on the Avachinsky Bay coast, along which the excavations were made (b); topographic profiles across the marine accumulative terrace, with multicolored lines showing the identified buried scarps (c), the sea is on the right (modified from [5]).

Several methods have been developed to determine the amplitudes of coseismic displacement [9, 5].

After coseismic subsidence, the erosion of the coast was activated. Amount of coastal retreat were determined by numerical simulation according to Bruun rule [10].

$$R = \frac{L}{B + h_*} S$$

where R – shoreline horizontal displacement (retreat); S – sea level change; L – cross-shore distance to the depth of closure h_*; B – berm height.

According to this rule, the coastal profile will be displaced landward or seaward, following to direction of the relative sea level change. The existence of accumulative beach is the result of the balance between wave energy, sediment supply (cross-shore, alongshore, from the river) and sea level. If there is a rapid rise in sea level, the balance is disturbed, that leads to a restructuring of the equilibrium beach and underwater profile. When the sea level rise, the shoreline retreat landward. If the sea level descends, the shore extends seaward. Estimating the volume of beach retreats due to coseismic displacement is an important aspect for studying coastal effects after the strongest earthquakes in subduction zone.

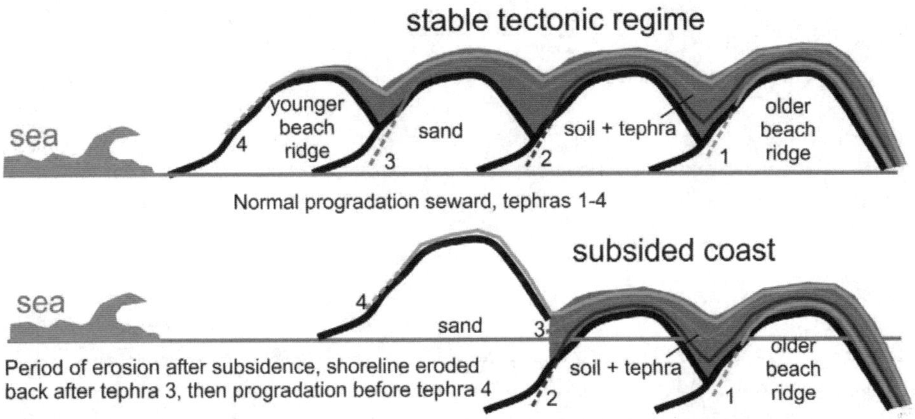

Figure 3. Method of dating paleo-beach ridges with tephra to estimate the age of buried erosional scarps/paleo earthquake (modified from [11]).

3. Results and discussing

All study sites on Avachinsky, Kronotsky and Shipunsky coasts are represented by marine accumulative terraces with beach ridges. Young beach ridges are higher than the old ones. Oldest beach ridges are not expressed in the relief and are identified either by vegetation or ground penetrating radar (GPR). As a rule, beach ridges are subdivided into series according to their height.

In places, where the height of the beach ridges changed, excavations were made to find and to describe buried scarps. They are geological evidences of coastal coseismic subsidence (associated with subduction-zone earthquakes). The scarps on the profiles are marked with multicolored lines (figure 2). Three seismic events accompanied by coastal coseismic subsidence were detected in the northern part of Avachinsky Bay during the past ~5 thousand years [5]. Five such events occurred at the coast of Kronotsky Bay and Shipunsky Peninsula during the past ~6 thousand years. [4]. The amplitudes of subsidence estimated from geological data did not exceed 2 meters and were usually within the range of 0.5–1 m. We modelled the shoreline retreat process and the amount of horizontal erosion for different amplitudes of subsidence. On the coast of Avachinsky Bay we obtained the following values: for 0.5 m surface subsidence, the value of coastal erosion was 72 m, for 1 m subsidence – 143 m, for 2 m – 287 m (table 1). South of Petropavlovsk-Kamchatsky, we found geological evidences of subsidence during one of the greatest historical tsunamigenic earthquake (1952 AD). The shoreline here still continues to erode. This is may be related either to insufficient amount of incoming sediment, or to the increase in the amplitude of subsidence to the south, or to both of these factors [12]. On the coast of Kronotsky Bay and Shipunsky Peninsula, the calculated coastal erosion was 24–55 m for 0.5 m subsidence, 70–111 m for 1 m subsidence, and 200–221 m for 2 m subsidence (table 1).

Table 1. Results of numerical simulation.

Amplitude of coseismic coastal subsidence	Shore retreat by Bruun rule		
	Avachinsky Bay	Shipunsky Peninsula	Kronotsky Bay
0.5	72	24	55
1	143	70	111
2	287	200	221

The data in the table show that the values of coastal retreat will be different for the same subsidence amplitudes in different sections. This may be due to both local features of the coast and different slope of the beach, as well as to the lack of versatility of the calculation model. The model does not take into account the features of morpho- and lithodynamics of the coast, in particular, the longshore sediment transport. In addition, in [13], standard coastal retreat trends were plotted for each section of Avachinsky Bay using the Bruun rule, which were subsequently compared with the results of coastal erosion calculations based on paleoseismological data from [5]. The comparison results are significantly different. The average coastal retreat, according to paleoseismological data, was 142 m, and according to the results of numerical simulation, 49 m [13]. Although in some areas the results of modelling and paleoseismological analysis coincide almost to 1 meter. We modified the Bruun model according to the coastal local features, based on geological and paleoseismological studies. The model is currently being finalized.

4. Conclusion

Shoreline displacements resulting from coseismic subsidence on Kamchatka (associated with subduction-zone earthquakes) have been simulated. The coastal retreats ranged from 24 to 287 m for typical in this region amplitudes of subsidence (0.5-2 m). These values can be used to predict the impact of the strongest earthquakes in the coasts of Kamchatka Peninsula. Modification the Bruun's rule of shoreline displacements as a result of relative sea level change for Kuril-Kamchatka region based on its verification with paleoseismological data is currently being developed.

On the coasts of the Kuril Islands, studies of the geological evidences of coseismic subsidence and numerical simulation of shoreline displacements have not yet been conducted. This task was set for the first time as a part of a new supported grant from the Russian Science Foundation (supervised by E.I. Gordeev). The first data were obtained during the 2021 field studies on Iturup Island. Detailed studies on Iturup Island are expected to be carry out in 2022. The study and modelling of shoreline changes during seismic events along the Kuril-Kamchatka subduction zone is very important for estimating the geomorphological impact of strongest earthquakes, especially for the areas with developing coastal infrastructure.

Acknowledgment

This study was supported by grant #21-17-00049 of the Russian Science Foundation, https://rscf.ru/project/21-17-00049, supervised by E.I. Gordeev and Russian Ministry of Education and Science (grant # 14.W03.31.0033 to N. Shapiro).

References

[1] Pinegina T K, Bazanova L I, Zelenin E A and Kozhurin A I 2015 *Proc. 5th Sci. and Tech. Conf. Problems of complex geophysical monitoring of the Russian Far East* Petropavlovsk-Kamchatsky, September 27–October 3, 2015 (Obninsk: GS RAS) ed V N Chebrov pp 373–376

[2] Pinegina T K, Bazanova L I, Zelenin E A and Kozhurin A I 2017 *Proc.6th Sci. and Tech. Conf. Problems of complex geophysical monitoring of the Russian Far East* Petropavlovsk-Kamchatsky, October 1–7, 2017 (Obninsk: FIC EGS RAS) ed D V Chebrov pp 235–239
[3] Pinegina T K, Bazanova L I, Zelenin E A, Khomchanovsky A L, Vydrin D S and Batanov F I 2019 *Proc.XXII All-Russian Sci. Conf. dedicated to the Day of the Volcanologist*, March 28–29, 2019 *Volcanism and related processes* (Petropavlovsk-Kamchatsky: IViS FEB RAS) pp 167–170
[4] Pinegina T K 2021 *Proc. XXIV Annual Sci. Conf. Dedicated to the Day of Volcanologists*, 29–30 March 2021 *Volcanism and related processes* (Petropavlovsk-Kamchatsky: Institute of Volcanology and Seismology Far East Branch of RAS) pp 102–104
[5] Pinegina T K, Bourgeois J, BazanovaL I, ZeleninE A, Krasheninnikov S P and Portnyagin M V 2020 *Quaternary Science Reviews* **233** 106171
[6] Lander A V and Pinegina T K 2017 *Proc. Sixth Sci. and Tech. Conf.* Petropavlovsk-Kamchatsky, October 1–7, 2017 *Problems of complex geophysical monitoring of the Russian Far East.* (Obninsk: FIC EGS RAS) ed D V Chebrov pp 221–225
[7] Pinegina T K et al 2018 *Pure and Applied Geophysics* **175** pp 1485–1506
[8] Gusev A A 2006 *Volcanology and seismology* **3** pp 39–42
[9] Pinegina T K 2014 *Spatial and temporal distribution of tsunamigenic earthquake sources in the Pacific and Bering Sea coasts of Kamchatka by paleotsunami sediments*: doctoral thesis abstract (Moscow) p 43
[10] Bruun P. 1988 *Journal of coastal Research* **5** pp 627–648
[11] Tatiana K Pinegina and Joanne Bourgeois 2020 Geological Records of Tsunamis and Other Extreme Waves *Elsevier* **34** Ed Max Engel et al. pp 745–759
[12] Batanov F I and Homchanovsky A L 2021 *Proc. XXIV Annual Sci. Conf. Dedicated to the Day of Volcanologists*, 29–30 March (Petropavlovsk-Kamchatsky: Institute of Volcanology of the Far East Branch of the RAS) pp 72–74
[13] Khomchanovsky A L 2020 *Proc. XXIII Annual Sci. Conf. dedicated to the Day of Volcanologist*, 29–30 March (Petropavlovsk-Kamchatsky: Institute of Volcanology of the Far East Branch of the RAS) pp 136–139

IOP Conf. Series: Earth and Environmental Science **946** (2021) 012002 doi:10.1088/1755-1315/946/1/012002

The first results of the seismic monitoring system of the Solntsevsky open pit coal mine area (on Sakhalin Island)

D V Kostylev[1,2*] and N V Boginskaya[1]

[1] Institute of Marine Geology and Geophysics Far Eastern Branch of Russian Academy of Science, Yuzhno-Sakhalinsk, 693022, Russia
[2]Sakhalin Branch, Geophysical Survey, Russian Academy of Sciences, Yuzhno-Sakhalinsk, 693010, Russia

E-mail: [*]d.kostylev@imgg.ru

Abstract. In 2021, work began as a part of the implementation of the decision of the protocol of the Sakhalin branch of the Russian Expert Council on earthquake prediction, seismic hazard and risk assessment dated October 6, 2020 on detailed monitoring in the area of active coal mining at the Solntsevsky open pit coal mine (Sakhalin Island). New points of seismic monitoring were installed directly in the area of the open pit coal mine. Integration of real-time data received from the points in real time into a unified seismic monitoring system in the Sakhalin Region was ensured. The results of registration of seismic events of various origins since the commissioning of the stations are presented. A significant increase in the accuracy of the determined epicenters and the possibilities of determining earthquakes and industrial explosions has been noted. The results of the monitoring system for studying the landslide process in the area of the open pit coal mine, as well as the probable factors that caused the landslide, are shown. The developed monitoring system allows for representative registration of seismic events with $ML \geq 0.8$ in the immediate vicinity of open pit coal mine, which makes it possible to control blasting operations with increased accuracy, as well as weak and possible induced seismicity formed as a result of a constant technogenic impact on the subsoil.

1. Introduction

The existence of induced seismicity around mining enterprises has been proven in many studies [1, 2]. The Uglegorsk earthquake with Mw = 4.8 ($I_0 = 5$) that occurred on Sakhalin Island on September 13, 2020, and the aftershocks that followed it were located in the area of active coal mining at the Solntsevsky open pit coal mine, which does not exclude the connection of the seismic process with technogenic seismicity [3]. The Solntsevsky open pit coal mine, located in the Uglegorsk district of the Sakhalin Region, is a key production asset of East Mining Company [4]. In addition, this area is saturated with numerous infrastructure facilities, and continues to develop – currently the construction of the largest coal conveyor in Russia with a length of 23 km is underway [5], which has an additional effect on the seismic regime and requires seismicity control. In connection with the above-mentioned earthquake, as well as the established change in the seismicity regime of the Solntsevsky open pit coal mine region [6], in accordance with the decision of the Sakhalin branch of the Russian Expert Council on earthquake prediction, a project for monitoring the Solntsevsky open pit coal mine was developed. This project took into account the author's experience in the development and creation of automatic points for instrumental seismological observations and their integration into a unified data collection system [7], as well as existing monitoring systems for similar coal deposits in the Kemerovo region [8].

2. Methods and results

The development of a complex monitoring system for the Solntsevsky coal deposit area envisages 2 stages. At the first stage, the existing seismic station "Uglegorsk" is retrofitted with equipment for complex geophysical observations, such as instruments for seismoacoustic monitoring and monitoring of the level of subsoil radon. In addition, at the first stage, additional autonomous stations are installed in the settlements of the region with the real-time transmission of information. In February 2021, the authors carried out reconnaissance work in the settlements of the region (measuring the level of seismic noise, the availability of power supply and communication systems, solving issues of equipment safety) and, based on the data obtained, two points were selected for placing additional stations. The were installed and commissioned at the end of April 2021. The stations were equipped with from Nanometrics broadband seismometers. The second stage envisages the installation of two more automatic stations equipped with short-period seismometers and is planned for implementation in 2022. A complete list of the equipment used, as well as the location of monitoring points in relation to the Solntsevsky open pit coal mine and other infrastructure facilities of the area is shown in figure 1.

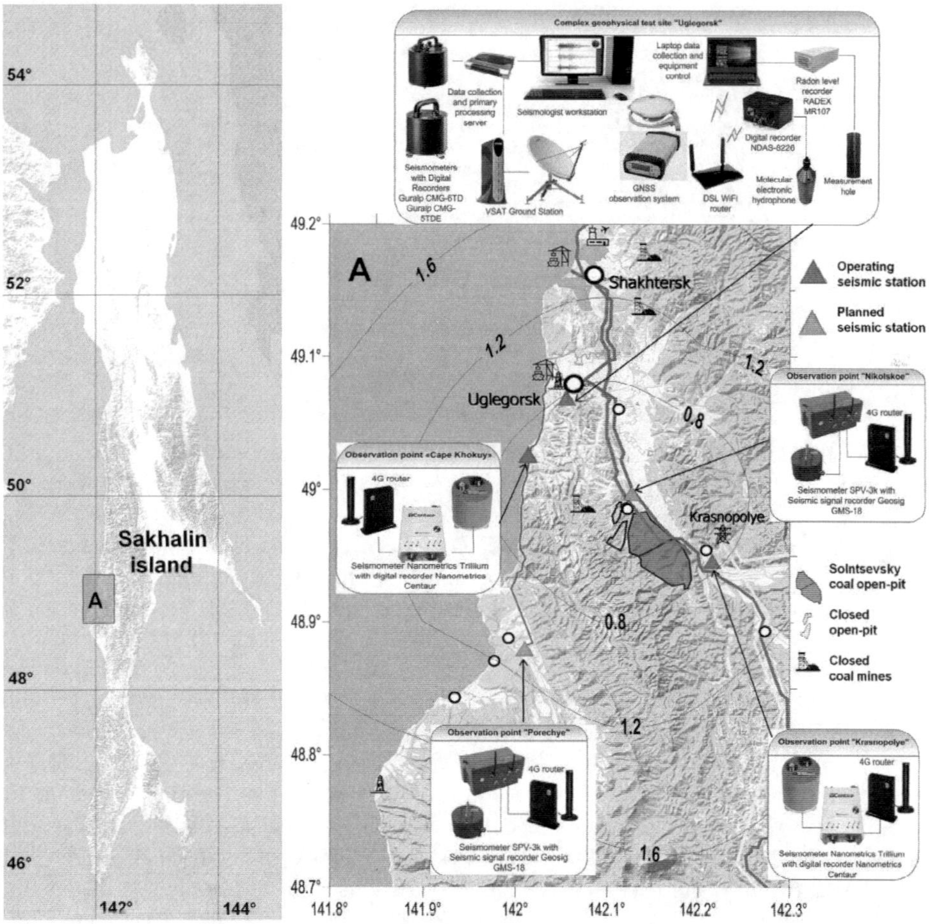

Figure 1. Monitoring area (polygon A). On the right are the locations of network stations and monitoring objects. Lines show infrastructure facilities: brown line – highways, yellow line – power transmission line (220 kV), lilac line – coal conveyor. The contour lines show the calculated registration capabilities of the network.

Based on the results of processing the records of waveforms obtained by means of new equipment in the period from April 22, 2021, to September 30, 2021, the authors created a catalog of seismic events in the area of the Solntsevsky open pit coal mine, including both earthquakes and industrial explosions produced at the Solntsevsky open pit coal mine. The total number of processed events is 118, of which 90 are industrial explosions and 28 are earthquakes. It is interesting to compare the results of processing events in the section area for the first four months of 2021 according to the seismological bulletins of the Uglegorsk station (in the period when the processing of events was carried out according to the records of one station) and the new results from April 22 to September 30 (figure 2).

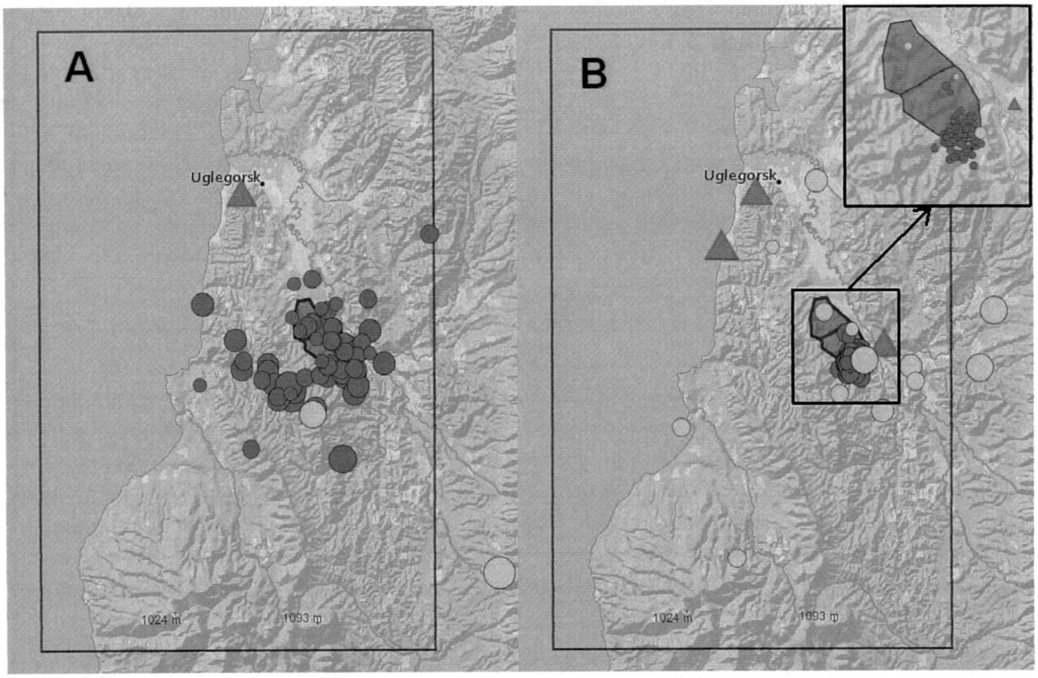

Figure 2. The position of the epicenters of the registered events before (A) and after (B) the installation of new stations. Yellow circles – earthquakes, purple circles – industrial explosions.

Obviously, the accuracy of determining the epicenters of explosions has significantly increased. The inset in figure 2B confirms that the main stripping work was carried out in the southern part of the open pit mine. The weak seismicity of the area in the first 4 months of 2021 is also noteworthy. This may be due to the really small number of earthquakes, but it is most likely that a number of earthquakes, when processed at one station, were identified as explosions. Thus, the commissioning of new stations helps to solve the problem of the determination of explosions and earthquakes.

The stations of the new network make it possible to register and assess not only earthquakes and explosions, but also other geophysical processes in the monitoring area, for example, such as mudflows and landslides typical for dumps of open quarries in coal mining areas on Sakhalin Island [9]. So, interesting results were obtained from the data of records of new stations during an environmental incident that occurred at the Solntsevsky open pit coal mine on the night of July 10 (local time) – a landslide from a temporary storage warehouse for overburden of an open pit coal mine, as a result of which, according to expert estimates, in about two to three million cubic meters of rock went down the river valley (figure 3). The authors analyzed seismic processes prior to the date of the landslide, since it is known that even small seismotectonic deformations can disturb the stability of slopes. It was found that the landslide was preceded by two explosions – on July 6 and 8 (highlighted in red in figure 3), and in the period between them (July 7), in the immediate vicinity of the landslide site, a seismic event with ML = 2.0 occurred. Using records from new stations, as well as from other stations of seismic networks in the Far East, the authors determined the focal mechanism of this earthquake, which corresponds to the type of seismic dislocation – strike-slip fault, which coincides with the mechanism of the 2020 Uglegorsk earthquake and is not typical for the natural seismicity of Sakhalin [6]. This fact once again indicates a possible technogenic cause of seismicity in 2020–2021 in the study area.

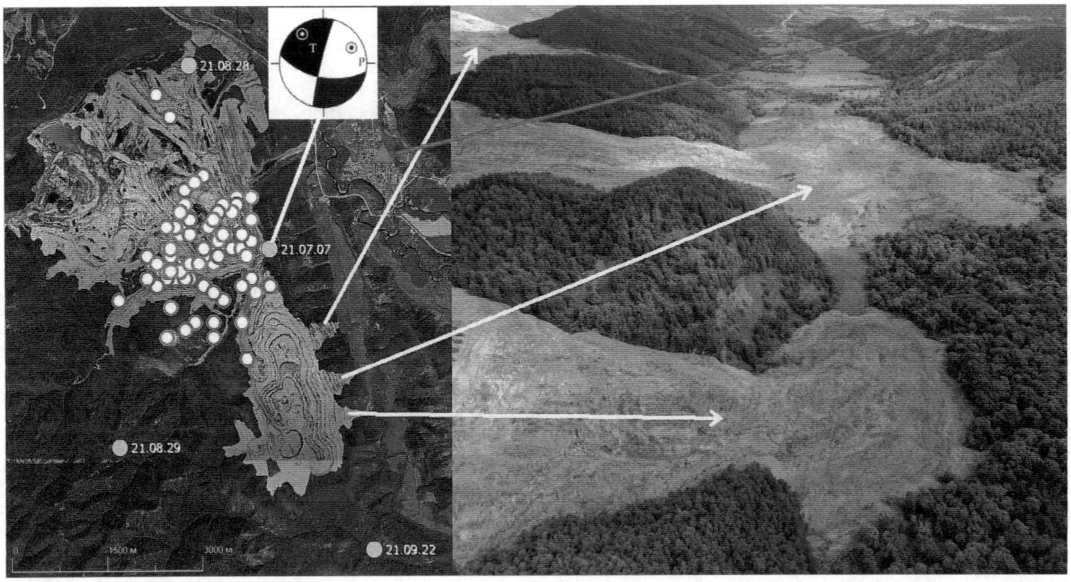

Figure 3. On the left is a satellite image of the Solntsevsky coal mine area (before the landslide). Purple outlines are explosions before the landslide, blue ones after. Yellow outlines – earthquakes in the period from April 22 to September 30. On the right are photographs after the landslide.

Using the techniques for identifying the landslides on seismograms [10], the time and duration of the landslide process were determined. Figure 4 shows waveforms from Krasnopolye station (UGL2) located 2 km from the landslide site and Cape Khokuy station (UGL3). As a result of the analysis of these waveforms, the authors identified a characteristic segment with a duration of about 10 minutes, with the start time at 12:09 UTC (or 23:09 local time), which can be considered a record of the landslide process (segments 1–3 and 2–4 in figure 4). The analysis of the records of the UGL2 and UGL3 stations was carried out in the low-frequency range (which is possible due to the characteristics of the equipment used) with a 0.01–0.1 Hz filter. Separately, the inset shows a 0–1 section with a 4–12 Hz filter. It can be seen, that for the Krasnopolye station (UGL2), a high-frequency segment with a duration of about 20 seconds can be distinguished, preceding the main (low-frequency) process. Additionally, the frequency composition of the landslide, as well as the earthquake and explosions that preceded it, was investigated (figure 4 below). On the left in Figure 4 (lower part) the Fourier spectra for the ground velocity at the Krasnopolye station of the segments 0–1 and 1–3 are shown according to filtered records. The figure shows that in the low-frequency (long-term) section, the fundamental harmonic is frequencies of 3–4 cHz (period of 25–30 seconds), and in the short (high-frequency) section – frequencies of 7–10 Hz. On the right, the figure shows the Fourier spectra for the unfiltered (initial) record of the 0–3 segment according to the Krasnopolye station data, as well as the spectra of the July 7 earthquake and the July 8 explosion. The spectral composition of the landslide record, as described in the works devoted to the identification of landslides on seismograms [10], has several local maxima. The spectra of the earthquake and explosion have values typical of such events and are higher in frequency.

IOP Conf. Series: Earth and Environmental Science **946** (2021) 012002 doi:10.1088/1755-1315/946/1/012002

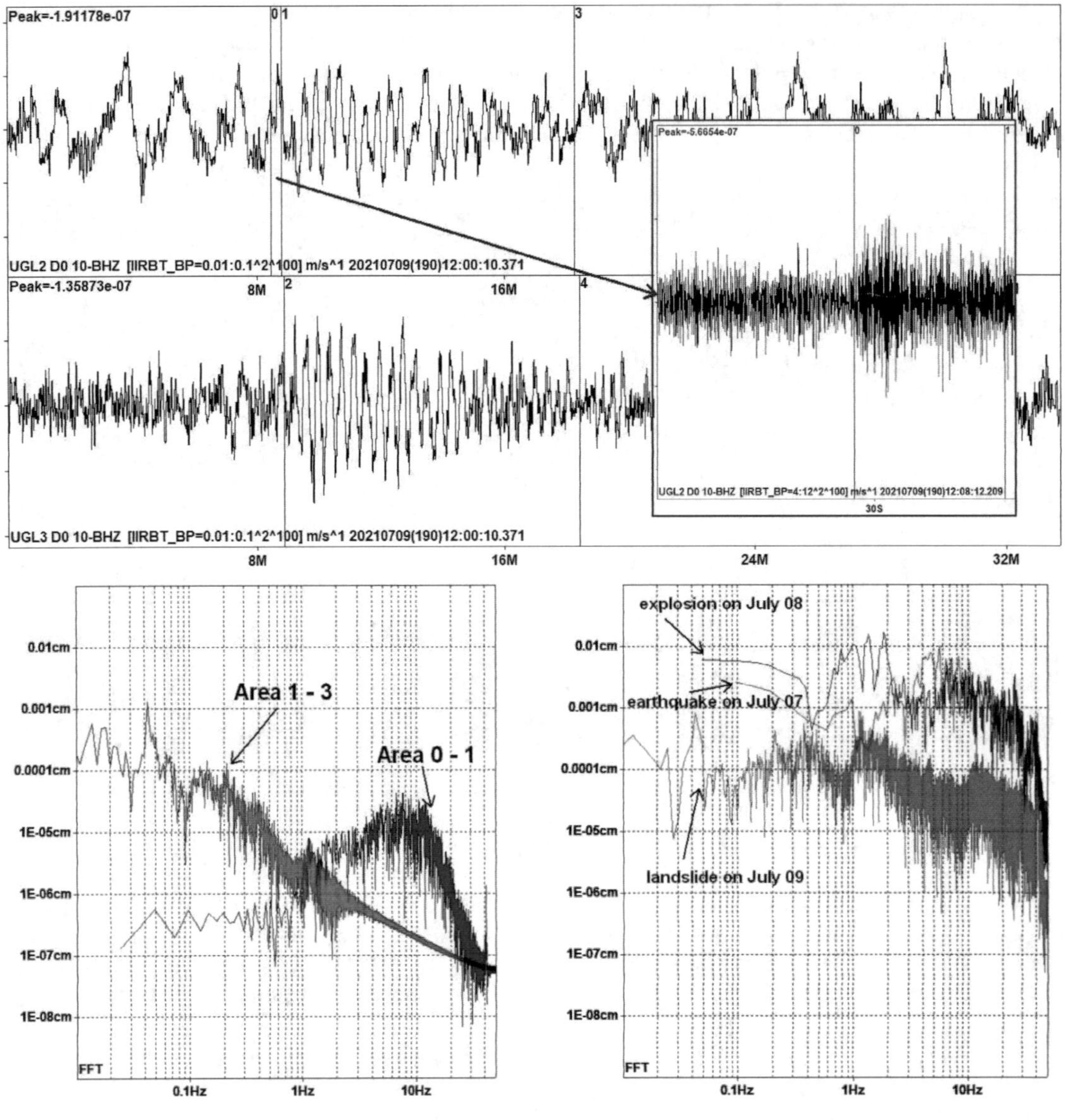

Figure 4. Waveforms according to data from Krasnopolye (UGL2) and Cape Hokuy (UGL3) stations of the landslide on July 9, 2021 (top). Fourier spectrum of soil velocity at the Krasnopolye station of filtered sections of landslide recording (bottom left) and joint analysis of the spectra of landslide, earthquake and explosion.

The results obtained confirm the correctness of the choice of equipment and the location of new stations, which makes it possible to study geophysical processes in a wide frequency range.

3. Conclusion

Currently, the developed monitoring system allows for representative registration of seismic events at a magnitude of ML = 0.5 in the immediate vicinity of open pit coal mine, which provides the very accurate control of weak as well as induced seismicity, formed under constant technogenic impact on the subsoil. The created monitoring network contributes to obtaining detailed data on the nature and features of the development of induced seismicity in the area of the Solntsevsky open pit coal mine, and also makes it possible to study the seismic effect of industrial explosions in the open pit coal mine. Integration of data coming from new points of seismic observation allows for prompt informing of potential stakeholders with the provision of accurate data on the location and strength of the seismic event and possible impacts on the infrastructure facilities of the monitoring area.

The work begun on the analysis of data obtained using a hydrophone and a device for registering the level of subsoil radon makes it possible to consider the created complex as an integrated monitoring system.

Acknowledgements

The work was supported by Ministry of Science and Higher Education of the Russian Federation. The data used in the work were obtained with large-scale research facilities «Seismic infrasound array for monitoring Arctic cryolitozone and continuous seismic monitoring of the Russian Federation, neighbouring territories and the world» [11].

References

[1] Adushkin V V and Turuntaev S B 2015 *Technogenic seismicity – induced and triggered* (Moscow: IDG RAS) p 364

[2] Emanov A F, Emanov A A, Fateev A V, Leskova E V, Shevkunova E V and Podkorytova V G 2014 *Journal of Mining Science* **50** 2 pp 224–228

[3] Semenova E P, Boginskaya N V and Kostylev D V 2020 *J. of Geosystems of transition zones* **4** 4 pp 474–485

[4] East Mining Company Ltd (*Electronic materials*) https://www.eastmining.ru/en/about-the-company/

[5] *East mining company Ltd, 2020 – start on the way to 20 million tonnes per year* 2021 *Russian Coal Journal* **3** 1140 pp 62–63

[6] Kostylev D, Boginskaya N and Zakupin A 2021 Seismic activity in the focus of the Uglegorsk earthquakes, Sakhalin Island in relevance to intensive development of coal deposits (*Preprint* rs-914003)

[7] Kostylev D V 2021 *Russian Journal of Seismology* **3** pp 41–53

[8] Emanov A A, Emanov A F and Fateev A V 2020 *J. Seismic Instruments* **56** pp 82–92

[9] Muzychenko L E and Kazakova E N 2020 *J. of Geosystems of Transition Zones* **4** 2 pp 359–368

[10] Sokolova I N and Shepelev O M 2005 *Landslide Record Identification on Seismograms* NNC RK Bulletin **2** 22 pp 165–168

[11] Geophysical Survey RAS (*Electronic materials*) http://www.gsras.ru/

IOP Conf. Series: Earth and Environmental Science **946** (2021) 012003 doi:10.1088/1755-1315/946/1/012003

Identification of disjunctive dislocations as one of the parameters for estimating a territory seismicity of North Sakhalin

O V Kuptsova[1*], V A Melkiy[2] and A A Verkhoturov[2]

[1]Sakhalin State University, Yuzhno-Sakhalinsk, 693001, Russia
[2]Institute of Marine Geology and Geophysics of Far Eastern Branch of the Russian Academy of Sciences, Yuzhno-Sakhalinsk, 693022, Russia

E-mail: *korsuncevaolesy@gmail.com

Abstract. The north of Sakhalin Island is characterized by frequent earthquakes and many disjunctive dislocations. One of the catastrophic earthquakes occurred in the settlement of Neftegorsk. The level of modern technologies makes it possible to track the change in environmental parameters accompanying dangerous natural processes with a high degree of certainty. The article proposes an interpretating technology for disjunctive dislocations detecting, which differs from the existing ones with the complex use and summation of satellite image data using image processing methods widely applied in "computer vision". The study aims to compile and describe a map of discontinuous faults of the village of Neftegorsk, located in the northern part of Sakhalin Island, using the developed decoding technology and geophysical data. Methods: Identification methods used in the work: 1) methods of primary image processing (correction, transformation, resolution change, cropping, visualization); 2) a set of "contextual" and "autonomous" methods of lineament analysis, with the help of which a series of images is processed (canny, erosion, Hough's algorithm); 3) methods for constructing maps of discontinuous faults (summation, sequential linking). Results. Using the developed technology, a map of discontinuous faults in the area of the Neftegorsk earthquake was constructed. The implementation of the technology makes it possible to provide the monitoring data on discontinuous faults to organizations that are engaged in seismic zoning, construction and operation of objects for various purposes, including mineral deposits.

1. Introduction

That is known that discontinuous faults reflecting deep geodynamic processes are genetically related to seismic processes. Elastic waves resulting from discharge of accumulated energy, are transmitted over long distances and cause the motion of blocks of the Earth's crust [1–4]. The faults, along which the blocks of the earth's crust are displaced separate these blocks and are places of stress discharge. When performing seismic zoning, it is necessary to analyse the distribution of earthquakes of study territory, in which the distance from a fault zone must be considered. These zones are characterized with the presence of linear elements recorded on satellite images [5, 6]. When decoding satellite images, groups of lineaments are revealed that indicate feathering cracks of deep active fault. The more cracks that support the fault, the deeper it is, since the number of cracks indicates geodynamic activity of the fault. In the areas of discontinuous faults, rocks are mainly characterized with good porosity, filtration, and permeability coefficient. Cracks are caused not only by geodynamic activity, but also by zones of cataclase, mylonitization, brecciation, crushing, slanting, and disintegration associated with them. These factors determine the state of vegetation growing in the fault zone, so these zones are clearly distinguishable in interpreted images [7–9].

2. Methods and materials

The materials for interpreting study territory were the data of satellite images of the Shuttle and Landsat-8 spacecraft [10–12]. The data were subjected to all necessary types of correction. The first level of processing corresponds to the materials obtained from the US Geological Survey website. Landsat8 data are single-channel GeoTIFF images. The materials of the Shuttle apparatus the images obtained using radar interferometric SRTM survey, which have passed procedure of filtering erroneous values.

The initial processing of the images was determined by the initial data needed for analysis in the LEFA program. Images that do not exceed size of 100 pixels are analysed faster. The format of materials uploaded to program must have *.tif extension and geographical reference. Images with different resolutions were introduced, cropping was performed to match the image of the studied area and its detailed analysis. The images were processed in the ArcGIS program [13] in order to obtain the materials with the resolution necessary for further analysis of the territory. The relief forms indicating the zones of discontinuous faults made it possible to identify SRTM materials. They were also processed further in the LEFA software package [14], in which automatic lineament analysis was performed. There are many methods of lineament analysis [15–17]. The results described in the article were obtained using Canny mathematical methods [18] and morphological erosion [19, 20]. The morphological erosion algorithm works on the principle of removing "excess" when comparing an image with a reference, so that their boundaries are determined in the original images [21]. The boundaries of the image determined by the Canny method are captured thanks to the matrix used in the method. The exact boundaries of the image are determined after processing by the double threshold method and the thinning method.

Each image of different resolutions was processed with different sensitivity lines: small, medium and many. The higher the sensitivity, the more lineaments are highlighted in the image. Collinear lines were identified using the Hough Transform [22, 23], which uses linear probabilistic parameters of known quantities of the equation $y = kx + b$. Determining which lines, according to probability theory, are extensions of each other, the program builds connected collinear lines.

Further integration of modified images and geophysical data was carried out in the QGIS geoinformation system [24]. The modified images were superimposed on each other, as a result of which, according to the greatest coincidence of the lines, considering the digital relief model and seismology data, faults were marked on the map.

The methods of various stages of processing materials to identify discontinuous faults allowed us to draw up a technological scheme for interpreting the faults. The technology includes 7 stages:

1) Search and input of materials suitable for solving the tasks of scientific research.
2) Pre-processing of materials.
   - definition of image boundaries using Canny and Erosion methods;
   - determination of stroke lines using automatic image analysis in the LEFA program;
   - determination of collinear lines indicating the probability of fault location using the Hough Transform;
   - map design, which includes the analysis of lineaments and geophysical data in the QGIS program;
   - map construction and printing.

As result of data transformation by means of the developed technology, a map of disjunctive disorders of the studied territory is created (figure 1).

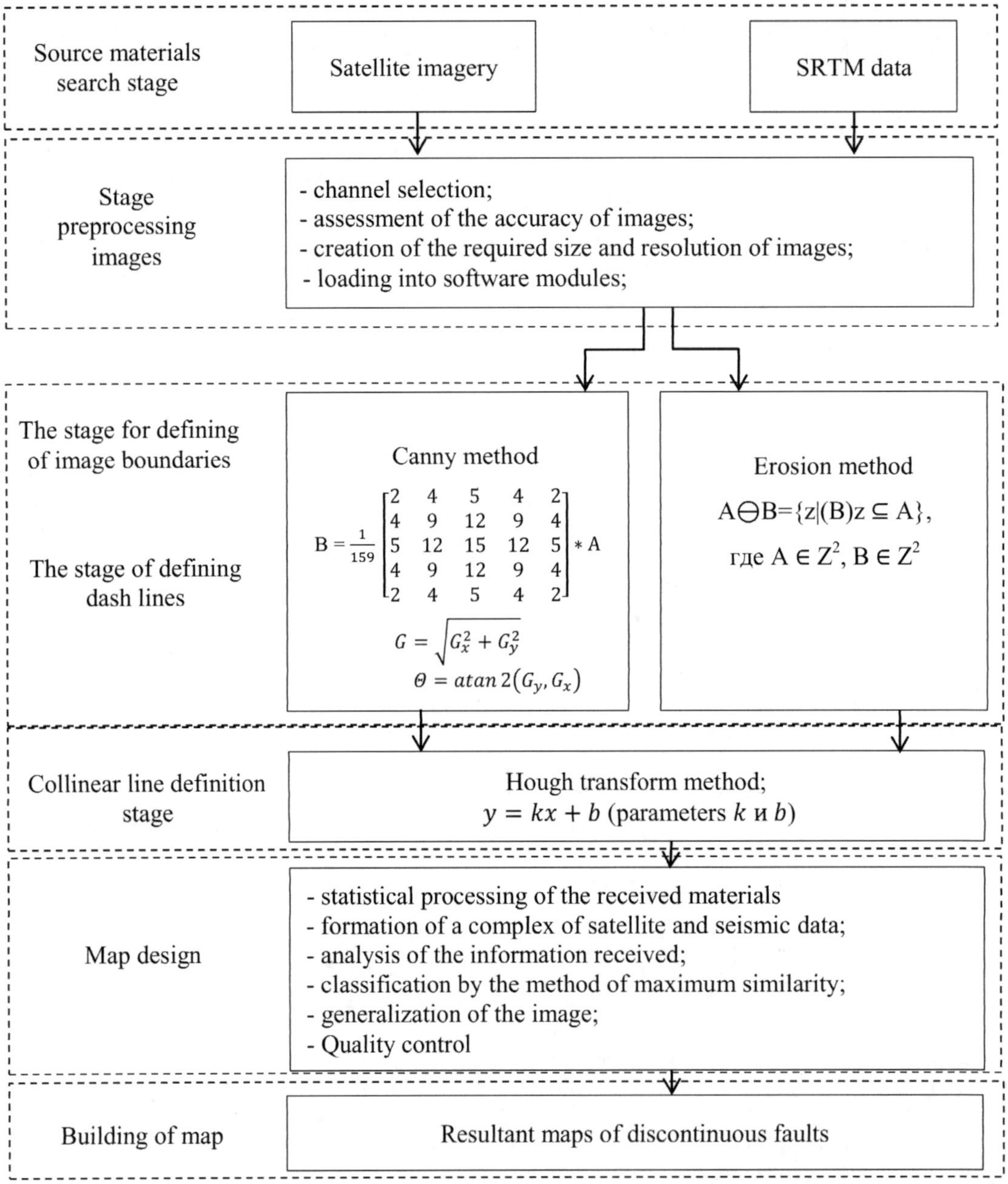

Figure 1. Technological scheme of interpreting the fault zones.

3. Results and discussion

When processing the materials of the study territory of North Sakhalin, zones of discontinuous faults were identified and characterized by increased number of cracks (figure 2).

The materials of the lineament analysis obtained after processing the original images show linear-striped textures that characterize the stress-strain state of the Earth's crust. This is due to the good permeability of rocks in fault zones. The obtained materials indicate a high mobility of the Earth's crust. The recorded lineaments are caused by a change in the stress-strain state of the Earth's crust. The integral effect of the images and seismic data makes it possible to identify the supposed fault structures by means of analysis.

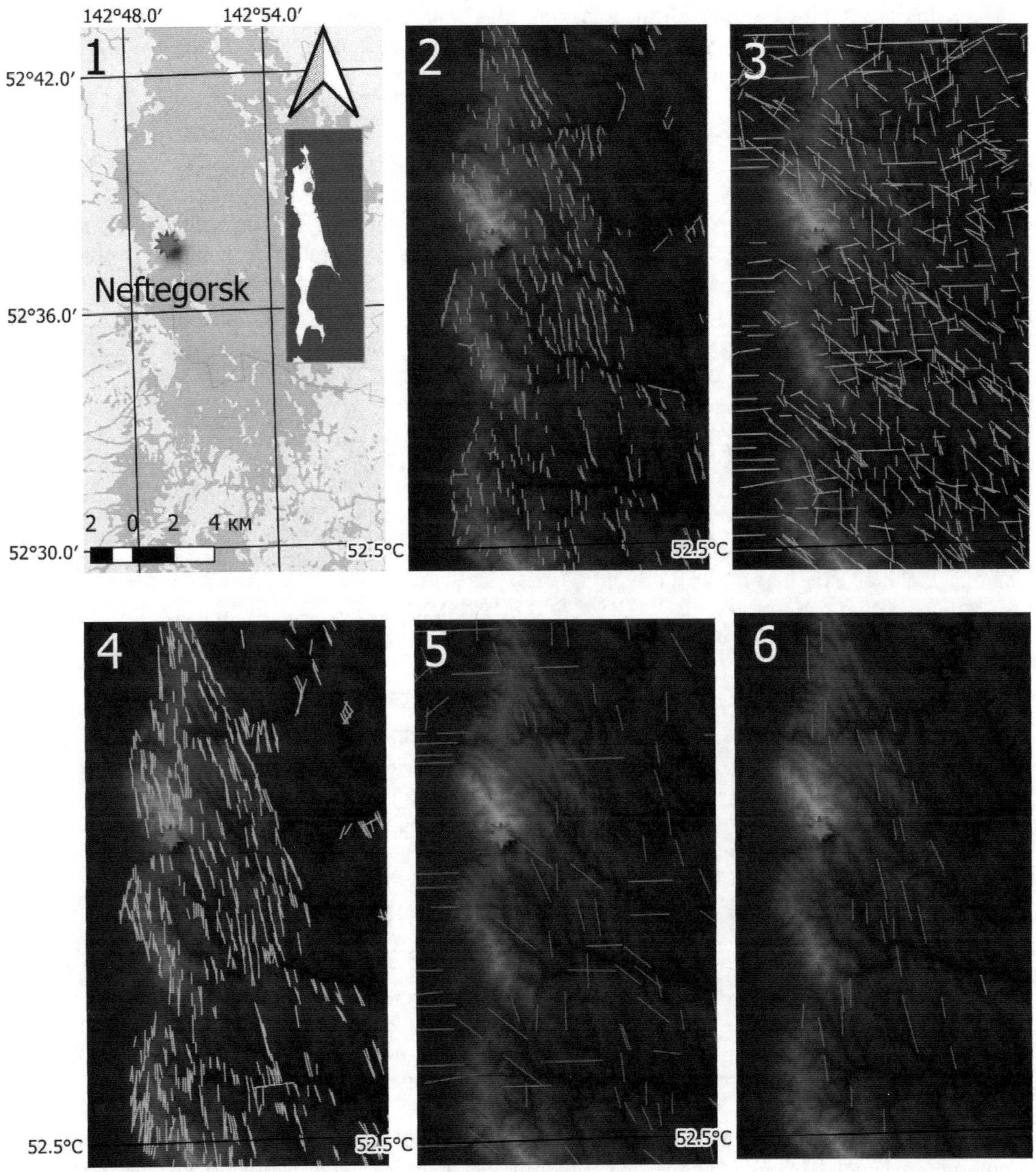

Figure 2. Images with lineaments highlighted into the LEFA program using various methods: 1 – map of Neftegorsk region; 2 – Canny many; 3 – Erosion medium; 4 – Canny medium; 5 – Erosion small; 6 – Canny small.

Earthquakes are located mainly near the faults. Figure 2 shows the areal distribution of the lineaments detected using the morphological erosion method and the local distribution obtained using the Canny method. Modified images with high sensitivity (many) give more detailed information about the Earth's surface. The average sensitivity of the lines (medium) draws our attention to some areas of the analysed territory. The low sensitivity of the lines highlights the most pronounced lineaments on the surface, corresponding to long-lasting discontinuous faults. Processed images and SRTM data in the shape file format are analysed in the Quantum GIS software module.

The materials obtained as a result of processing the information about the location of earthquakes and discontinuous faults allow us to judge the nature of the interdependence of seismic activity and fracturing of rock massifs in the fault zones (figure 3). Seismic activity in the fault zone creates structural strength inhomogeneities in the rocks composing it. When selecting lineaments using automated interpretation, the personal component of the interpreter, its subjectivity, is minimized. Software products allow you to see and display what the human eye could miss and not notice. After the lineaments are selected by program, it is time to analyse and forming map in QGIS, where, integrating all the data obtained, certain lines, that form regional and local structural patterns of the Earth's crust, are selected. Then the epicenter points of the earthquakes with the intensity above four points over the last 100 years are added to the map [11], after which the mutual location of point objects (earthquake epicenters) and linear elements is analyzed.

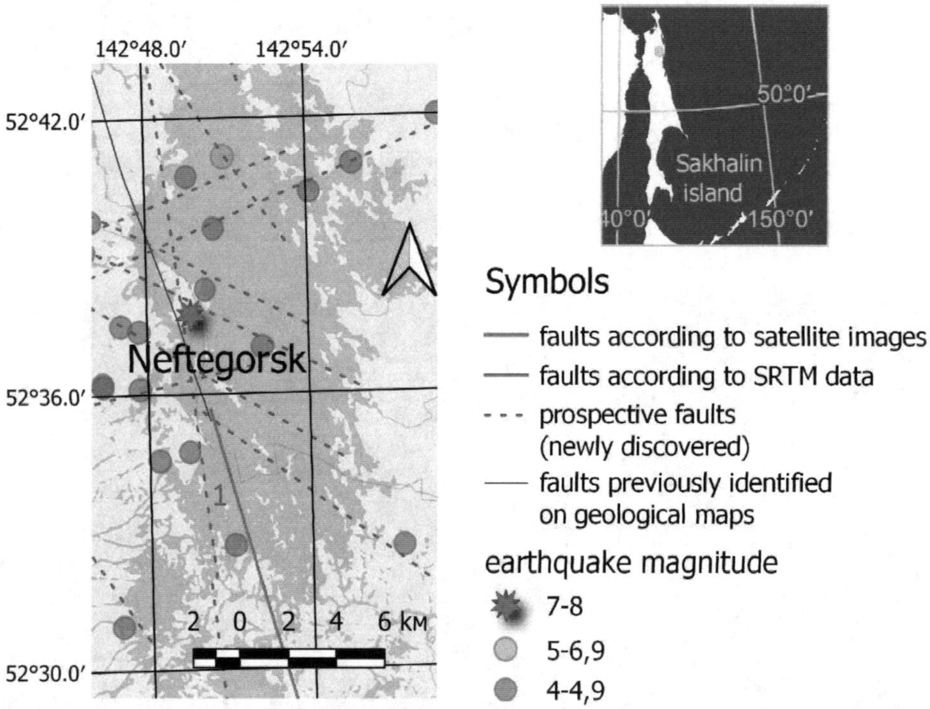

Figure 3. Discontinuities in the area of Neftegorsk earthquake revealed by lineament analysis data and position of epicenters: 1 – Hokkaido-Sakhalin fault (Verkhnepiltunsky segment).

It can be noted on the resulting map, that earthquakes are clearly confined to the zones of discontinuous faults. The areas of the marked earthquakes are characterized with increased stress of the Earth's crust, as evidenced by numerous cracks in fault zones. The western part of the study area is characterized by seismic activity confined to the zone of the Hokkaido-Sakhalin lithospheric fault, previously identified by numerous researchers [25, 26].

Analysis of results has shown that about 90 % of earthquakes are confined to the fault zones and are located at 0 to 3 kilometers from them (table 1).

Movements in the fault zones can lead to both material losses and human casualties. Therefore, the zones of discontinuous faults should not be considered as a territory for the construction of important objects of national economic purpose in order to minimize the risks associated with them. To prevent situations like the one that occurred with Neftegorsk village, a more thorough study of the zones near faults and their comprehensive analysis are necessary [27–29].

Table 1. Distance from the faults to earthquake epicenters in the north of Sakhalin Island.

Fault name	Percentage of earthquakes on Sakhalin Island, confined to the faults, with magnitude greater than 3 for the period 1900–2021, %	The distance from the faults to earthquake epicenters is less than 3 km	Fraction of earthquakes lying near the fault, %
Piltun fault	4.6	0–1	63
		1–2	6
		2–3	6
		3–4	25
Hokkaido-Sakhalin fault (Upper-Piltun segment)	20.9	0–1	92
		1–2	2
		2–3	4
		3–4	2

4. Conclusion

The results obtained deserve special attention, since in addition to the previously known zones, new areas of discontinuous faults have been identified. A more thorough study of this area is needed using large-scale images, as well as a number of geophysical observations. The research materials allow us to propose a fundamentally new approach to the study of discontinuous faults based on space monitoring of the study territory.

References

[1] Aplonov S A 2001 *Geodynamics* (St. Petersburg: Publishing House of St. Petersburg State University) p 360
[2] Zverev A T 2013 *Engineering geodynamics: a textbook for students of higher educational institutions* (Moscow: MIIGAiK Publishing House) p 324
[3] Parnachev V P 2011 *Fundamentals of geodynamic analysis: a textbook* (Tomsk: Publ. House of NTL) p 308
[4] Koronovsky N V 2017 *General geology: textbook* (Moscow: INFRA-M) 2nd ed p 474
[5] Bondur V G and Zverev A T 2007 Mechanisms of formation of lineaments recorded on satellite images when monitoring earthquake-prone territories *Issledovanie Zemli iz kosmosa* **1** pp 47–56
[6] Bondur V G, Zverev A T, Gaponova E V and Winter A L 2012 Investigation from space of the precursor cyclicity in the preparation of earthquakes, manifested in the dynamics of lineament systems. *Issledovanie Zemli iz kosmosa* **1** pp 3–30
[7] Trifonov V G 1985 Features of the development of active faults *Geotectonics* **19** 2 pp 16–26
[8] Trifonov V G 2001 Living faults of the Earth's crust *Soros educational magazine* **7** 7 pp 66–74
[9] Zhmakin E Ya and Davydova E G 2010 Faults of Earth's crust and vegetation features in zone of their influence on territory of the Kaluga region *Bull. of the Kaluga University* **3** pp 57–60
[10] USGS EROS Archive. Digital Elevation – Shuttle Radar Topography Mission (SRTM) Void Filled *GloVis Website USGS* (*Electronic Materials*) https://glovis.usgs.gov/aP
[11] *United States Geological Survey* (*Electronic Materials*) https://doi.org/earthexplorer.usgs.gov/

[12] Sentinel Missions. *Website European Space Agency* (*Electronic Materials*) https://sentinel.esa.int/web/sentinel/missions/sentinel-2

[13] Resources for ArcMap. *ESRI.com* (*Electronic Materials*) https://www.esri.com/en-us/arcgis/products/arcgis-desktop/resources

[14] Shiverev S L 2018 LEFA software: an automatized structural analysis of remote sensing imagery in MATLAB environment *Uspekhi sovremennogo estestvoznaniya [Advances in Current Natural sciences]* **10** pp 138–143

[15] Kuptsova O V 2021 Methodology for detecting disjunctive disorders based on remote sensing data using lineament analysis. *Monitoring. Science and Technology* **1** 47 pp 84–91

[16] Zverev A V and Zverev A T 2015 Application of automated lineament analysis of satellite images in the search for oil and gas fields, prediction of earthquakes, slope processes and ways of underground water migration *Izvestia of higher educational institutions. Geology and exploration* **6** pp 14–20

[17] Kuptsova O V, Verkhoturov A A and Melkiy V A 2021 *Proc. Int. Conf. InterCarto. InterGIS 27. GI support for sustainable development of territories* Mapping of faults on territory of the North Sakhalin plain by remote sensing data (Moscow: MSU, Faculty of Geography) **27** part 1 pp 317–329

[18] Canny J F A 1986 A Computational Approach to Edge Detection. *IEEE transactions on pattern analysis and machine intelligence*, pami-8, 6, NOV679–698 (*Electronic Materials*) http://perso.limsi.fr/vezien/PAPIERSACS/canny1986.pdf

[19] Shovengerdt R A 2010 *Remote sensing. Models and methods of image processing* (Moscow: Tekhnosfera) p 560

[20] Kashnitskiy A V, Balashov I V, Lupyan E A, Tolpin V A and Uvarov I A 2015 Creating tools for remote processing of satellite data in modern information systems *Current problems of remote sensing of the Earth from space* **12** 1 pp 156–170

[21] Galamhos C, Matas J and Kittler J 1999 Progressive probabilistic Hough Transform for line detection *IEEE Computer Society Conference on Computer Vision and Pattern Recognition* pp 554–560

[22] Hobbs W H 1904 Lineaments of the Atlantic border region. *Bull. Geol. Soc. Amer.* **15** pp 483–506

[23] Shahtahmassebi A, Yang N, Wang K, Moore N and Zhangquan Sh 2013 Review of shadow detection and de-shadowing methods in remote sensing *Chinese Geographical Science* **23** 4 pp 403–420

[24] The QGIS Line Direction Histogram Plugin. *Håvard Tveite. Created using Sphinx 1.6.7.* (*Electronic Materials*) http://plugins.qgis.org/plugins/LineDirectionHistogram/

[25] Kharakhinov V V, Galtsev-Bezyuk S D and Tereshchenkov A A 1984 Faults of Sakhalin *Pacific Geology* 2 pp 77–86

[26] Lomtev V L and Zherdeva O A 2015 To the seismotectonics of Sakhalin: new approaches. Geology and minerals of the World Ocean 3 pp 56–68

[27] Dolgopolov D V 2020 Use of earth remote sensing data for formation of geodata space of pipeline transport. *Vestnik of the Siberian State University of Geosystems and Technologies* **25** 3 pp 151–159

[28] Dolgopolov D V, Nikonov D V, Poluyanova A V and Melkiy V A 2019 Possibilities of visual interpretation of trunk pipelines and infrastructure facilities using satellite images of high and ultra-high spatial resolution *Vestnik of the Siberian State University of Geosystems and Technologies* **24** 3 pp 65–81

[29] Karpik A P., Seredovich V A, Dubrovskij A V, Kim E L and Malygina O I 2012 *Proc. Int. Sci. Interexpo GEO-Siberia 2012* Analysis of natural and man-made features of the emergency geospace **3** (Novosibirsk: SSGA Publ) pp 178–184

IOP Conf. Series: Earth and Environmental Science **946** (2021) 012004 doi:10.1088/1755-1315/946/1/012004

Diagnostics of changes in the stress-strain state of the massif after earthquakes by geodetic methods

A A Panzhin* **and D A Koptiakov**

Institute of Mining, Ural Branch of the Russian Academy of Sciences, Yekaterinburg, 620144, Russia

E-mail: *panzhin@igduran.ru

Abstract. The results of diagnostics of changes in the stress-strain state of the massif during the earthquake in the region of Katav-Ivanovsk in September 2018 are presented. The data were obtained as a result of a study of regional geodynamics using permanent stations of the Global Navigation Satellite System (GNSS) of the Urals. The numerical values of the amplitudes of changes in coordinates, the distribution of displacements and deformations of the rock mass have been determined.

1. Introduction

The modern seismicity of the Ural region is mainly characterized by a large number of shallow-focus events with a magnitude of 2–3 [1, 2]. In September 2018, a series of earthquakes with magnitudes 4.2–4.5 occurred in the South Urals. The epicenter of the earthquakes was 7 km to the northwest of the city of Katav-Ivanovsk, Chelyabinsk Region, at a depth of 10 km (figure 1).

Previously, geodynamic research in the Urals was carried out mainly by geophysical methods. Deformation studies of the geodynamics of the Northern and Middle Urals by geodetic methods, according to GPS data, were first carried out under the leadership of V.I. Utkin (IGF UB RAS) in 2009–2010. When redetermining the coordinates of points of geodetic networks, the changes in the stress-strain state (SSS) of the massif was determined. Based on the results of the work performed, the following conclusions were drawn [3]:

- the Ufa ledge of the East European Platform is a tectonic formation, which continues to slowly move eastward in modern times wedging into the Ural structure;
- the movement of the Ufa ledge will inevitably occur in the future and lead to the accumulation of elastic stresses at the boundaries of the ledge, which can cause sufficiently strong earthquakes when discharging;
- it was noted the need to organize a comprehensive detailed geodynamic monitoring in the area of the Ufa ledge of the East European Platform in order to predict the possible next major tectonic event.

However, later, the monitoring of the SSS of the Urals has not been organized and studies have not been carried out by the Institute of Geophysics of the Ural Branch of the Russian Academy of Sciences by the number of reasons. Nevertheless, the Institute of Mining of the Ural Branch of the Russian Academy of Sciences has been researching regional geodynamics for several years using the initial data of permanent stations of the GNSS of the Northern, Middle and Southern Urals [4]. At the same time, periodic, four times a year, calculations of their spatial coordinates from IGS points in the ITRF-2014 system, determination of annual displacement rates, construction of displacement and deformation fields based on speed differences are performed.

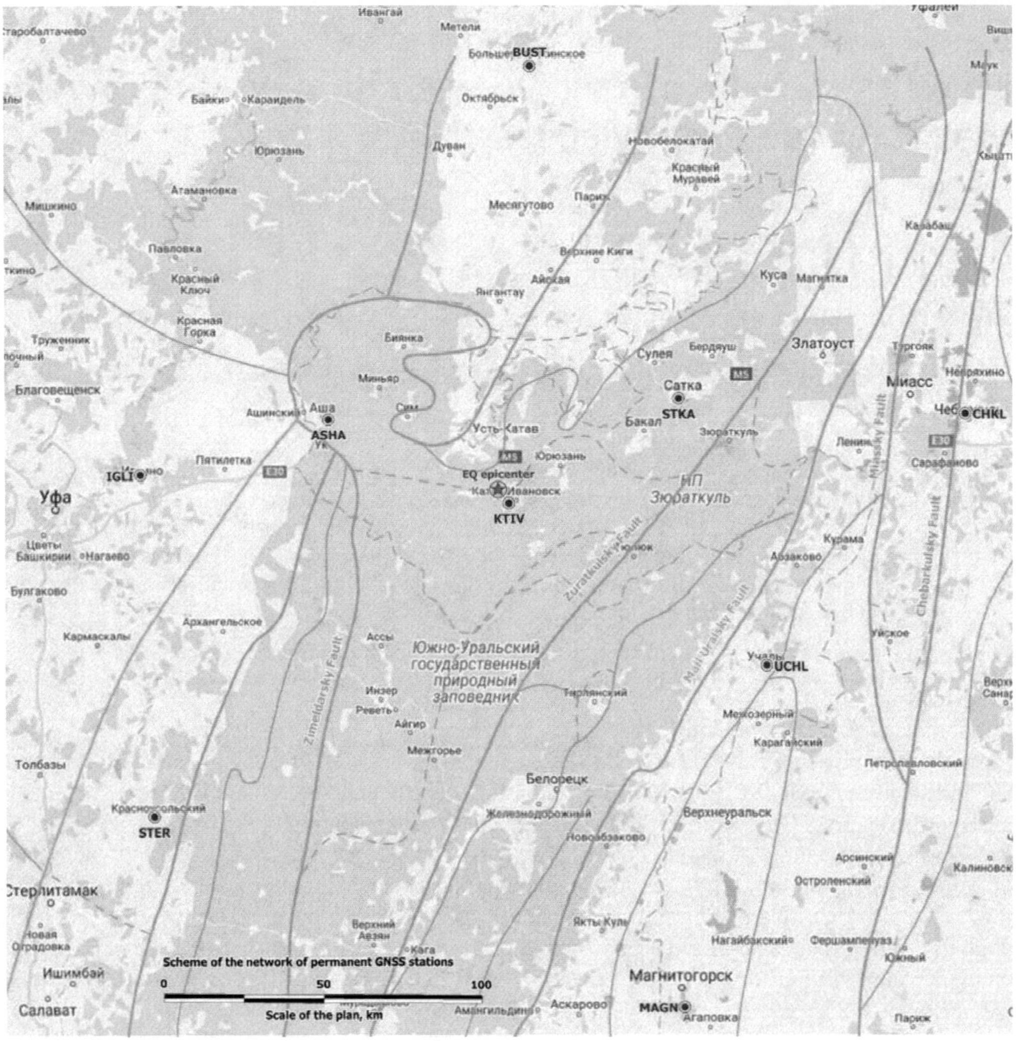

Figure 1. Scheme of the network of permanent GNSS stations in the Southern Urals in the area of the earthquake epicenter.

2. Research methods

In connection with the earthquake that occurred in September 2018, an experiment was carried out to diagnose changes in the stress-strain state of the rock mass in the area of Katav-Ivanovsk. The dimensions of the study area were 280x250 km. The experiment involved 9 permanent GNSS stations of the Southern Urals, using the data accumulated by the stations in the RINEX format. Cameral processing was carried out in the Bernese Software (by the Precise Point Positioning (PPP) method) and Waypoint GrafNet (by the Double Difference (DD) method) with the determination of the spatial coordinates of the points for each daily series.

A sample of the initial data was made for the period from 15.08.2018 to 15.10.2018, to record movements and deformations before, during, and after the earthquake.

The experiment program included:

- determination of the absolute coordinates of points and their changes along the coordinate axes daily, for 61 days, by referencing them from 10–12 reference points of IGS in the ITRF-2014 system;
- processing and adjustment of the GNSS network, in order to study the trend movements by comparing the spatial coordinates of points obtained in various series of monitoring measurements.

3. Research results

As a result, numerical values were determined:

- daily amplitudes of changes in coordinates along three coordinate axes, amplitude and trend components before an earthquake, between series of earthquakes and after ones (figure 2).
- distribution of horizontal displacements and deformations of the rock mass in the area – in the form of displacements to the east with amplitudes of 7–10 mm (figure 3);
- distribution of horizontal deformations of the rock mass in the area – extension and compression with an amplitude of up to $2.4 \cdot 10^{-7}$ (figure 4).

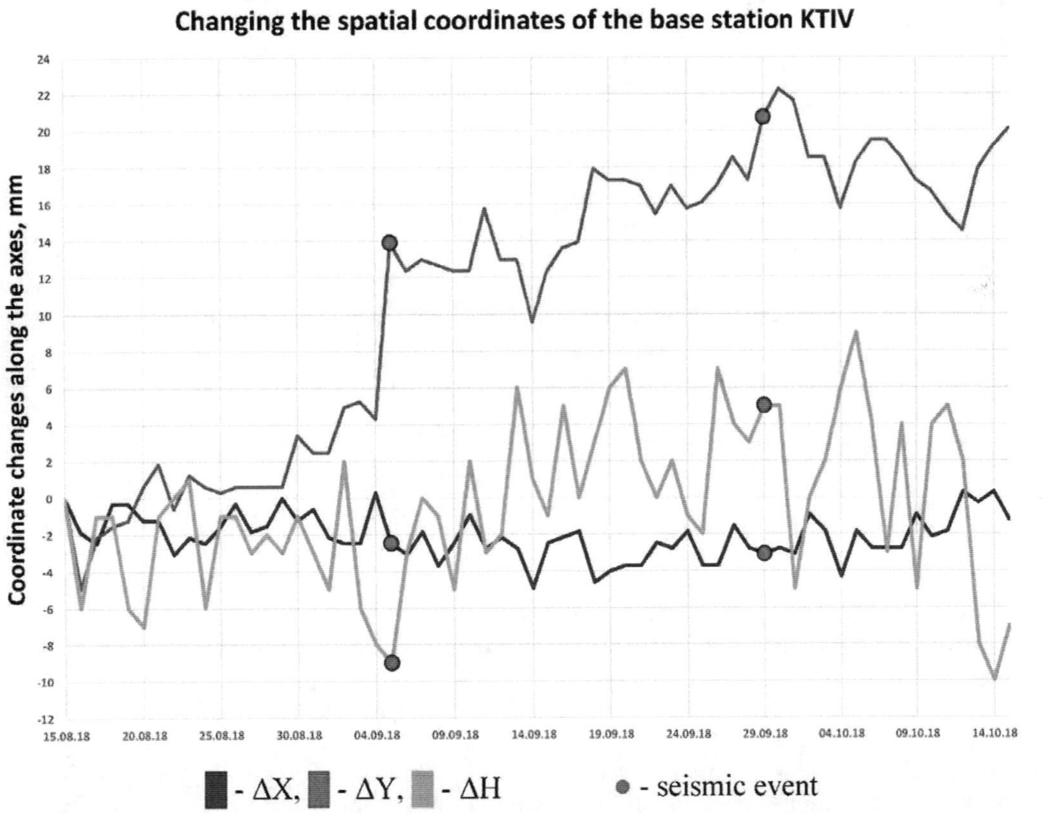

Figure 2. Amplitudes of changes in the spatial coordinates of the GNSS network point KTIV before and after the earthquake.

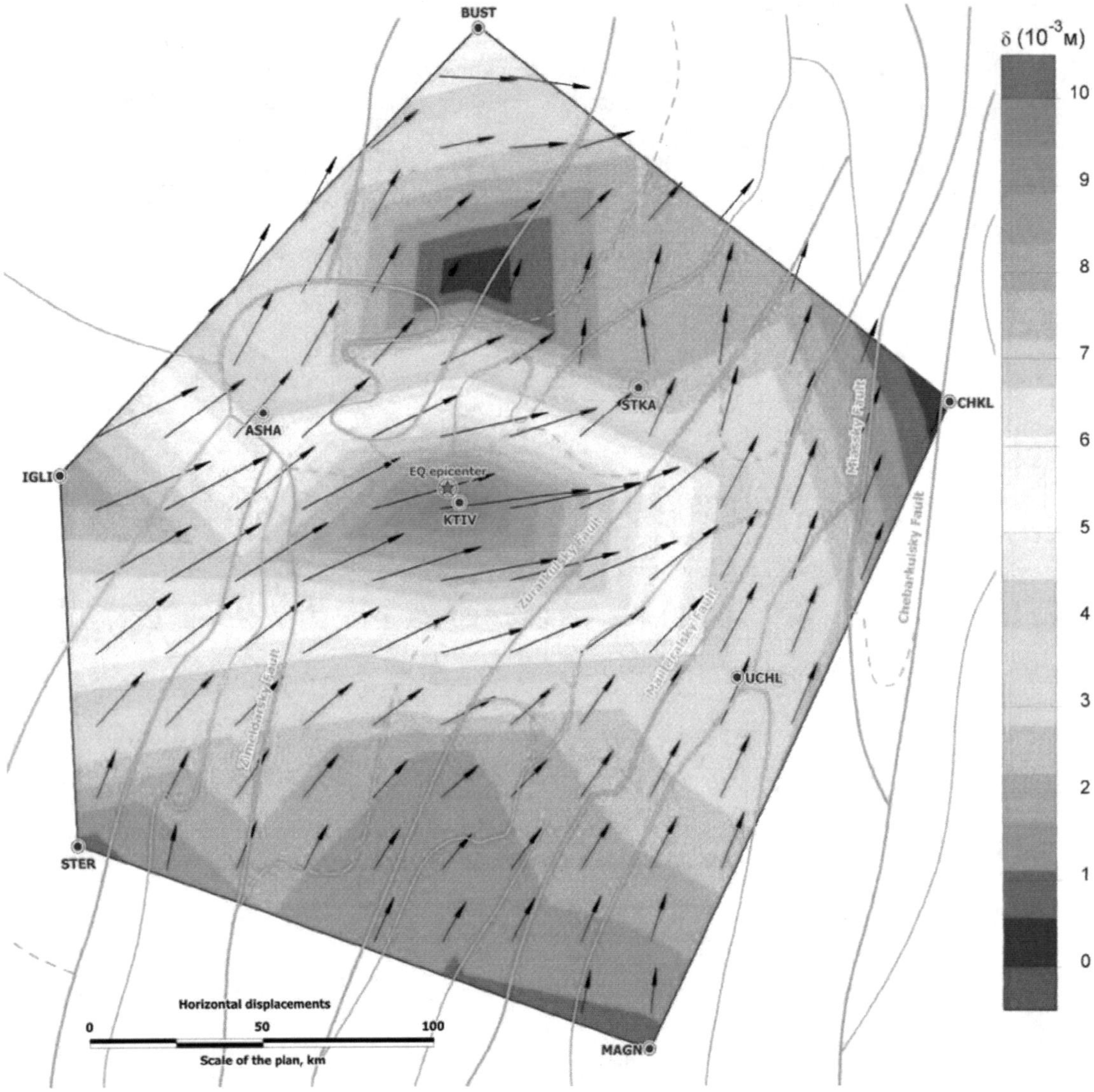

Figure 3. Horizontal displacements to the east along the axis of IGLI–KTIV stations with amplitudes of 7–10 mm.

To construct the field of horizontal displacements and deformations, the finite element method was used, as the initial data were the values and directions of the displacement vectors measured at the GNSS points.

Vertical movements were also recorded, which are manifested in the form of a uniform inclination: uplifts in the southwestern part, subsidence in the northeast. Tensile deformations were noted in the southwestern and western parts of the site at azimuths of 135° and 45°, with compressive deformations prevailing in the eastern part of the study area.

Also, according to the measurement results, the azimuth diagrams of displacements were built for all possible ΔN, ΔE, ΔH, 2D, 3D between the points of the GNSS network. Both correspondence and inconsistency have been established for the main directions of the prevailing orientations of the faults in the Ural region.

Figure 4. Horizontal deformations of rock mass.

4. Research prospects

One of the promising areas of research of modern geodynamic movements of the upper part of the Earth's crust is the method of radar interferometry.

The studies were carried out on the territory of Kuzbass, where the earthquake of magnitude 4.9 occurred on August 12, 2021, at a depth of 10 km, not far from the cities of Prokopyevsk and Kiselevsk, as a result of the development of coal deposits.

A test study conducted used C-SAR images from the Sentinel-1 mission. The mission is a group of two satellites in a near-polar sun-synchronous circular orbit, performing C-band synthetic aperture radar (C-SAR) imagery in 4 data acquisition modes [5]. Scenes were selected for research and pairs were compiled covering the area of interest in the area of the earthquake epicenter. As a result, the seismic deformation field, located at the epicenter of the event, was identified. Figures 5a and 5b show the fields of coseismic deformations obtained from interferograms. The shape, location and values of the deformation fields are different. This is due to the difference in dates, satellite courses and angles of incidence of the interferogram axis.

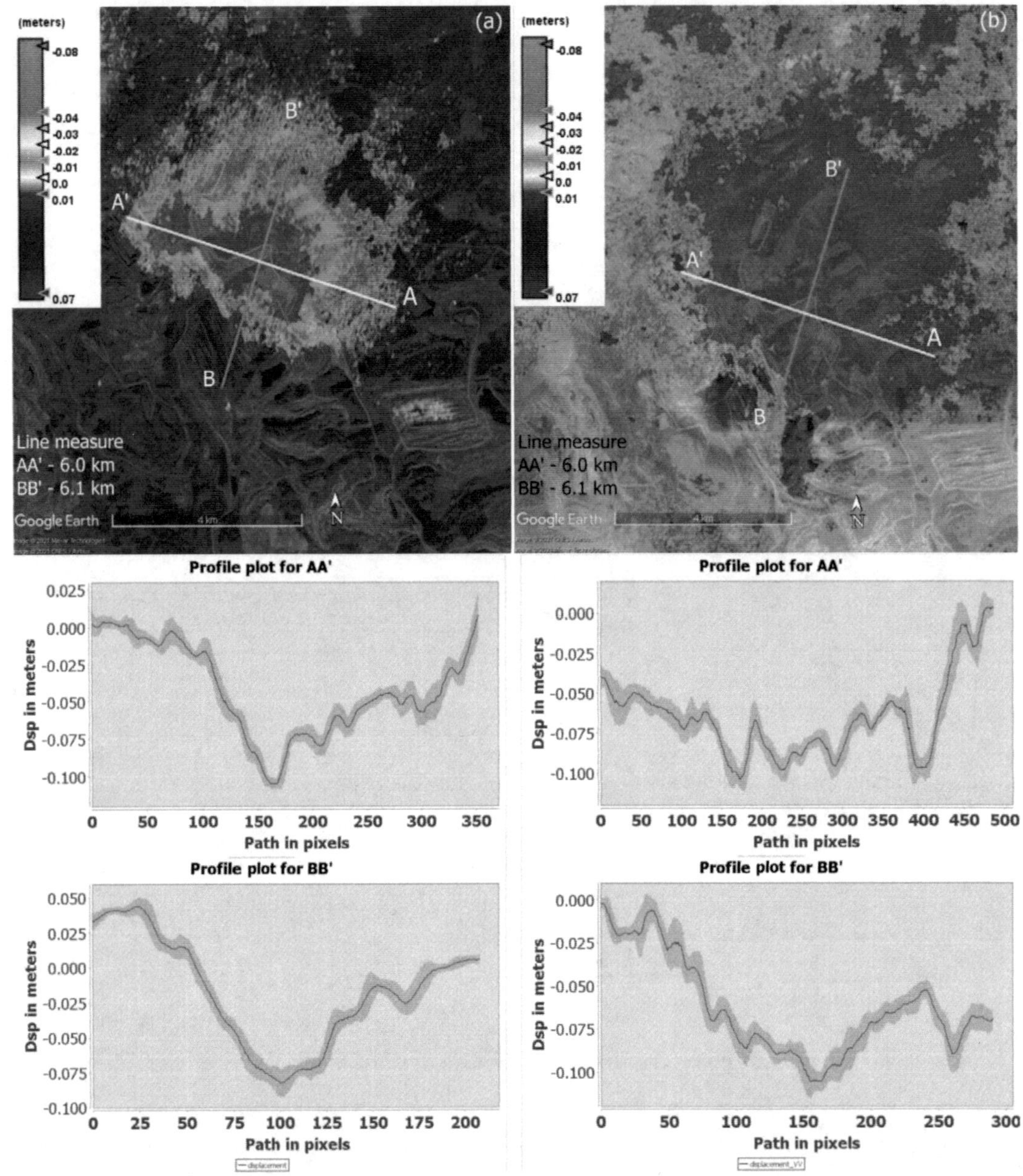

Figure 5. Vertical displacements obtained from two pairs of images (a) and (b) with profile plots for AA' and BB'.

5. Conclusion

As a result of the studies, it has been confirmed that at present in the Southern Urals there is a slow movement to the east of the Ufa ledge of the East European Platform. The result of such a movement is the accumulation of elastic deformations, which can cause earthquakes when discharging.

Further directions of the research are determination of new data on the rates of displacement of points of the GNSS network of the area located in the zone of changes in the stress-strain state of the massif during an earthquake in the vicinity of Katav-Ivanovsk, for the period 2018–2019 and further years, during the period of stabilization of the geodynamic situation.

One of the directions of research is the establishment of the main characteristics of the vector field of modern movements [6], in particular divergence [7], which will determine the distribution patterns of its characteristics and identify the sources of the formation and flow of deformation processes.

Also, within the framework of the indicated experimental studies, work continues on the creation of a database of modern geodynamic movements of the upper part of the Earth's crust in the Ural region [8]. The initial data for filling the database are spatial displacements and their velocities, determined from the results of processing geodetic measurements of more than 60 GNSS points located in the Northern, Middle and Southern Urals.

Acknowledgements
The study was supported by State Contract No. 075-00581-19-00. Topic No. 0328-2019-0005.

References

[1] Kononenko I I, Khalevin N I, Blumin M A et al 1990 *Modern geodynamics of the Urals* (Sverdlovsk: UB RAS) 91 p

[2] Gulyaev A N 2016 Seismicity of the Urals and adjacent territories *Izv. vuzov. Gornyi Zhurnal* 4 pp 45–49

[3] Utkin V I, Belousova A A, Tyagunov D S et al 2010 Research of geodynamics of the Northern and Middle Urals according to GPS data *Reports of the Academy of Sciences* **431** 2 pp 246–251

[4] Panzhin A A 2015 Study of geodynamic movements CORS to substantiate the methodology for controlling the displacement process in the fields of the Ural region *Bull. of the Nosov Magnitogorsk State Technical University* **49** 1 pp 22–26

[5] William H P, Saul A T, Willaim T V and Brian P F 1992 *Numerical Recipes: The Art of Scientific Computing* (New York: Cambridge University Press) 1262 p

[6] Sashurin A D and Panzhin A A 2020 Current problems and objectives in geomechanics *Mining information and Analytical Bull.* 3–1 pp 188–198

[7] Kolmogoron V G, Mazurov B T and Panzhin A A 2018 Algorithm for estimating the divergence of vector fields of movements of the earth's surface from geodetic data *Geodesy and Cartography* **79** 10 pp 46–53

[8] Panzhin A A and Panzhina N A 2019 Evaluation of the stability of control points as a basis for geodynamic monitoring *Izv. vuzov. Gornyi Zhurnal* 6 pp 31–40

2000 Uglegorsk earthquake

A Yu Polets

Institute of Marine Geology and Geophysics Far Eastern Branch of Russian Academy of Science, Yuzhno-Sakhalinsk, 693022, Russia

E-mail: aypolec@gmail.com

Abstract. The paper presents the results of waveform inversion of the Mw 6.8 August 4 (5), 2000 Uglegorsk earthquake (Sakhalin Island, Russia). The detailed rupture process of the 2000 Uglegorsk earthquake was simulated using the waveform inversion method. The average parameters were calculated for both nodal planes. Waveform inversion was carried out on the basis of Global Seismographic Network (GSN) data. Only P-waves from BHZ channels of all stations from the GSN were used. The simulated source parameters included a double-couple source, the scalar seismic moment, the source time function, and the slip directions. The performed studies made it possible to investigate the features of the rupture development and the amplitude of displacements along the east and west-dipping nodal planes of the August 4 (5), 2000 Uglegorsk earthquake. The obtained P-slip model for the 2000 Uglegorsk earthquake source area is in good agreement with the surface manifestations of the rupture according to the field geology data and the results of geodetic inversion.

1. Introduction

A strong Mw 6.8 earthquake occurred on Sakhalin Island (48.786^0 N, 142.246^0 E) on August 4, 2000, at 21:13 pm (August 5 at 8:13 am, local time) and caused a swarm of earthquakes. The magnitude of the largest aftershock was MLH 5.5.

The earthquake occurred in a sparsely populated area of the Uglegorsk district, under the Western Sakhalin mountain range. The event caused significant damage and destruction of buildings and structures in the settlements located close to the epicentre. By lucky chance, there were no casualties. The earthquake was named Uglegorsk or Uglegorsk-Ainu earthquake [1].

The earthquake has confirmed that the Uglegorsk district is one of the most earthquake-prone areas of the Sakhalin Region. Before the devastating Neftegorsk earthquake (the Mw 7.0 Neftegorsk earthquake occurred in Northern Sakhalin, killed almost 2000 people [2]), this part of the island, like the entire Sakhalin Region, belonged to the moderate seismic zone.

Some strong earthquakes occurred earlier in the area: 1924 Lesogorsk-Uglegorsk earthquake (epicentral intensity I_0 8 – 9); 1928 Lesogorsk earthquake (I_0 7 – 8), 1931 Lesogorsk earthquake (I_0 6), 1956 Uglegorsk earthquake (I_0 6 – 7), 1970 Lesogorsk earthquake (I0 6), 1973 Uglegorsk earthquake (I_0 7). There are only 6 earthquakes with Mw > 5 for the period 1924 – 2000. Four of them were realized outside the island, in the Tatar Strait, and only two on the island (1924 Lesogorsk-Uglegorsk and 1970 Lesogorsk earthquakes). All earthquakes within the island, including 2000 Uglegorsk earthquake, are characterized by the high magnitudes – Mw 6.7 (1924 Lesogorsk-Uglegorsk earthquake), Mw 7.0 (1970 Lesogorsk earthquake), Mw 6.8

(2000 Uglegorsk earthquake). It should also be noted that the 2000 Uglegorsk earthquake was localized much to the south of all previous events [3].

On September 13, 2020 (20 years after the 2000 Uglegorsk event), an earthquake with a moment magnitude Mw = 4.8 occurred in the study area. The focal mechanism of the earthquake is a reverse fault type (table 1). The epicentral intensity was I_0=5 according to MSK-64. The seismic process lasted for about 2 days, the most of the aftershocks were recorded during the first 7 hours [4]. The epicenters of the earthquakes are located within the Western Sakhalin fault structure.

It should be noted that in the seismic history of the region there are irregular phases of intensification and decline of seismic activity, replacing each other in about 20 years [1]. However, today it is difficult to say whether this earthquake is associated with a new phase of intensification of seismic activity or not, more research is needed in this area.

Present study aims to simulate the rupture process of the 2000 Uglegorsk earthquake using a waveform inversion method [5].

2. The tectonic position and the main 2000 Uglegorsk earthquake fault parameters

The tectonic position of the 2000 Uglegorsk earthquake is determined by three main orographic elements: the Western Sakhalin Mountains (represented by the Kamyshev Ridge), the Uglegorsk-Ainu Plateau near the western border of the Kamyshev Ridge, the Lomanon volcanic massif, separated from the plateau by one of the main branches of the Western Sakhalin Ridge. The 2000 Uglegorsk earthquake occurred under the horizontal compressive sublatitudinal stresses. The focal mechanism of the earthquake is a reverse fault type (table 1).

Table 1. The earthquake focal mechanism parameters for Uglegorsk 2000 and 2020 events (according to the https://www.globalcmt.org/).

	T		N		P		Nodal planes						Focal mechanism solution
							NP1			NP2			
	PL	AZM	PL	AZM	PL	AZM	STK	DP	SLIP	STK	DP	SLIP	
Mw 6.8 Uglegorsk earthquake	69	134	177	353	13	259	328	36	60	183	60	110	
Mw 4.8 Uglegorsk earthquake	60	345	29	151	6	244	3	47	132	130	57	55	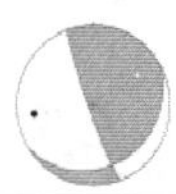

The tectonic position of this reverse in the structure of the northwestern part of the transition zone from the Asian continent to the Pacific Ocean is the following: on the one hand, the earthquake is located near the western boundary of the Sakhalin fold area with the deep-water Japan Basin which is pinching out to the north. Therefore, the concentration of compression stresses can be associated with the Japan Basin expansion. On the other hand, the fault orientation coincides with the strike of numerous echeloned region folds, which can be considered as a manifestation of shear deformations, then the formation of the fault should be associated with the release of stresses on the reverse fault in the meridionally oriented right-side shearing zone [6].

The fragments of the 2000 Uglegorsk earthquake fault (later they were named Ainu fault) were observed in three areas. During the initial observation of the epicentral zone, the fault fragments were located nearby Krasnova Mountains in two areas spaced about 5 km from each other. In the first one, north-east of the mountain, the fragment of seismic rupture is represented by a reverse fault. The fault plane has the north-northwest strike and dips to the west. The value of the vertical surface displacement was 70–80 cm (the western side of the fault was lifted). Approximately 300–400 m to the south, the value of the vertical surface displacement sharply decreased and it does not exceed 10 cm. The size of the vertical cut of the road bed for each of them is 20–30 cm (the western wings are raised). The total vertical displacement of the Earth's surface along the fault fragments was about 60 cm.

In another area, at the eastern foot of Krasnova Mountain, two more fault fragments were discovered. The fault fragments differ sharply from each other. They have different morphokinematic properties and the values of vertical displacements. The eastern fault fragment, striking in the north-east direction, is similar to those discussed above. The value of the vertical displacement (the western fault side was raised) was 40 cm. Another western fault fragments with a total length of about 1 km is located at an insignificant distance to the west of the considered seismic fault. The western seismic rupture with the vertical plane of the displacer had a northeastern strike and a maximum amplitude of the vertical displacement of 2.5 meters (the eastern side was uplifted). At another point, this fault fragment had a vertical displacement of 2.2 m [1, 6].

The total length of the observed fault fragments was at least 5 km. It should be noted, that signs of horizontal displacements were not found.

Initially, it was impossible to distinguish exactly which of the nodal planes was a fault plane and the west dipping plane was chosen as the fault plane. Additional study of the earthquake has shown that the east-dipping plane is preferred since it fits the data significantly better, it is more consistent with the aftershock distribution and the location of the surface rupture, and the resulting geodetic moment is closer to the seismically determined value [7, 8].

3. A waveform inversion method

Recent developments in waveform inversion method for inferring earthquake source behaviour, together with the establishment of high-quality digital seismograph networks and advances in computer technology, have led to great progress in understanding earthquake source behaviour. Using waveform inversion methods, the detailed rupture process of the earthquake can be retrieved from observed data. Also moment tensors for many moderate and large earthquakes can be determined in a routine way [9].

The waveform inversion method was used to construct a source model and study the features of the development of the seismic process in the source area of the Uglegorsk earthquake [5]. The method allows us to determine the source mechanism and estimate its parameters, the orientation of the seismic rupture in the source area, the form of the source time function, the value of the seismic moment, the velocity of rupture propagation, etc. The method allows calculating synthetic seismograms for the current source model, comparing them with the observed waveforms and calculating corrections to the model in order to bring the synthetic seismograms closer to the observed ones.

The inverse problem solution is to compare the observed and calculated waveforms to obtain the best match. Source data – broadband seismograms of body waves recorded by the stations of the GSN (http://www.iris.edu/). The working data include seismic records, characteristics of seismic channels, coordinates and names of stations, as well as information about the event (coordinates, time, magnitude).

4. The results of waveform inversion for the 2000 Uglegorsk earthquake

The azimuthal network of the stations in the source area was inhomogeneous. Only P-waves recorded by broadband stations of the GSN in the range of epicentral distances from 20 to 100°, on the vertical channels (BHZ) were used. The boundaries of the aftershock area were taken as the boundaries of the August 4 (5), 2000 Uglegorsk earthquake. In the process of waveforms selecting one or several records that significantly differ in amplitude and oscillation forms at closely spaced stations were removed. In the final analysis, 69 P-wave seismograms were selected (figure 1). The deviation between real and synthetic waveforms for both nodal planes was about 0.4. The resulting P-slip models for two nodal planes of the 2000 Uglegorsk earthquake are shown in figure 2.

The waveform inversion was performed using a grid of 10x5 nodes (along strike and dip, respectively), with a grid step of 8 km. The arrows show the displacement direction, the color reflects their magnitude.

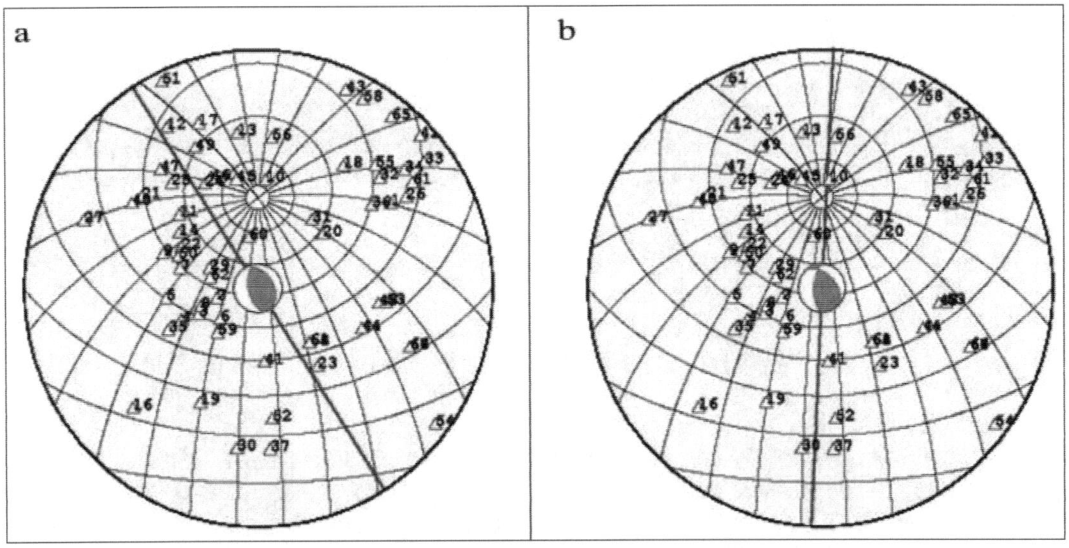

Figure 1. Station distribution used for P-slip inversion of the August 4(5), 2000 Uglegorsk earthquake: (a) the east-dipping nodal plane; (b) the west-dipping nodal plane

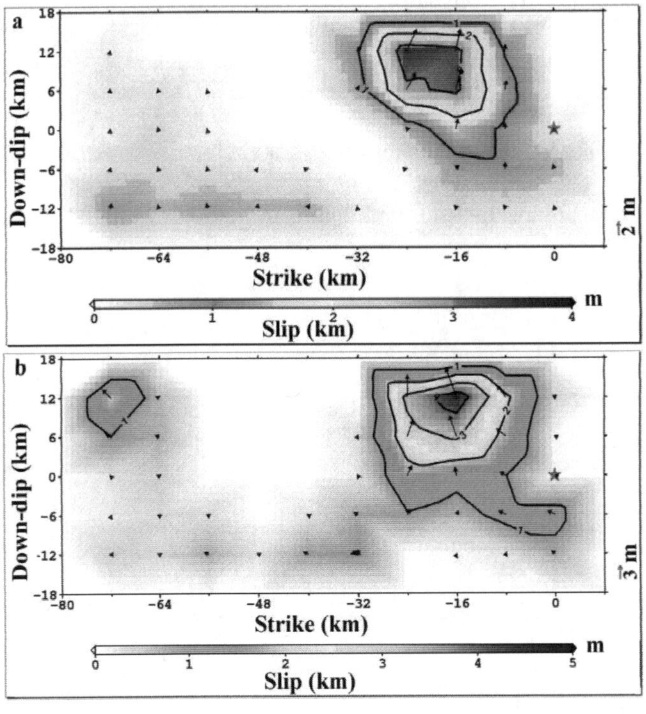

The average parameters were calculated for both planes. Thus, for the east-dipping nodal plane (STK 328 DP 36 RAKE 73), the average displacement was about 2 m (the maximum displacement was about 4 m), the rupture time was about T = 28 s, and the seismic moment was $M_0 = 3.1\times10^{19}$ N·m (Mw = 6.93). The time source function has peaks at the 8th and 12th seconds. For the west-dipping nodal plane (STK 183 DP 60 RAKE 119), the average displacement was about 3 m (the maximum displacement – 5 m), the seismic moment was $M_0 = 3.6\times 10^{19}$ N·m (Mw = 6.97).

Figure 2. Horizontal axis is a distance along the strike and vertical axis is a distance along the dip of the fault planes. Red star marks the hypocentre location. (a) the east-dipping plane; (b) the west-dipping plane.

5. Conclusion

Direct observations of the source area of a strong earthquake are possible only in the cases, when the rupture reaches the surface. Therefore, the indirect remote sensing methods in various combinations and modifications are the most acceptable. The carried out waveform inversion made it possible to simulate the development of the seismic process of the August 4 (5), 2000 Uglegorsk earthquake source zone and to study the features of the rupture development and the amplitude of displacements. The simulated source parameters included a double-couple source, the scalar seismic moment, the source time function, and the slip directions. The obtained dislocation model for the 2000 Uglegorsk earthquake source area is in good agreement with the surface manifestations of the rupture according to the field geology data and the results of geodetic inversion.

References

[1] Koff G L, Bulgakov R F, Ivashchenko A I et al 2000 *Preliminary assessment of the August 4 (5), 2000 Uglegorsk earthquake consequences on Sakhalin Island* (Petropavlovsk-Kamchatsky: Operativnaya Poligrafiya) p 66

[2] Ivashchenko A I, Kim Ch U, Oscorbin L S, Poplavskaya L N, Poplavsky A A, Burymskaya R N, Mikhailova T G, Vasilenko N F and Streltsov M I 1997 The Neftegorsk, Sakhalin Island, earthquake of 27 May 1995 *J. Island Arc.* **6** 3 pp 288–302

[3] Poplavskaya L N et al 2006 August 4, 2000 Uglegorsk-Ainu earthquake, MLH = 7.0, I0 = 8–9 (Sakhalin) *Earthquakes of Northern Eurasia in 2000* eds A V Nikolaev and F F Aptikaev (Obninsk: GS RAS) pp 265–284

[4] Semenova E P, Boginskaya N V and Kostylev D V 2020 Uglegorsk earthquake on September 13 2020 (Sakhalin Island): preconditions for the occurrence and results of observations in the epicentral zone *J. Geosystems of Transition Zones* **4** 4 pp 474–485

[5] Kikuchi M, Kanamori H and Satake K 1993 *J. Geophys. Res.* **98** pp 15797–808

[6] Ivaschenko A I et al 2001 *The 1st Russian-Japanese seminar, Khabarovsk, September 26–29, 2000: Problems of geodynamics and earthquake prediction* The August 4 (5) 2000 Earthquake on Sakhalin Isand ed F G Korchagin (Khabarovsk: ITIG FEB RAS) pp 109–125

[7] Kogan M G, Scholz C H, Bürgmann R, Vasilenko N F, Ivashchenko A I, Kim Ch U, King R W, Frolov D I, Egorov S G and Steblov G M 2003 The 2000 Mw 6.8 Uglegorsk earthquake and regional plate boundary deformation of Sakhalin from geodetic data *J. Geophys. Res. Let.* **30** 3 p 1102

[8] Prytkov A S and Vasilenko N F 2006 The dislocation model of the 2000 Uglegorsk earthquake source (Sakhalin Island) *J. of Pacific Geolog.* **25** 6 pp 115–122

[9] Yoshida S 1995 Waveform inversion methods for the earthquake source *J. Phys. Earth* **43** pp 183–209

IOP Conf. Series: Earth and Environmental Science **946** (2021) 012006 doi:10.1088/1755-1315/946/1/012006

Study of the tectonic structure and modern geodynamics of the nickel-copper-sulfide Kun-Manyo deposit at the stage of its field the development

M I Potapchuk[1], **I Yu Rasskazov**[2], **V I Usikov**[1] **and V G Kryukov**[1]

[1]Institute of Mining of the Far Eastern Branch of the Russian Academy of Sciences – Separate Division of Khabarovsk FEB RAS, Khabarovsk, Russia
[2]Federal State Budgetary Institution of Science Khabarovsk Federal Research Center of the Far East Branch of the Russian Academy of Sciences, Khabarovsk, Russia

E-mail: *potapchuk-igd@mail.ru

Abstract. Currently, the Kun-Manyo nickel-copper-sulfide deposit in the north of the Khabarovsk Krai is being prepared for development, with part of the reserves expected to be mined by underground mining. To justify the rational order of opening and excavation of sub-ore reserves it is necessary to have objective information on rock mass condition, which can be received as a result of complex geodynamic and geomechanical investigations.

The established geodynamic position of the deposit, determined by its location at the junction of actively interacting large tectonic elements of the Euro-Asian tectonic plate – the tectonic stress of the North Asian craton and the Amur plate, as well as within the modern Olekmo-Stanovo seismic zone, has allowed the massif of the field area to be classified as tectonically stressed.

An analysis of the data of the GPS-observation points on the territory of the Russian part of the Amur tectonic plate, the results of calculations of the vector field of velocities of modern movements of points, made within the framework of the ITRF – 2000 coordinate system, as well as the results of in-situ geomechanical studies of rock massifs of the Amur Plate's rock-bump hazardous deposits, have made it possible to establish the current tectonic regime of the deposit area – a region of intense modern compression with a predicted intensity of more than 50 MPa. By methods of morphometric analysis and remote sensing, it has been found that the relative relief excesses were significant (700–1000 m), which may lead to an uncompensated horizontal component of geostatic stress. The most extended lineaments of the relief have predominantly southeasterly extension. The identified features of the tectonic structure and regional neotectonics have made it possible to determine the most probable direction and magnitude of the current main horizontal compression, which could be further used in solving various geomechanical problems in the exploitation of the field.

1. Introduction

Currently, the Kun-Manyo nickel-copper-sulfide deposit is being prepared for the development involving open pit and underground methods. The experience of developing ore deposits shows that, as a rule, the transition to underground development is associated with deterioration of mining and geological and geomechanical conditions. Significant heterogeneity of natural stress fields, predetermined by the complexity and peculiarities of the tectonic and geological structure of the deposits, is further enhanced by the technogenic impact on the rock mass as a result of mining operations. Redistribution of initial stresses in space and time, as well as concentration thereof in some areas, is one of the main causes of dangerous failure of massif areas in the dynamic form [1–3].

To ensure the safety of underground mining operations and reduce the risk of dangerous geodynamic phenomena in the mining enterprise, the organization of geomechanical observations in the continuous geophysical monitoring mode, and linking them with the regional and local neotectonic position of the operation object will be required here. As experience shows, the study of neotectonic processes and tectonic structure, which are associated with natural and anthropogenic seismicity, is of great importance [4–6].

2. Materials and Methods

The Kurumkanskoye ore field includes 8 deposits – independent areas, including the Malyi Kurumkan, Sobolevskaya, and Kubuk deposits. In plan view, the ore field (up to 2–3 km wide) is a linearly stretched in south-east direction 31 km long area consisting of numerous sheetlike and lenticular intrusions, less frequently dikes of mafic-ultramafic rocks with disseminated nickel-copper sulfide mineralization. The mining depth is determined by the base stratification depth and will be (by the maximum depth of the wall) from 20–25 to 150–200 m for different areas. A total of 121 thousand tonnes (C_1 category) and 184 thousand tonnes (C_2 category) of nickel reserves have been reported at the deposit.

Assessment of the tectonic structure and modern geodynamics of the Kun-Manyo field, which influence the choice of the stripping pattern and the most rational parameters of the underground development system, was carried out using a set of methods, including geostructural analysis and geodynamic zoning of the field. Geodynamic zoning of the field area with emphasizing of active faults and determining the nature of deformation of block rock massifs was carried out using morphometric analysis and Earth remote sensing (ERS) [7, 8]. The SRTM30 and GTOPO90 digital elevation models (DEMs) were visualized in QGIS software and presented with elevation wash images in WGS84 geographic projection [9, 10]. The DEM interpretation is based on the concept of tectonic flows [11–13], adapted for the East Asian region [9].

3. Results and Discussion

At the current stage of exploration, the Kun-Manyo nickel-copper-sulfide deposit is the only prospective property of Stanovaya copper-nickel province for industrial exploitation. According to Stepanov V.A.'s conclusions [14], platinum-copper-nickel potential of Northeast Russia can be significantly increased by further exploration of the North Baikal and Stanovaya provinces, which are located near the southern margin of the North Asian craton (figure 1).

Stanovaya province stretches along the southeastern margin of the mentioned craton for 1 300 km, with a width of about 250–300 km. According to gravimetric data at depth, it corresponds to a commensurate giant stratified basite-hyperbasite abyssalite, manifested by a zonal maximum of the gravity field, with a pronounced centroclinal occurrence in the west and east [15]. The Kun-Manyo deposit is located on the eastern margin of abyssalite, in the marginal southwestern part of the Kun-Manyo Early Archean gabbro-anorthosite massif. The Stanovaya province covers the southeastern part of the Aldan Shield and coincides in direction with the transition zone of active interaction of the North Asian craton with the Amur lithospheric plate [16]. The current activity of the zone is expressed by the presence of the Olekmo-Stanovaya seismotectonic zone extending along the southern boundary of the craton. Due to variations in the thickness of the transition zone, where the modern northern boundary of the Amur tectonic plate is located, researchers' opinions on its position differ significantly. The authors of the paper believe that it is most important for geomechanical studies that this boundary runs along seismically active faults. The copper-nickel mineralization of the described province belongs to the sulfide-platinum-copper-nickel formation and is genetically related to different-age basite-ultrabasite complexes – from Late Archean to Early Cretaceous ones.

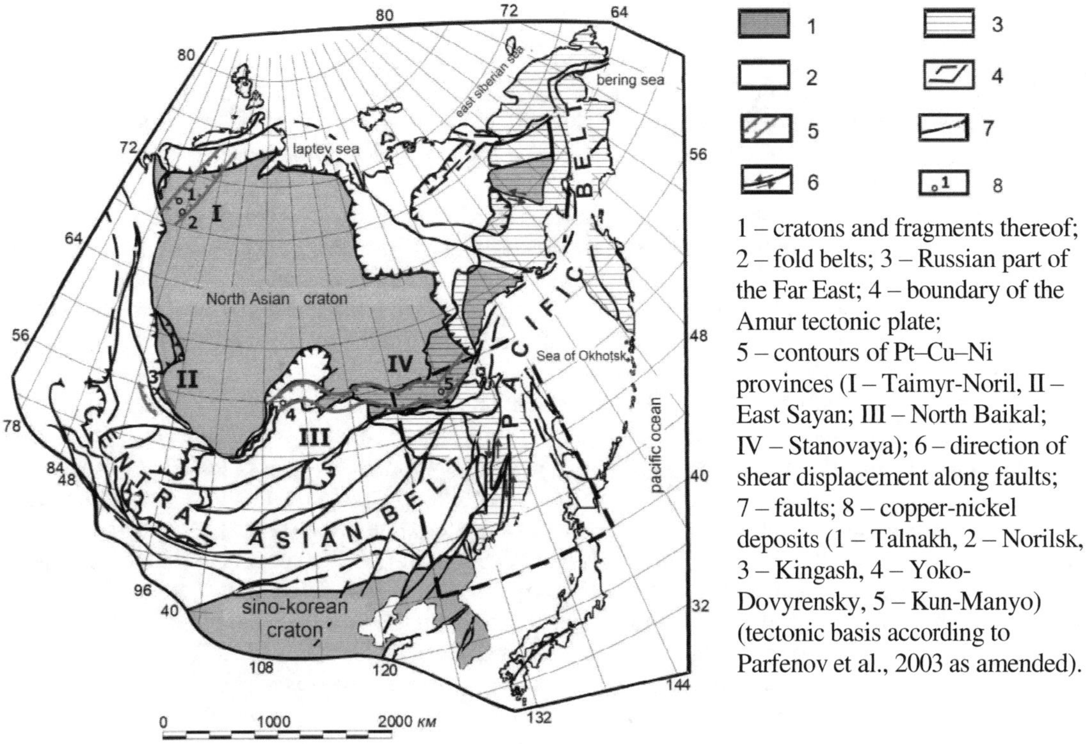

Figure 1. Geotectonic position of Pt – Cu –Ni provinces of Eastern Russia (according to [14]).

The Early Proterozoic Kun-Manyo gabbro-norite-pyroxenite-peridotite complex is the most productive within the Kurumkan ore field, which includes the Kun-Manyo deposit [17]. Nickel-bearing intrusions are dominated by hornblende, olivine, and plagioclase websterites. These are small bodies of strata and plate shape with thickness from the first meters to 63 m. They can be traced in the northwest direction at a distance of up to 40 km. Gentle bedding and dip to the northeast and north predominate.

Structurally, rocks of the Kunmaniev Complex are confined to the junction area of the Dzhagdinsky and Tukskaya blocks of the Aldan Shield crystalline basement. Numerous ancient tectonic faults have been repeatedly activated, creating complex structures such as flake thrusts. Ore-bearing intrusions are located in such structures in tiers. As a rule, they are sharply discordant towards the folded structures of the crystalline basement [14].

Thus, the geodynamic setting of the ore field, including the Kun-Manyo deposit, determined by its location relative to large structural elements, is quite complex.

For the geomechanical analysis of the data set, the features of the regional stress field, which is being formed in the upper part of the Earth's crust section in the recent period, are of the greatest interest. It is here that modern natural geodynamic processes take place, with which the mining engineering system, being formed for the mining of sub-ore reserves at the Maly Kurumkan deposit site, will interact. In this regard, the geometry of the rock massif zoning scheme by stability categories should, if possible, be similar to the current structural-tectonic situation in the territory and not contain elements of past geological epochs, which were not renewed in the new time. It is known that the best indicator of neotectonic processes in Eastern Russia is the relief. The mentioned natural surface better reflects modern deformations of the Earth's crust. It is also important that it is uniformly studied by both terrestrial instrumental and remote methods, and there are well-developed and tested methods for the morphotectonic analysis of the relief (including the method of studying the upper crust

IOP Conf. Series: Earth and Environmental Science **946** (2021) 012006 doi:10.1088/1755-1315/946/1/012006

stratification based on the analysis of 3D models of the relief). To date, numerous materials have been published [18–21, etc.], which indicate that the region's lithosphere is under stress. However, it is possible to differentiate this territory by the degree of change in the internal state of the upper part of the Earth's crust only by additional identification of different-scale zones (or regions) of "extension – compression" based on a set of indirect features.

The data from a network of GPS observation sites on the territory of the Russian part of the Amur tectonic plate [22–26], as well as the results of calculations of the vector field of current velocities of displacements carried out within the framework of ITRF-2000 coordinate system, have allowed us to determine the general nature and average velocities (about 30 mm/yr) of the upper crustal masses movements of the territory, the differential movements of local blocks in specific parts thereof, as well as the general direction of contemporary mass drift, which for the whole area is stable (to the south-east) and is used as an indicator to determine the regional direction of maximum contemporary compression of the upper parts of the Earth's crust.

The geomechanical information model of the Amur lithospheric plate, developed on the basis of an integrated analysis of multi-scale data, including the results of morphometric analysis, numerical modeling and instrumental measurements of the stress state, was used to predict the state of the upper part of the Earth's crust within the study area [27]. According to the geodynamic position, the Kun-Manyo deposit is located within a local area of intense modern compression with a predicted level of 50 MPa or more at depths of 500–700 m. Only Antaeus, one of the most rock-bump hazardous deposits of the territory, characterized, in addition, by changes in the direction of the main horizontal compression with depth and severity of other signs of the high-stress state of the massif of host rocks, is located in such a situation. Judging by the materials of geomechanical study of the Antaeus deposit [5], which is located in a similar zone, the ratio of the main stresses in the Kun-Manyo field is expected to be as follows – $\sigma_1:\sigma_2:\sigma_3 = 2.5:1.4:1.0$.

In addition, the field is localized within the northern end of a major Border Fault, which divides the Amur Plate into first-order blocks and is clearly expressed by the regional gravity field gradient. This fault controls the morphology of the largest Meso-Cenozoic depression. The orientation of the long axis of the Kun-Manyo ore field is orthogonal to the direction of the fault.

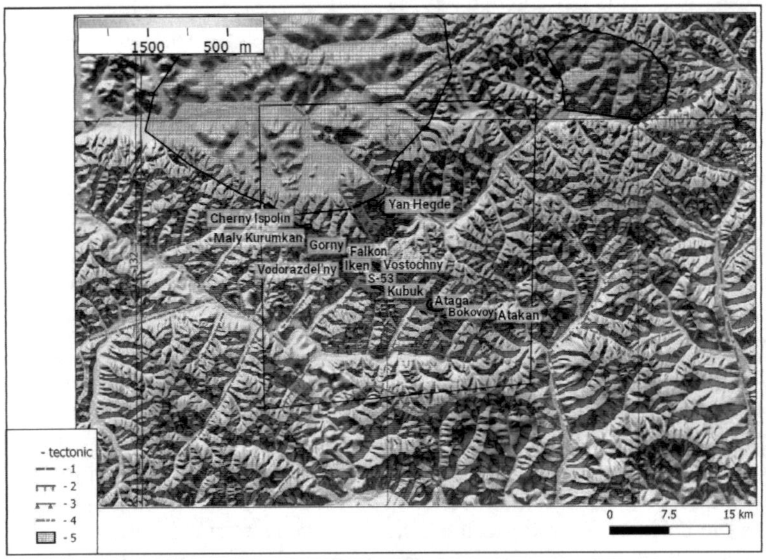

1 – currently active tectonic faults; 2 – line of hanging wall rupture from the lying reverse thrust; 3 – trailing edge of the hanging wall of the reverse thrust; 4 – faults of unclear kinematics; 5 – rigid tectonic block with more weakly dissected relief as compared to the adjacent areas.

Figure 2. Morphotectonic scheme of Kun-Manyo ore district and adjacent areas based on DEM visualized in the form of a color relief map.

The study of the regional neotectonic features of the area including the Kun-Manyo field revealed that the relative elevations are significant (700–1000 m), which in turn may lead to the appearance of an uncompensated horizontal component of geostatic stress. The lineaments of the relief have considerable variation in direction, although, a south-eastern extension can be distinguished among the longest, having an average azimuth of approximately 119°. Figure 2 shows that a very high degree of topographic dissection can be observed over most of the area. All this testifies to a frequent change in the direction of lateral movement of the upper crust layers, complicated by chaotic movements of tectonic flakes relative to each other.

Hard rock massifs are identified in the considered area and beyond, which are confidently identified by the smoothed landforms of low dissection. The massif located in the northern part of the plate has a complex composition of rocks: granitoids, gneisses, diafluorites, tuffs, and metagabbro.

All considered similar objects outside the area are, in one way or another, associated with granites, i.e. are inside the contour of the outcrop massif, or entirely contain it. A small massif of Proterozoic granites, or rather its eroded part, is also located inside the northern rigid block contour. The thrust front is mapped along its southern contact.

Based on the interpretation, two main cycles of horizontal displacements of the tectonic flow in the considered area can be distinguished. The first cycle is a southward submeridional drift. This resulted in the formation of sublatitudinal and submeridional lineaments. In some places, they are disturbed by diagonal structures formed as a result of the second cycle of displacements in the southeasterly direction, which continues to the present day. As already mentioned, everything is compounded by the chaotic oscillation of separate tectonic elements.

4. Conclusion

The ore field of Kun-Manyo deposit is located in a complex geodynamic environment, which is determined by its location at the junction of actively interacting large tectonic elements of the Euro-Asian tectonic plate: North Asian craton and the Amur plate (which began to form as an independent tectonic element in the Jurassic), as well as within the modern Olekmo-Stanovoi seismic zone. The geodynamic position of the Kun-Manyo deposit determines its association with the local area of intense modern compression with a predicted intensity of more than 50 MPa.

Based on the results of morphometric analysis of the relief according to satellite geodesy data, it has been found that the field area, in general, is characterized by a compression regime and the vector of the modern main horizontal compression at the site is directed to the SE. The direction of flow and horizontal tectonic stresses determined from the relief comfortably coincides with GPS data [28]. Since the mineralization is concentrated in a mountain range oriented towards modern movement, if the deposit is developed by underground mining, a deviation of the main direction of the horizontal compression vector to the south can be expected at lower elevations.

References

[1] Rasskazov M I, Rasskazova A V, Potapchuk M I and Teryoshkin A A 2020 *Proc. of the 14th Int. Congress on rock mechanics and rock engineering* Rock Mechanics for Natural Resources and Infrastructure Development pp 793–799

[2] Potapchuk M I, Rasskazov I Yu, Potapchuk G M, Sidlyar A V and M I Rasskazov 2011 *Problems of integrated development of georesources* (Khabarovsk) pp 110–118

[3] Rasskazov I Yu, Potapchuk M I, Osadchiy S P and Potapchuk G M 2010 Mining Information and Analytical *Bull. Sci. and Tech. J.* **7** pp 137–145

[4] Oparin V N, Sashurin A D and Kulakov G I 2008 *Modern geodynamics of the upper lithosphere rock massif: origins, parameters, impact on subsoil facilities* (Novosibirsk: SO RAN Publishing House) p 449

[5] Rasskazov I Yu 2008 *Control and Management of Mining Pressure at Mines in the Far East Region* (M.: Gornaya Kniga Publishing House) p 329

[6] Mclnikov N N, Kozyrev Λ Λ and Panin V I 2015 *Geoecology and Engineering Geology* 5 pp 425–433

[7] Petukhov I M and Batugina I M 1999 *Geodynamics of the subsurface* (M.: Nedra Communication Ltd) p 256

[8] Rasskazov I Yu, Saksin B G and Petrov V A 2014 *Izv. Physics of the Solid Earth* **50** 3 pp 444–452

[9] Usikov V I 2011 *Pacific Geology* **6** pp 14–33
[10] Rybas O V and G Z Gilmanova 2011 *Exploring Earth from Space* **6** pp 45–52
[11] Leonov M G 2013 *Lateral tectonic flows in the Earth's lithosphere* (Moscow: GEOS) 604 p 318
[12] Patalaha E I, Lukienko A M and Gonchar V V 1995 *Tectonic flows as a basis for understanding geological structures* (Kiev: Fenix) p 160
[13] Smirnov A V 2007 *Geodynamics of the Earth's mobile belts formation* (Ekaterinburg: Institute of Geology and Geochemistry, Ural Branch of RAS) pp 285–289
[14] Stepanov V A 2013 *Regional geology and metallogeny* **56** pp 78–87
[15] Mishin V M, Sokolov E P and Andreev A P 2013 *Geology and Mineral Resources of the North-East of Russia* (Yakutsk) pp 25–29
[16] Malyshev Yu F, Podgorny V Ya, Shevchenko B F and Romanovsky N P **2007** *Pacific Geology* **2** pp 3–18
[17] Prikhodko V S, Perestoronin A N and Guryanov V A 2010 *Vestnik ONZ RAS* **2**
[18] Levy K G, Sherman S I and Sankov V A 2007 *Map of modern geodynamics of Asia* (Irkutsk: IZK SB RAS)
[19] Saksin B G, Rasskazov I Yu and Shevchenko B F 2015 *FTPRPI* **2** pp 53–66
[20] Zabrodin V Yu, Rybas O V and Gilmanova G Z 2015 *Fault tectonics of the Russian Far East mainland* (Vladivostok: Dalnauka) p 132
[21] Nazarova L A, Nazarov L A and Dyadkov P G 2002 *FTPRPI* **5** pp 3–9
[22] Bykov V G and Bormotov V A 2009 *Vestnik ONZ RAS* (Vladivostok) p 93
[23] Gusev G S and Imaev L P 2014 *Exploration and protection of subsoil* **12** pp 23–35
[24] Timofeev V Yu, Gornov P Yu and Ardyukov D G 2008 *Pacific Geology* **27** 4 pp 39–49
[25] Ashurkov S V, Sankov V A and Miroshnichenko A I 2011 *Geology and Geophysics* **2** pp 299–311
[26] Serov M A and Zhizherin V S 2012 Irkutsk IZK SB RAS pp 116–118
[27] Oparin V N, Sashurin A D and Kulakov G I 2008 *Modern geodynamics of the upper lithosphere rock massif: origins, parameters, impact on subsoil facilities* (Novosibirsk: IGD SB RAS Novosibirsk Publishing House SB RAS) p 449
[28] Ashurkov S V et al 2016 *Geology and Geophysics* **57** 11 pp 2059–70

Neotectonics and geodynamics of the Streltsovsky ore region according to the analysis of digital elevation models

I Yu Rasskazov, V I Usikov and Yu V Fedotova*

Mining Institute of the FEB RAS, Khabarovsk Federal Research Center, Khabarovsk, 680000, Russia

E-mail: *fjulia@mail.ru

Abstract In the case of underground mining of mineral deposits under difficult mining and geological conditions and at great depths, one of the most urgent problems is the prevention of dangerous geodynamic events. The use of digital elevation models significantly expands the possibilities of preliminary assessment of the potential hazard of their manifestation. This article discusses the results of an assessment of the neotectonics and geodynamics of the Streltsovsky ore region carried out using digital elevation models. The interpretation of the obtained data is based on the concept of lateral tectonic flows adapted for the Far Eastern region of Russia. Digital elevation models based on the GTOPO30 made it possible to identify and evaluate the drift direction of three tectonic flows in the region. The direction determined by the relief coincides satisfactorily with the GPS direction.

1. Introduction

In the case of the mineral deposits underground mining under difficult mining and geological conditions and at great depths, one of the most difficult problems is the prevention of such dangerous geodynamic events as rockburst and mining-induced earthquakes. Assessment of these risks and hazard prediction should be carried out at all stages of deposit development, starting from the earliest.

The modern relief of the Earth's surface is determined by a compromise of the oppositely directed impact of endogenous and exogenous processes on a given territory at a given moment of geological time [1]. It contains, in a hidden form, information about neotectonic, geodynamic phenomena and the stress-strain state of rock mass in the study area. Therefore, the study of relief forms and their distribution over the area is an urgent task. The purpose of the research is to identify the neotectonics and geodynamics of the Streltsovsky ore region based on the analysis of digital elevation models.

2. Research method

Geodynamic zoning of ore areas is carried out at the initial stage of deposit development using lineament and morphostructural analysis of the Earth's surface from satellite images and topographic maps. With the development of mining operations, the amount of information about the rock mass increases significantly and the created models of geodynamic zoning are refined and supplemented.

In the Mining Institute of the FEB RAS geodynamic zoning of ore areas was supplemented and modified through the use of digital elevation models (DEM), which are based on elevation matrices. DEM is a data set for various studies that include morphometric, neotectonic, hydrological, etc. studies [2]. The GIS platform is ideally suited for morphometric analysis because of its capabilities in processing and quantifying topographic data [3].

The main sources of information for digital terrain modeling are large-scale topographic maps, materials of field instrumental surveys, as well as data from remote sensing of the Earth (RSE). Each of the data sources has its advantages and disadvantages, but in general, it should be noted the tendency of the growing role of RSE data in the analysis of the relief of a particular territory.

The use of GIS technologies and RSE data makes it possible to significantly simplify not only obtaining quantitative relief data, but also to conduct modeling on them. In many GIS software products known today, there are methods for constructing DEM, in addition, global DEM have been developed, such as ASTER and SRTM, which are freely available on the Internet. A number of studies conducted by Farr [4] has shown that the digital flight altitude of the SRTM shuttle radar topography is much better than that of ASTER, provided that relatively accurate data is needed to perform morphometric analysis. SRTM data is recommended because of their high vertical and horizontal accuracy [5]. The most accurate representation of the relief allows you to reduce risks when solving various tasks, to develop measures for the safe use of land. The DEM is created on the basis of digital data from the air and gives complete information about the coordinates of points on the terrain, elevation marks, water edges, etc. in triangulation or matrix form.

To conduct research at the Streltsovsky ore field, the publicly available SRTM30 (GTOPO30) matrices were loaded from the NASA portal [6] in increments of 30'' (~1 km) and SRTM03, respectively 3'' (~100 m) [7]. The advantage of this approach is that the work is carried out with a "clean" relief, without the influence of details in the form of landscape features, infrastructure facilities, especially linear – communication routes, energy lines, and, also, it becomes possible to visualize and interpret the DEM in different versions [8, 9]. When using color, another visual element is added to visually display the elevation of the relief. The use of an anaglyphic map with the help of special glasses allows you to work with a stereo image.

During the research of the Streltsovsky ore field, a methodical technique was applied using the latter option, but it is unsuitable for the presentation of reporting and illustrative graphics. Therefore, the final thematic maps and plans have, as a basis, an image either in shades of gray or color.

The interpretation of the obtained data is based on the concept of lateral tectonic flows adapted for the Far Eastern region of Russia [10–12]. According to this concept, in the upper part of the lithosphere, stratification occurs into separate tectonic layers that move in a horizontal direction. Due to the fact that the horizontal dimensions of the layers are many times larger than the vertical ones (capacities), they are split into separate scales, which forms ensembles. As a result, a peculiar wave pattern is formed, consisting of structures of crowding and pulling apart, which appear in relief in the form of alternating mountain ridges – intermountain depressions, elongated parallel to each other. The extension of these structures is almost perpendicular to the direction of drift of the tectonic flow. Often the stream is broken by stratiform faults according to the direction of movement of the plate blocks limited by them, which are expressed in relief in the form of canyon-like valleys, dissecting both ridges and intermountain depressions.

3. Results and discussions

The GTOPO30-based DEM made it possible to identify and estimate the drift direction of three tectonic flows in the region shown in figure 1. The direction determined by the relief satisfactorily coincides with the GPS direction (see figure 2) [13, 14].

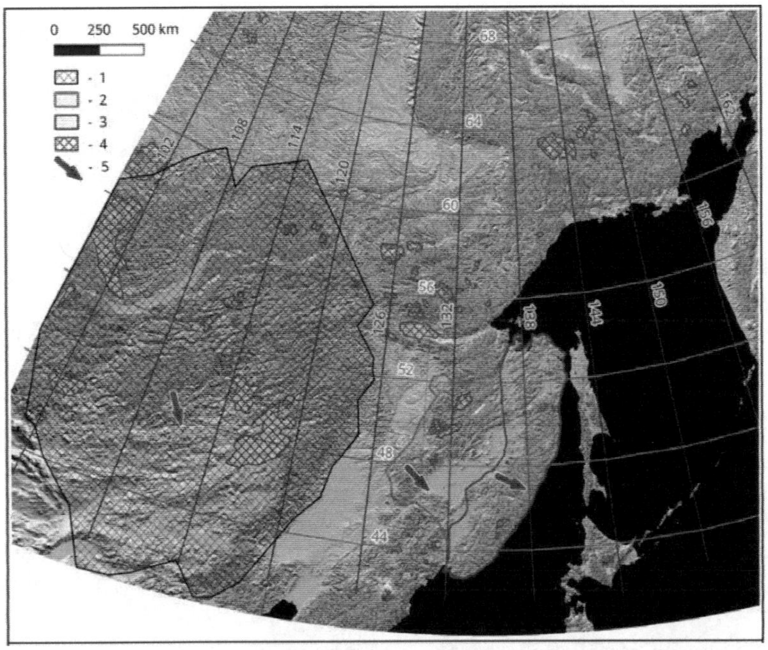

Tectonic flows: 1 – Baikal; 2 – Amur; 3 – Sikhote-Alin; 4 – contours of rigid flows; 5 – directions of flow drift. The basis is a shaded relief map built according to the DEM using the GTOPO30 matrix

Figure 1. Map tectonic flows and rigid arrays in the folded belts of the Far Eastern region of Russia.

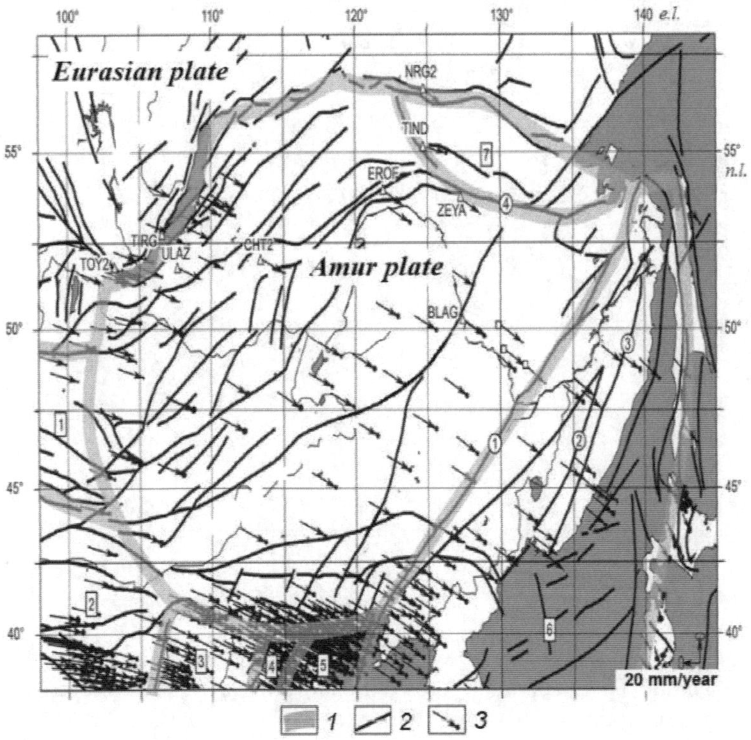

1 – boundaries of blocks and plates; 2– faults, according to [15] with changes; – velocity vectors and error ellipses in the 95 % confidence interval. Triangles are permanent and semi-permanent GPS points. Squares are GPS points of periodic measurements. The geoblocks are indicated by numbers in squares: 1 – Mongolian, 2 – Alyoshan, 3 – Ordoz, 4 – Taiheng, 5 – North China Plain, 6 – Japanese-Korean, 7 – Stanovoy. The numbers in the circles indicate the faults: 1 – Elan-Egunsky, 2 – Arsenevsky, 3 – Central Sikhote-Alinsky, 4 – South Tukuringrsky.

Figure 2. Map modern horizontal velocities according to GPS data in the global reference coordinate system ITRF 2008 according to [13].

The analysis of the DEM made it possible to note one previously unknown natural phenomenon, which probably sheds light upon the origin of the abnormal stress-strain state of mountain massifs, provoking their tendency to dangerous geodynamic events during the development of deposits. Within the mentioned tectonic flows, accompanied by a high degree of dissection of the relief, plateaus of various shapes are observed, where this dissection is much lower. Such structures are confidently registered by anaglyphic relief maps, maps with shadow washing in the color version of the elevation horizontal map. They usually have forms close to isometric, often accompanied by a decrease in the density of small-focus earthquakes within their limits.

It is logical to represent such objects as rigid mountain massifs, lowly destructured by discontinuous disturbances, drifting in streams. Their low disturbance indicates a rare discharge of horizontal stresses and, conversely, their concentration and accumulation. Accordingly, the probability of their discharge increases in case of violation of massifs by mining operations and provoking dangerous geodynamic events. The model of rigid arrays is most clearly manifested in deposits with an increased risk of rockbursts and mining-induced earthquakes in the Dalnegorsky ore district (Primorsky Krai) [16].

Figure 3 shows an overview map of the relief of the Baikal region and Transbaikalia in shades of gray, based on the GTOPO30 DEM. Morphostructures in the form of plateaus with smoothed relief, probably displaying rigid arrays, are confidently identified on the territory. Streltsovsky ore district is located in the convergence zone of three such structures, possibly exposed parts of one shield-shaped mountain massif. As we can see from the data shown in figure 2, these directions correspond to the instrumental GPS measurements given in [13, 14].

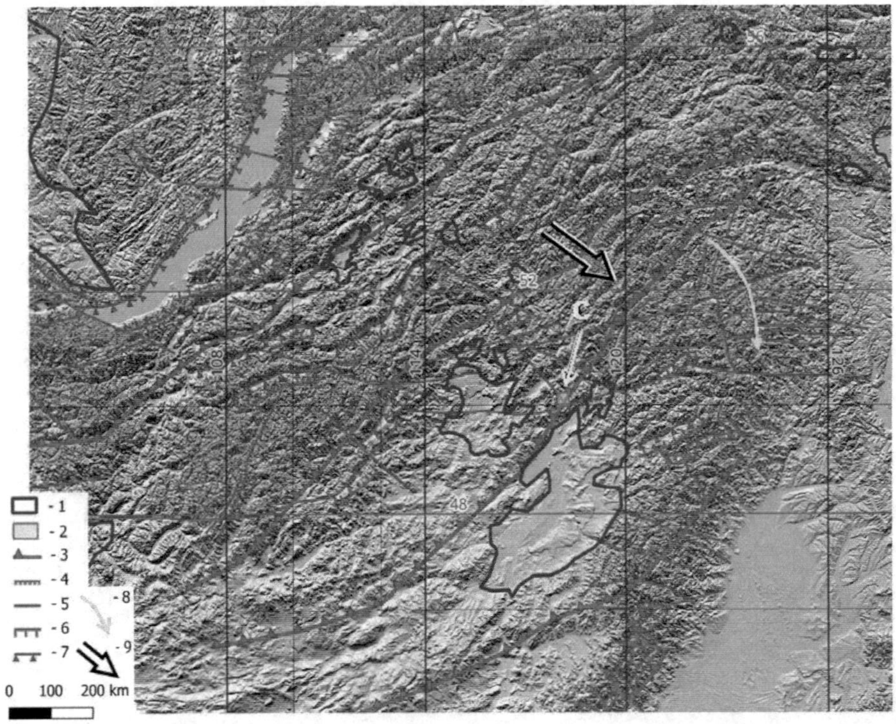

1 – contours of rigid massifs in plan; 2 – Streltsovsky ore field; 3 – thrust fronts; 4 – faults; 5 – tectonic disturbances of various kinematics; 6 – separation lines of hanging wings of reversed thrusts; 7 – rear edges of hanging wings of reversed thrusts; 8 – direction of reversal of the eastern part of the flow; 9 – general direction of drift of the tectonic flow. The yellow arrow (C) indicates the Streltsovsky ore district.

Figure 3. Morphotectonic scheme of the Baikal region and Transbaikalia, based on the GTOPO30 matrix.

On the map, one can note a characteristic feature for this, and probably other flows. Local motion anomalies are often superimposed on the general drift. For example, attention is drawn to the southeastern part of the flows, near which the study area is located. Here you can observe the clockwise reversal of allochthonous masses. This is evidenced by the change in the stretch of the lineaments reflecting the position of the elongated relief elements. The center of rotation is located near the Streltsovsky ore field. Probably, there is a rigid protrusion of autochthon here – a structure that exerts the maximum inhibitory effect on the flow.

Figure 4 shows the morphotectonic scheme of the Streltsovsky ore field, based on the analysis of the SRTM 03 DEM on a larger scale (in detail close to 1:100 000). The assumed main flow direction (as well as compression) is shown by an arrow on the plan.

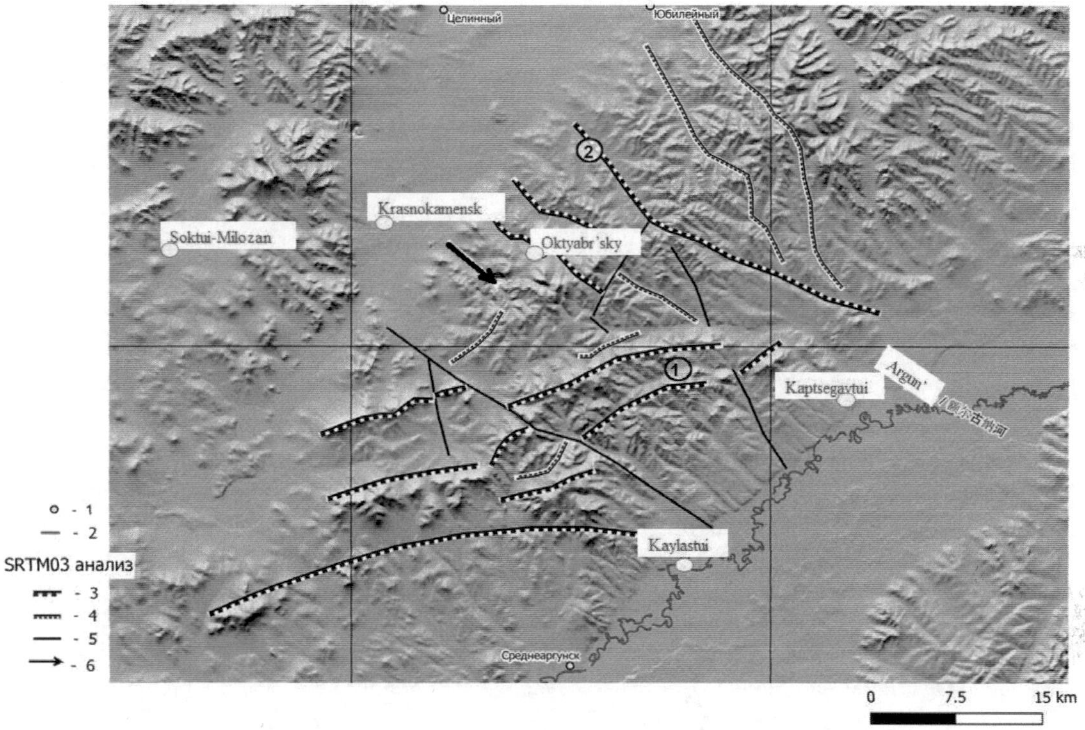

1 – settlements; 2 – hidrografic network; 3 – upthrow faults, bergstrichs are directed towards the upthrow wing; 4 – faults, bergstrichs are directed towards the lowered wing; 5 – discontinuous disturbances of uncertain kinematics: 6 – general direction of the tectonic flow.

Figure 4. Morphotectonic scheme of the Streltsovsky ore field. The basis is the SRTM 03 height matrix.

The details of the morphotectonic scheme suggest that the decisive influence on the geodynamic situation of this area is exerted by upthrow faults, primarily the step series 1. Series 2 appears to be identified by the anomaly of the flow drift. The ratio of the average modern heights of the lowered and upthrow wings indicates that these structures are currently active.

According to [17], most of the main ore regions of the Far East are confined to the global fault zone. At the same time, it is believed that within the eastern edge of the Eurasian continent, probably in the Mesozoic, a left-sided fault has been formed, during the long development of which the faults have been repeatedly activated. As a result, numerous fault dislocations of various types are recorded within the global zone. These are extended faults of the north–northeastern strike, which limit narrow tectonic blocks and lenses, as well as obliquely oriented upthrow faults, thrust faults and upthrow–

thrust faults. The article [8] presents the results of research to identify the most common patterns of formation of regional natural stress fields of the Amur geoblock. And in the book by Rasskazov [18], data on the stressed state of rock massifs of a number of ore deposits in the Far Eastern region are given. According to the author, in the upper part of the Earth's crust at the depts from 300 to 800 m and more, unequal stress fields act with a predominance of subhorizontal compressive stresses oriented in the direction from sublatitudinal to southwestern. The results of the conducted studies have shown compliance with the conclusions presented in these publications.

4. Conclusion

The analysis of the DEM made it possible to note one previously unknown natural phenomenon, which probably sheds light upon the origin of the abnormal stress-strain state of mountain massifs, provoking their tendency to dangerous geodynamic events during the development of deposits. Within the mentioned tectonic flows, accompanied by a high degree of dissection of the relief, plateaus of various shapes are observed, where this dissection is much lower. Such structures are confidently registered by anaglyphic relief maps, maps with shadow washing in the color version of the elevation horizontal map.

Due to the complex interaction of the displacement vectors of individual parts of the flow, a complex field of abnormal compression stresses is also formed. This explains the increased risks of dangerous geodynamic phenomena in the deposits of the region.

Acknowledgments
The research was carried out with the financial support of RFBR grant No. 20-55-53028 GFEN_a.

References

[1] Ufimtsev G V 2008 *Mountains of the Earth (climatic types and phenomena of the latest orogeny* (Moscow: Scientific world) p 352
[2] Kanhaiya S et al 2018 Morphometric analysis, bed-load sediments and weathering intensity in the Khurar River Basin, Central India *J. Geol. J.*
[3] Altaf F, Meraj G and Romshoo A S 2013 Morphometric analysis to infer the hydrological behaviour of Lidder Watershed, Western Himalaya, India. *J. of Geography* pp 1–14
[4] Farr T et al 2007 The shuttle radar topography mission (*Rev Geophys* 45:RG2004)
[5] Kaushik P and Ghosh P 2015 3D DEM delineation of Chambal river basin from SRTM data using remote sensing and GIS technology *Int. J. Remote Sens. Geosci.* **4** 4 pp 1–6
[6] (*Electronic materials*) https://dds.cr.usgs.go
[7] Index of /srtm/version2_1/SRTM30 (*Electronic materials*) https://dds.cr.usgs.gov/srtm/version2_1/ (srtm03 v1,2.1, srtm30)
[8] Rasskazov I Yu et al 2014 Modern stress-strain state of the upper levels of the crust Amur lithospheric plate *J. Phys. of the solid Earth* **3** pp 144–153
[9] Usykov V I 2011 3D models of relief and structure of the upper part of the earth's crust of the Amur region *Pacific geology* **6** 14–33
[10] *Lateral tectonic flow in the lithosphere of the Earth* 2013 ed M G Leonov (Moscow: GEOS) **604** p 318
[11] Patalakha E I, Lukienko A M and Gonchar V V 1995 Tectonic flows as a basis for the understanding of geological structures (Kiev: Phoenix) p 160
[12] Usykov V I 2011 *Proc. of the All-Russian Conf., September 12–15, 2011, Khabarovsk* Dynamics and structure of tectonic flows Analysis of 3d relief models ed by A N Didenko, Yu F Manilov (Khabarovsk: Federal State Budgetary Institution of Science Kosygin Institute of Tectonics and Geophysics FEB RAS) pp 328–331
[13] Ashurkov S V et al 2016 Modern deformations of the Amur Plate and surrounding structures according to GPS measurements *J. Geology and Geophysics* **57** 11 pp 2059–70

[14] Shestakov N V et al 2011 Present tectonics of the southeast of Russia as seen from GPS observations *J. Geophys. J. Int.* **184** 2 pp 529–540

[15] Bouysse P 2009 Geological map of the world at 1:50,000,000, 3rd ed Commission for the Geological Map of the World (Paris)

[16] Rasskazov I Yu, Usikov V I and Saksin B G 2019 *Materials of the All-Russian Conf. with Int. part. Tectonics, deep structure and mineralogy of East Asia* The use of remote sensing methods in the study of modern geodynamics of the Sikhote-Alinsky orogenic belt and a comprehensive assessment of the VAT of the mountain massif of the Nikolaevsky deposit (Khabarovsk: Federal State Budgetary Institution of Science Kosygin Institute of Tectonics and Geophysics FEB RAS) pp 300–302

[17] *Geodynamics, magmatism and metallogeny of the East of Russia* 2006 ed A N Khanchuk (Vladivostok Dalnauka) in 2 books p 981

[18] Rasskazov I Yu 2008 *Control and management of mining pressure in the mines of the Far Eastern region* (Moscow Publishing House "Gornaya kniga") p 329

Types of strong earthquake precursor behavior obtained from world and regional catalogs of earthquakes

M V Rodkin[1,2,*], **M Yu Andreeva**[2] **and E V Liperovskaya**[3]

[1]Institute of Earthquake Prediction Theory and Mathematical Geophysics, Russian Academy of Sciences, Moscow, 117997, Russia
[2]Institute of Marine Geology and Geophysics, Far Eastern Branch of the Russian Academy of Science, Yuzhno-Sakhalinsk, 693022, Russia
[3] Schmidt Institute of Physics of the Earth of the Russian Academy of Sciences, Moscow, 123242, Russia

E-mail: *rodkin@mitp.ru

Abstract. Typical features of seismicity inherent to the period prior to the strong earthquake occurrence are examined using the regional and world-wide earthquake catalogs by the method of construction of the Generalized vicinity of a large earthquake (GVLE). These features are foreshock power-law cascade, the weak long-term background increase in seismic activity, and the definite clustering of the main events. Worth to note, that all these features appear to indicate the loss of a stability of a system, without any reference to the physical mechanism of development of the instability. Thus, these features are expected to occur in connection with the bifurcation points in systems of other physical nature, examples of such behavior are given. Besides, a tendency of a decrease in earthquakes depth in the close GVLE vicinity was found. This precursor feature has a specific character and evidences in favor of a fluid involvement in the mechanism of earthquake occurrence.

1. Introduction

The However, even the mostly supplied with statistical data case of large earthquake occurrence is far from the whole understanding and the possibility of strong event forecasting. The study of strong earthquake precursor behavior meets serious difficulties because of a very strong turbulent random component in the seismic process, that results in a great diversity in regimes of foreshock and aftershock sequences of individual strong earthquakes. One meets essential difficulties if tries to identify typical features of the foreshock and aftershock behavior upon the background of a large random component of the seismic regime. The variations that are observed in different fore-and aftershock sequences are so great that the very existence of common regularities in foreshocks and aftershock behavior such as the power-law foreshock cascade and the occurrence of the aftershock process following the Omori law have been questioned [17; et al.]. And although the existence of such patterns had been confirmed [2; et al.], the details of foreshocks' behavior remain unclear.

The existence of a set of typical common features in the foreshock–aftershock process was confirmed, and some details of these processes were revealed or verified by the method of construction and analysis of the generalized vicinity of a large earthquake (GVLE) [10–13]. The idea of this method is to carry out a scaled aggregation of data from a large number of individual foreshock and aftershock sequences into a generalized foreshock–aftershock sequence. The method of data combination for a large number of events, which is usually referred to as the method of epoch superposition, is not new. As an example, [9] made use of this method to perform the statistically rigorous identification of the tendency of a decrease in b-values before large earthquakes. The approach used in the present study differs from those of the other authors in the use of larger (up to 1000) number of individual foreshock and aftershock sequences and in scaling of the spatial vicinity by the size of the main shock.

Application of the GVLE method to the regional data [14, 15] had met an inevitable decrease in data volume, but it gave a possibility to reveal some new features of the precursor regime in seismicity. This way a noticeable clustering of the strong and main events (after aftershocks removing) was found. The brief discussion of the regularities in the precursor seismic behavior revealed by GVLE method is given below.

2. Complex of precursors found in GVLE

In the procedure of the GVLE construction the events that occurred near the N strongest events of the used catalog are selected (below, when using the world GCMT and ISC catalogs, the vicinities of 500 strongest events are summed up). Each event in the GVLE, in addition to its parameters, is characterized by the time interval of its occurrence before (or after) the corresponding large earthquake, and by the distance from the hypocenter of the corresponding large event normalized to the source size of the corresponding large event.

The source size R (in kilometers) is calculated here as a function of the magnitude M of the corresponding large event following the relation [18]:

$$R = 10^{0.5M - 1.9}. \tag{1}$$

Further, the events are sorted by time before (or after) the moment of the corresponding large event. As the result of this procedure, the number of fore- and aftershocks in the GVLE increases sharply, and the influence of random variations of the seismic regime decreases. The experience of the GVLE construction shows, the details of the procedure, such as the specific form of relation (1), are not so essential.

The GVLE that displays several important regularities in precursor behavior is shown in figure 1 (earthquakes with depth H <70 km and Mw $\geq$ 5.4 from the global GCMT catalog are used). First of all, the power-law sequence of growth in foreshock number can be seen clearly. Besides, at distances up to a couple of source size of the main event, the weak long-term increase in seismic activity can be seen; this growth in earthquake number begins up to several (maybe up to 5–10) years before the moment of the generalized large earthquake (GLE) event. In figure 1 it can also be seen that the source area of a large event is surrounded by a zone of a decreased seismicity. This decrease in seismic activity occurs at a distance of about 4–6 source sizes around the GLE event and is seen during one to several days (see figure 1).

Besides the change in fore- and aftershocks number the set of anomalies of a few different characteristics of the seismic regime (such as b-value, mean earthquake magnitude value, mean earthquakes' depth and some others) were found in GVLE. All these anomalies increase in their amplitude A approaching the GLE time moment as minus logarithm of time Δt before (or after) the GLE time moment

$$A = a + b\, log(\Delta t), \tag{2}$$

where a, b – coefficients, $b < 0$.

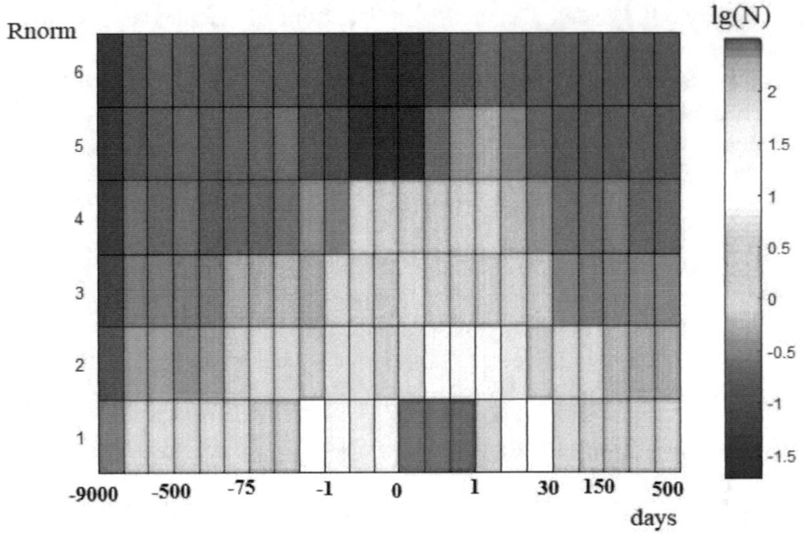

Figure. 1. Spatiotemporal density of number of events N in vicinity of generalized large earthquake. The spatial distance is normed by the source size of the corresponding main earthquake. X-axis – time before and after the GLE moment, Δt=0, days, earthquakes with $Mw \geq 5.4$, $H \leq 70$ km are used.

As an example, figure 2 shows the plots for the current mean magnitude Mw in the GVLE. It can be seen that mean Mw value increases in foreshock sequence approaching the GLE event time moment according relation (1).

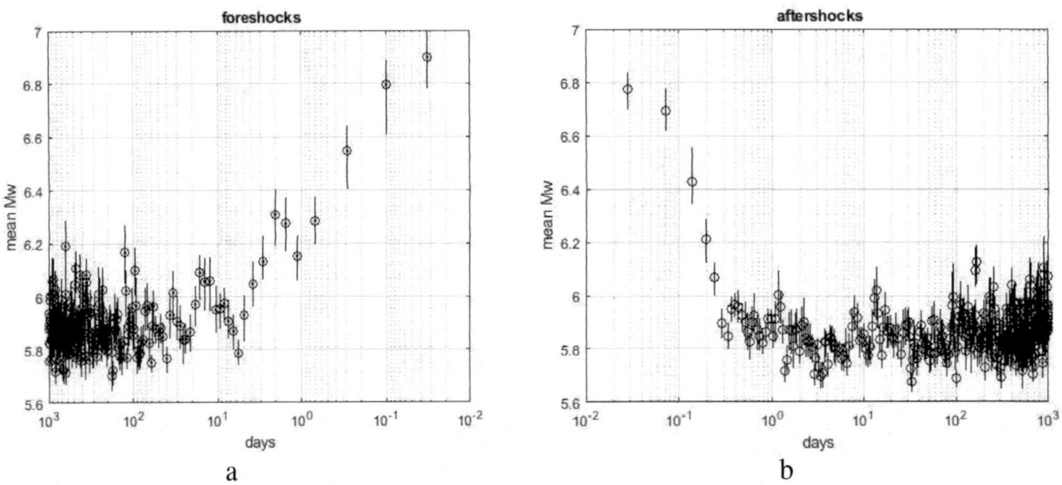

Figure. 2. The mean earthquake magnitude ($Mw \geq 5.4$, $H \leq 70$ km) in the GVLE, foreshocks (a) and aftershocks (b), GCMT catalog (1976–2019).

Figure 2 shows a linear increase in foreshocks' mean magnitude with Δt decrease (and increase in |log (dt)|) and a power-law increase in the average energy of foreshocks with the approach of the moment of the generalized main earthquake, respectively. We emphasize that this precursor anomaly can be interpreted almost as a tautology – "a sign of an increase in the probability of a strong earthquake is an increase in the average amplitude of current earthquakes."

Applying the GVLE method to regional data allows new results to be obtained. In this case, growth of foreshock activity as a rule became to be complicated by a number of strong bursts of seismicity arising near the moment of a generalized strong earthquake (GLE event). Figure 3 shows the distributions of time intervals *dt* between events for real main events and for a virtual random in time catalog with the same number of events. ISC catalog is used. The program by V.B.Smirnov of aftershock removing by Molchan-Dmitrieva algorithm [8] is applied. It can be seen that for both regions, the subduction zone of South America and the Mid-Atlantic Ridge zone, the probability of observed small time intervals between events is significantly higher than would be expected for a sequence of independent events. Note that this anomaly will not be so prominent with a linear scale along the y axis. Note also, that the anomaly is more pronounced for the subduction zone than for the mid-ocean ridge zone, an area of rock submelting. This tendency also appears to be valid for other similar seismogenic zones (for example, it has been identified for the Kuril-Kamchatka zone [14]). We suggest that when analyzing global seismicity, this clustering effect has been masked by overlapping data from different seismogenic zones.

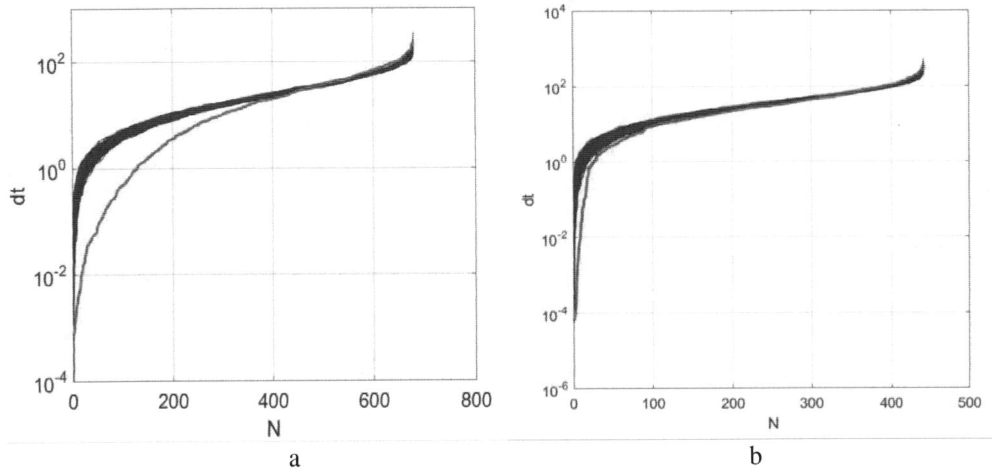

Figure. 3. Distribution of time intervals *dt* between successive main events for the subduction zone of South America (a) and for the region of the Mid-Atlantic Ridge (b). Plots for real main events (red line) and for 500 random realizations for the same number of events and the same time intervals (blue lines); *N* – number of intervals between events ordered by *dt* value.

Should be emphasized, that all the above precursor anomalies – signs of an increase in the probability of occurrence of a strong earthquake – are nonspecific. They can be expected to arise in development of bifurcations in systems of very different physical nature.

An anomaly specific for the case of earthquakes is shown in figure 4. This is an anomaly of a decrease in the average depth of earthquakes as the moment of the generalized main event approaches (H <70 km). This anomaly can be explained by the breakthroughs of the low-density deep fluid into the overlying horizons of the lithosphere because of growth of fracturing in the large earthquake focal zone. The development of such an anomaly testifies in favor of the widely spread model of existence of fluid in the source zones of many earthquakes [3, 16; and many others].

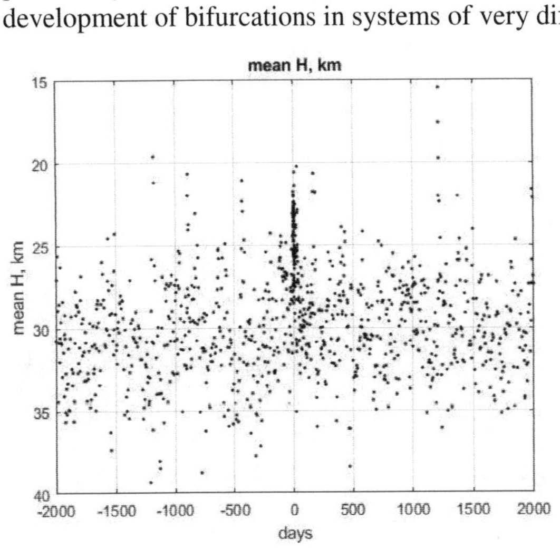

Figure. 4. Change in the mean depth of focus in the GVLE, ISC world catalog [13].

3. Conclusion

A complex of precursor anomalies has been identified by the GVLE method using data of world and regional earthquake catalogs. The list of identified anomalies includes a weak background increase in seismic activity, the development of a foreshock power-law cascade, increasing in the average magnitude of earthquakes, the "shadow" of precursor seismic activity, clustering of strong (and main) earthquakes and some other anomalies [10–13]. The clustering was examined in more detail for this region (Sakhalin and Kuril Islands); the presence of the clustering effect was shown both for the strongest earthquakes [14] and for a larger set of main events after the removal of aftershocks [15].

The majority of the precursor anomalies are nonspecific; their development can be expected in systems of very different nature. Indeed, based on a similar approach in [4, 5], a change in the administration of the White House (USA) was predicted. An analogue of clustering of strong (and main) earthquakes can be seen in the events of the Arab Spring 2010–2011.

Besides nonspecific predictive anomalies, a specific predictor of a decrease in the average depth of earthquakes was revealed by GVLE method. This effect is associated with the existence of the deep fluid in the source areas of earthquakes and with the breakthrough of this low-density fluid into the upper horizons of the lithosphere in connection with growth of fracturing during the preparation and realization of strong earthquakes.

The complex of nonspecific and specific precursor anomalies identified in the GVLE can be used as a basis for the development of earthquake prediction algorithms. With regard to the earthquake prediction problem in this region (Sakhalin and the Kuril Islands), this approach can be used in the future as an addition or alternative to the used earthquake prediction methods based among other methods on the method of calculating self-developing processes [5, 6] and the load/unload response ratio method [20]. The effect of developing of the self-developing process appears to be very similar with the foreshock cascade, the precursor anomaly concerned with the change in load/unload response ratio can be added into the list of precursor anomalies found in the GVLE and described briefly above.

Acknowledgments

This work was carried out for a state assignment of the Institute of Earthquake Prediction Theory and Mathematical Geophysics, Russian Academy of Sciences (topic AAAAA19-119011490129-0), Schmidt Institute of Physics of the Earth of the Russian Academy of Sciences (topic AAAA-A19-119110690110-6) and of the Institute of Marine Geology and Geophysics Far East Branch of the Russian Academy of Sciences (topic AAAA-A18-118012290125-2.2), and was supported by the Russian Foundation for Basic Research, project no. 19-05-00466.

References

[1] Bak P 1996 *How Nature Works* (New York: Springer-Verlag) p 212
[2] Bufe Ch and Perkins D 2005 *Bulletin of the Seismological Society of America* **95** 3 pp 833–843
[3] Gratier J-P, Favreau P, Renard F and Pili E 2002 *Earth Planets Space* **54** pp 1139–46
[4] Keilis-Borok V I, Lichtman A J 1981 *Proc. of the National Acad. of Sci.* **78** pp 7230−34
[5] Lichtman A J 2020 *Predicting the next president: The Keys to the White House* (Rowman & Littlefield Publishers) p 226
[6] Malyshev A I and Malysheva L K 2018 *Geosystems of Transition Zones* **2** 3 pp 141–153
[7] Malyshev A I, Malysheva L K 2021 Geosystems of Transition Zones **5** 2 pp 84–112
[8] Molchan G and Dmitrieva O 1992 *Geophys.J. Int.* (Moscow) **109** pp 501–516
[9] Molchan G M, Kronrod T L and Nekrasova A K 1999 *Physics of the Earth and Planetary Interiors* **111** 3–4 pp 229–240
[10] Rodkin M V 2008 *Journal of Volcanology and Seismology* (Moscow) **2** 6 pp 435–445
[11] Rodkin M V 2012 *Extreme Events and Natural Hazards: The Complexity Perspective Geophysical Monograph Series 196* (Berlin: Springer) pp 27–39
[12] Rodkin M V and Tikhonov I N 2016 *Physics Chemistry Earth* (Moscow) **95** pp 73–84

[13] Rodkin M V 2020 *Journal of Volcanology and Seismology* (Moscow) **14** 1 pp 58–69

[14] Rodkin M V, Andreeva M Yu and Grigorieva OO 2020 *Journal of Volcanology and Seismology* (Moscow) **14** pp 410–419

[15] Rodkin M V, Andreeva M Yu and Liperovskaya EV 2021 *Materials of the 13th Int. school-conf. Problems of Geocosmos March* 24–27, 2021 (St. Petersburg, Russia) pp 279–288

[16] Role of water in earthquake generation 2001 *Special Issue Bull. Earth. Res. Inst.* **76** 3–4

[17] Romashkova L L, Kosobokov V G 2001 *Journal of Volcanology and Seismology* (Moscow) **32** pp 162–189

[18] Sobolev G A 1993 *Osnovy prognoza zemletryasenii (Principles of Earthquake Prediction)* (Moscow: Nauka, Russia) p 313

[19] Sornette D 2000 *Critical phenomena in natural sciences* (Berlin: Springer) p 434

[20] Zakupin A S and Boginskaya N V 2021 *Geosystems of Transition Zones* **5** 1 pp 27–45

Lithospheric plate tectonics and mass extinctions of biological species

V V Snakin

Institute of Fundamental Problems of Biology of the Russian Academy of Sciences, Moscow, 142290, Russia

E-mail: snakin@mail.ru

Abstract. The merging of lithospheric plates and the formation of supercontinents are considered to be the main causes of global species extinctions within the Earth's biosphere. Under those conditions, the factor of geographic isolation is diminished and interspecies competition is accelerated, allowing for the survival of the best-adapted species. The divergence of lithospheric plates triggers a new spurt of speciation that surpasses the previous one, as it involves the participation of the winning species.

1. Introduction

The development of life on our planet, over the billions of years of the biosphere's existence, has been under the continuous influence of geological processes, both catastrophic in character (e.g., volcanism, earthquakes, etc.) and the much slower but continuously manifesting processes, such as the horizontal and vertical movements of the Earth's crust. Such movements cause changes in climate, relief, and ecological conditions, affecting also the factor of inter-species isolation.

One of the most interesting evolutionary problems, which has not been solved thus far, is the clarification of the causes of global mass extinctions of biological species, which took place regularly in the evolution of life on the Earth.

The present work aims to analyze the relationship between the slow movements of the Earth's crust altering the degree of geographical isolation of biological species and mass extinctions, when paleontologists note the disappearance of sometimes more than half of the species in the biosphere.

2. The Global Dynamics of Biodiversity

In the process of biospheric evolution, the number of species has continuously increased: from the Cambrian to the Silurian, the number of marine animal families increased from five to 400 over the course of 200 million years. By the Quaternary, the number of marine animal families alone had increased to 750. At the same time, the dynamics of biodiversity, as well as the whole evolutionary process, were characterized by abrupt starts: there were several large bursts of speciation and extinction, when whole families, orders, and even types of organisms appeared or disappeared. Modern science has identified several mass extinctions that occurred over the past 500 million years.

The best-known, though not the greatest of these, the Cretaceous-Paleogene extinction (~65 million years ago) took out more than half of all the species, including the dinosaurs. During the "great" Permian extinction (~250 million years ago), approximately 90 percent of all the species went extinct.

2.1. Why, what for, and how do extinctions occur?

There are no unequivocal answers as yet to these basic questions that arise when analyzing the dynamics of biodiversity. According to Charles Darwin [1], we needn't marvel at the fact of extinction, for the extinction of species and entire groups of species, which has played such an outstanding role in the history of the organic world, is an almost inevitable consequence of the principle of natural selection. According to V.A. Krasilov [2], "extinction is a means of regulating diversity in variable conditions so as to reduce the competition".

From evolutionary positions, the cause of species extinction is considered to be their constant improvement for the purposes of maximizing their use of the environmental resources, even as the environment itself is altered by new species, in their own favor. In the process of the biosphere's evolution, the number of species increased, while the biosphere expanded into areas previously unoccupied by life, involving new substances in its growing orbit of activity, and utilizing solar energy and chemical compounds more and more efficiently. Academician M.A. Fedonkin [3] noted: "For the biosphere, extinctions were as good as it is good for a species to have aging individuals die naturally from old age. In both cases, carriers of faulty hereditary information that might hamper evolution are cut off".

In the period accessible to the paleontological record, five mass extinctions are distinguished, as outlined in table 1. Work [4] contains a proposal to add a sixth extinction to this list, that being in the Carnian pluvial period (~233 million years ago). In work [5], the modern period associated with human activity is proposed as the sixth mass extinction.

Table 1. Major mass extinctions and the geodynamic situation (after [6, 7]).

Major mass extinction (scale of the extinction)	When it took place (millions of years ago, or "m.y.a.")	Supposed causes of the mass extinction
Ordovician-Silurian mass extinction (~85 % of all the living species eliminated)	Paleozoic Era, End-Ordovician (~440–445 m.y.a.)	- The onset of glacial cycles on the Earth, with corresponding changes in the sea level; - Changes in atmospheric and oceanic chemistry related to the rise of the Appalachian Mountains
Late Devonian mass extinction (~80 % of all the living species eliminated)	Paleozoic Era, End-Devonian (~383–359 m.y.a.)	- Climate change, possibly related to the diversification of land plants; - Decrease in oxygen levels in the deep ocean
Permian-Triassic mass extinction (~ 96 % of all the living species eliminated)	Paleozoic Era, End-Permian (~ 250–252 m.y.a.)	- Volcanic activity; - Climate change; - Decrease in oxygen levels in the deep ocean; - Changes in atmospheric chemistry; - Changes in oceanic chemistry and circulation
Triassic-Jurassic mass extinction (~50 % of all the living species eliminated)	Mesozoic Era, End-Triassic (~ 200 m.y.a.)	- Volcanic activity
Cretaceous-Paleogene (K-T) Mass Extinction (~75 % of all the living species eliminated)	Mesozoic Era, End-Cretaceous (~66 m.y.a.)	- Asteroid impact; - Volcanic activity; - Climate change; - Changes in atmospheric and oceanic chemistry

The causes of all the just-mentioned mass extinctions are attributed to absolutely different phenomena of predominantly catastrophic nature. The most popular of them are volcanism, trap magmatism, cosmic catastrophes (e.g., collisions with asteroids), change of magnetic poles, climate change, etc.

Another cause of one of these extinctions (the Late Devonian) has been recently proposed, that being the influence of a supernova explosion and the correspondingly powerful ionizing radiation on the ozone layer of the planet [8].

The just-listed heterogeneous causes are attributed to different cases of mass extinctions of species, and none of them happens to be universal.

2.2. The peculiarities of mass extinctions

An analysis of the nature of mass extinctions allows us to note two characteristic features:

1. after a sharp decrease in the diversity of biological species in the course of their extinction, biodiversity recovers at the next stage of evolution, subsequently exceeding the pre-crisis level;
2. the duration of mass extinctions, sometimes estimated at millions of years (the Late Devonian extinction, according to estimates in work [9], lasted approximately thirteen million years), a figure incommensurate with the typical durations of catastrophic events and their aftermath (those, as a rule, constitute several years).

It remains an open question why the events indicated in table 1 have caused mass extinctions only some of the time, although they have occurred much more frequently. Giant eruptions of super-volcanoes had occurred regularly; the Yellowstone super-volcano erupted relatively frequently: 640,000 years ago, and before then, 1.3 and 2.1 million years ago.

In our view, since mass extinctions have been repeated many times in the history of the biosphere, there must be a single major cause of global fluctuations in the number of species. Without downplaying the role of catastrophic events in the decline of many species, we try to look at mass extinctions from an evolutionary perspective, taking into account the horizontal and vertical movements of the lithospheric plates. These movements lead to the unification and disconnection of continents and islands and, consequently, to changes in the geographical isolation of biological species. In our earlier works, [10] and [11], we have already put forward some considerations in this regard.

3. The Proposed Mechanism of Mass Extinctions

We propose that the relatively slow movements of the lithospheric plates, due to which continents and islands periodically unite and separate, and large continental associations, up to a single super-continent like Pangea (figure 1), Monogea, Rodinia, etc. are formed, are the main cause of mass extinctions.

The movements of lithospheric plates substantially affect the isolation factor that plays a very important role in speciation. Species populations find themselves either united over vast expanses or isolated on separate continents and islands.

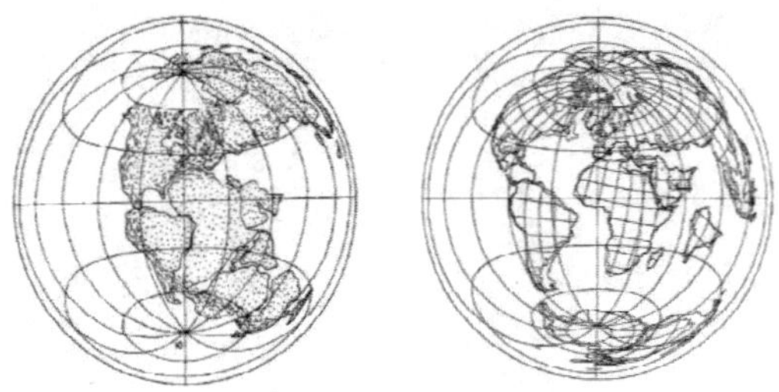

Figure 1. Wegener's reconstruction of the Pangea super-continent (approximately 200 m.y.a.) and its separation (approximately 60 m.y.a.) (after [12]).

It is important that the periodic merging and divergence of the lithospheric plates had a regular character. One piece of evidence for this periodic merging of the lithospheric plates, with the formation of the super-continents like Pangea, Gondwana, Rodinia, Columbia etc., is represented in figure 2.

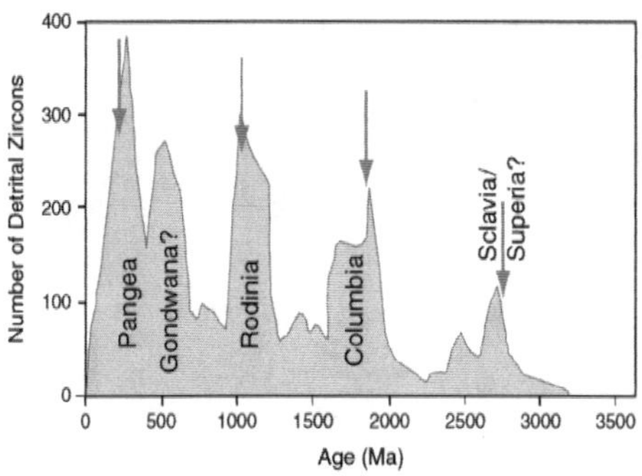

Figure 2. Comparison of orogenic peaks (arrows) observed in radiometric data with spectra of U–Pb detrital zircon crystallization ages and proposed times of super-continent assembly [13, 14].

3.1. Ecological consequences of the unification of lithospheric plates

In ecological terms, the unification of lithospheric plates leads to the removal of geographical barriers and to a reduction of geographical isolation of species.

This involves not only change in the climatic and geomorphological conditions, but also the unification of several species occupying the same ecological niche, who come together in the unified space. As a result, interspecific competition increases and the rule of competitive exclusion comes into effect (the rule of competitive exclusion, or the Voltterra-Gauze principle, was formulated in 1926 by Vito Voltterra, based on his study of a mathematical model of the dynamics of two populations competing for the same food resource. In 1931–1935, George F. Gause showed by experiments on protozoa how competitive exclusion of one species by another occurs. The discussion of the competitive exclusion rule played an important role in the development of the *ecological niche* concept and the ecological and geographic model of speciation, as well as in the evaluation of interspecific competition as a factor maintaining the structure of communities [15]). According to this rule, two or more species cannot coexist stably in a limited space if they occupy the same ecological niche. The process of competition always proceeds towards complete displacement of one species by another. Species less adapted to the environmental conditions then die out.

Another important factor of continental unification is a reduction in the number of ecological niches, due to a decrease in the diversity of climate and other geographical conditions, which also reduces the species' prospects for continued existence.

Hence, the unification of lithospheric plates entails the association of species in similar ecological niches, and the reduction of their diversity trigger the mechanism of interspecific competition. In the further process of species extinction, a variety of factors take on significance, including physiological characteristics and the degree of adaptation to food resources and to altered climatic and other geographical features. Theoretically, for instance, the union of six continental plates could result in up to six different species sharing the same ecological niche, five of which are evolutionarily doomed to extinction, owing to interspecific competition. Similar processes take place in the oceanic expanses, where several geographically isolated oceans and seas are replaced by a single world ocean (e.g., the Tethys). The area of the life-saturated shelf zone then changes dramatically.

As the continents drift apart, the surviving, most evolutionarily developed, or adapted, species, during continued drift in the conditions of geographical isolation, give rise to a new round of evolution, surpassing the previous one. Following this, continental convergence under new geographical conditions leads to a new cycle of extinction and subsequent increase in biodiversity when the continents diverge. Thus, the unification of continents and the accompanying extinction contribute to the preservation and further evolution of the most adapted species, which, during the subsequent separation of the continents, have given rise to new and even more promising species.

A comparison of the timing of mass extinctions and collisions of lithospheric plates (table 2) shows that almost all of the five mass extinctions under consideration have occurred at the time of significant lithospheric plate unions up to the formation of super-continents.

Table 2. Lithospheric plate collisions during mass extinctions.

Major mass extinction	The associations of the continents
Ordovician-Silurian	The formation of Lavrusia (Euramerica), a Paleozoic super-continent, as a result of the collision of the North American (ancient continent of Laurentia) and East European (ancient continent of Baltica) platforms during the Caledonian orogenesis (500–400 m.y.a.)
Late Devonian	In the Devonian, the northern continents formed a single large continent, Atlantia, to the east of which was Asia. Gondwana continues to exist
Permian-Triassic	In the second half of the Carboniferous and the Permian (~300–180 m.y.a.), the Pangea formed. This period is long and occurred in several stages
Triassic-Jurassic	The collision of the continents that broke away from Gondwana with parts of Eurasia
Cretaceous-Paleogene (K-T)	In the Paleogene (~65 m.y.a.) the territory of Eurasia expanded: the Zondian Archipelago joined the continent, the Balkan Peninsula was united with Asia Minor, Europe joined Africa in the area of modern Gibraltar, and in the north-west the conjunction with North America occurred

4. Vertical Tectonic Movements and Biodiversity

Similarly to the horizontal movements of lithospheric plates, upward and downward movements of the Earth's crust lead to transgression and regression of the seas or, in other words, to processes of unification and decoupling of territories, with concurrent changes in the factor of geographical isolation and in biodiversity.

The situation with Late Pleistocene megafauna extinction 20–24 thousand years ago (about 36 % of large mammals died out in Northern Eurasia, in North America – about 72 %) may be explained by the periodic formation of "Eurasia-America" supercontinent due to the periodic formation and disappearance of the Bering Isthmus. It is known that, over the past 3 million years, the territory of Beringia has risen and gone underwater about six times [16]. Eurasia and North America separated for the last time 10–11 thousand years ago.

Before that, the isthmus existed for 15–18 thousand years. Of the large herbivores, the mammoth and the woolly rhinoceros became extinct at that time; of the large carnivores, the giant sloths, saber-toothed lions, and cave bears became extinct.

5. Globalization as a Factor in the Extinction of Species

The proposed mechanism of mass extinction is also applicable to explaining the current situation with biodiversity on the planet. The processes of globalization, which are actively developing thanks to human activity, generally act similarly to the effects of unification of continents, reducing the effect of geographical isolation. The intensification of both accidental and deliberate transport of animals and plants with the help of humans (the processes of invasion and introduction) leads to competitive struggle between species that find themselves in the same ecological niche, with the inevitable extinction of the less adapted. There is a steady increase in the number of invasive species (figure 3), which undoubtedly contributes to the extinction of native species. In the context of globalization, island populations are particularly endangered.

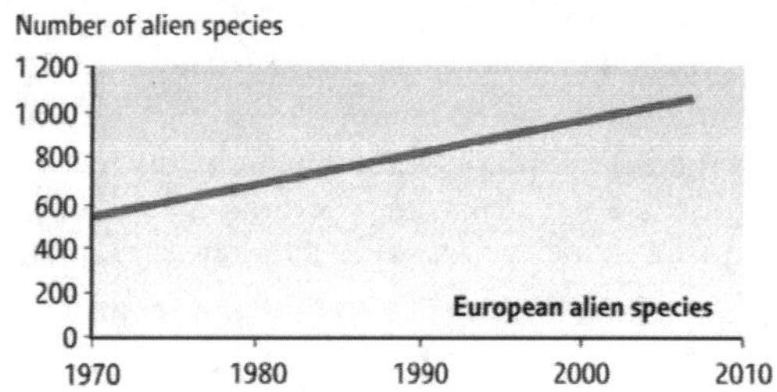

Figure 3. Numerical dynamics of invasive species in Europe [17].

6. Conclusion

The tectonic movements of the Earth's crust, which cause periodic unification and separation of the lithospheric plates and consequently change the degree of geographical isolation of species, are proposed to be the main factor triggering the process of mass extinctions of species that has repeatedly occurred in the history of the biosphere.

The extinction mechanism is realized by virtue of increased interspecific competition and the death of the less adapted species as the diversity of available ecological niches decreases, and different species find themselves in competition over any given ecological niche, according to the rule of competitive exclusion. This occurs when lithospheric plates merge and especially when super-continents come into being. When lithospheric plates disconnect once again, a new progressive leap in speciation occurs, as the most competitive among the surviving species now take part in the process.

The removal of the isolation factor also explains the phenomenon of the Late Pleistocene extinction of megafauna in Northern Eurasia and North America, as a result of sea transgression and regression and of the periodic unification of the Old and the New Worlds.

Like the unification of the lithospheric plates, human activity has greatly reduced the degree of species isolation. In the period of globalization, migration processes have significantly accelerated due to species introduction and invasion, which inevitably entail an increase in interspecific competition, a decrease in biodiversity, and may ultimately become the basis of a new mass extinction.

Catastrophic natural processes referred to in the literature as causes of mass extinctions undoubtedly played and continue to play a major role in the reduction of the number of species inhabiting the affected regions, but they cannot fully explain the regularity of global failures in the dynamics of biodiversity and their duration. Such cycles of mass extinction during continental unification, followed by spurts of speciation after continental separation, constitute evolutionary progress for the biosphere, as the most competitive of the surviving species take part in the new round of speciation.

Further clarification of periods of mass extinction and periods of unification and divergence of lithospheric plates will make it possible to create a more accurate model of biodiversity dynamics, which must ultimately be taken into account in the development of conservation policy.

References

[1] Darvin Ch 1859 *The origin of species by means of natural selection, or the preservation of favoured races in the struggle for life* (London: John Murray) 574 p
[2] Krassilov V A 1992 *Nature Conservation: Principles, Problems and Priorities* (Moscow: VNIIPriroda) 173 p
[3] Fedonkin M A 1991 Biosphere: the fourth dimension *Priroda* **9** pp 10–18
[4] Dal Corso J et al 2018 Multiple negative carbon-isotope excursions during the Carnian Pluvial Episode (Late Triassic) *Earth-Science Rev.* **185** pp 732–750
[5] Ceballos G et al 2015 Accelerated modern human–induced species losses: Entering the sixth mass extinction *Science Advances* **1 5** pp 1–5
[6] Greshko M and National Geographic staff 2019 What are mass extinctions, and what causes them? *IOP Publishing PhysicsWeb* https://www.nationalgeographic.com/science/prehistoric-world/mass-extinction/
[7] Scoville H 2020 The 5 Major Mass Extinctions *IOP Publishing PhysicsWeb*
[8] Brian D Fields et al 2020 Supernova triggers for end-Devonian extinctions *PNAS*
[9] Grzegorz Racki 2020 A volcanic scenario for the Frasnian–Famennian major biotic crisis and other Late Devonian global changes: More answers than questions*? Global and Planetary Change* **189** pp 103–174
[10] Snakin V V 2016a Geographic isolation of species as a factor of global biodiversity dynamics *Zhizn' Zemli* **38** pp 52–61
[11] Snakin V V 2016b. Mass Extinctions in Biosphere History: Another Hypothesis *Izvestiya RAN Ser. Geograph* **5** pp 82–90

[12] Sorokhtin O G, Chilingarian G V and Sorokhtin N O 2011 Evolution of Earth and its Climate. Birth, Life and Death of Earth *Developments in Earth & Environmental Sciences* (Elsevier) **10** 353 p
[13] Meert J G 2012 What's in a name? The Columbia (Paleopangaea/Nuna) supercontinent *Gondwana Research* **21** pp 987–993
[14] Nance R D Murphy J B and Santosh M 2014 The supercontinent cycle: A retrospective essay *Gondwana Research* **25** pp 4–29
[15] Snakin V V 2020 Ecology, global environmental processes and biosphere evolution: Encyclopedic dictionary (Moscow University Press) 528 p
[16] Lorenzen E D et al 2011. Species-specific responses of Late Quaternary megafauna to climate and humans *Nature* **479** pp 359–364
[17] Global Biodiversity Outlook 3 2010 (Montreal: Secretariat of the Convention on Biological Diversity) 95 p

IOP Conf. Series: Earth and Environmental Science **946** (2021) 012010 doi:10.1088/1755-1315/946/1/012010

Approaches to assessing the effectiveness of earthquake predictions using the LURR (load-unload response ratio) method on Sakhalin

A S Zakupin

Institute of Marine Geology and Geophysics Far Eastern Branch of Russian Academy of Science, Yuzhno-Sakhalinsk, 693022, Russia

E-mail: a.zakupin@imgg.ru

Abstract. A retrospective analysis of the seismicity of Sakhalin from 1997 to 2019 was performed to demonstrate the possibilities of the LURR technique recently in previous our work. The following results were obtained: 84 % of earthquakes (16 out of 19, with $M \geq 5$) are predicted, 25% alarms (4 out of 15 predicted areas) were false. This paper proposes an analytical dependence to describe the forecast effectiveness (Ke) for this research. The extremes of Ke were found at the value of the alarm period of 12 and 24 months. At the same time, Ke is significantly higher for the alarm period of 24 months and decreases after a two-year alarm period. Another way to prove the results obtained is the random spatio-temporal distributions of the predicted objects (19 earthquakes with $M \geq 5$). 10 such random sets have been assigned to 15 predicted areas, the result shows a significant advantage of a real sample over random ones, and also practically confirms the reliability of the algorithm for using the LURR technique. The methodology and results of this work can serve as practical recommendations for working with the LURR method for seismologists.

1. Introduction

In 2021, the paper [1] was published in the journal "Geosystems of Transition Zones", in which the LURR (Load-Unload Response Ratio) [2] earthquake prediction method was retrospectively applied for Sakhalin seismicity. For the first time for Sakhalin, such a detailed analysis was made for a continuous time period (1997–2019) and on the territory of the entire island. It is very important that after a number of works on this topic [3–5] the authors have systematized the materials, bringing all the calculations to a single database of seismological data according to the catalog of the Sakhalin branch of the Geophysical Survey of the Russian Academy of Sciences. It should be noted that this study aroused a great interest not only among the predictors, a theoretical review of the LURR method, for example, was published almost immediately [6]. So, in the study [1], the Sakhalin territory was scanned with calculated areas in the form of circles with a radius of one degree, but unlike previous works, this was done with a high resolution (number of areas). The entire territory was covered by such zones in increments of 0.5 degrees in latitude and longitude, and the grid was detailed to 0.1 degrees in the three most dangerous seismic zones. A total of 323 calculated samples were processed on the territory of the island (of which 119 were basic and 204 were detailed). The number of earthquakes satisfactory for calculation contained 36 main and all detailed samples. In total, 77 anomalies in the LURR parameter were revealed in 36 main areas over 22 years (1997–2019). Additional calculations (step is 0.1°) helped to identify 5 more anomalies. After identifying the

IOP Conf. Series: Earth and Environmental Science **946** (2021) 012010 doi:10.1088/1755-1315/946/1/012010

anomalies of 2019 (it is impossible to retrospectively close the alarm period for them), 69 anomalies were grouped as follows. The calculated areas with anomalies that appeared with small time difference from each other were combined into the forecast zone. The beginning of the alarm period was determined for each forecast zone (from the moment of appearance of the first anomaly in the group). In total, 15 forecast zones have been established on the island for 22 years, which differ in the number of calculated areas and the area of the covered territory. For example, figure 1 shows a large alarm zone of 12 calculated areas and a small one of 2 calculated areas.

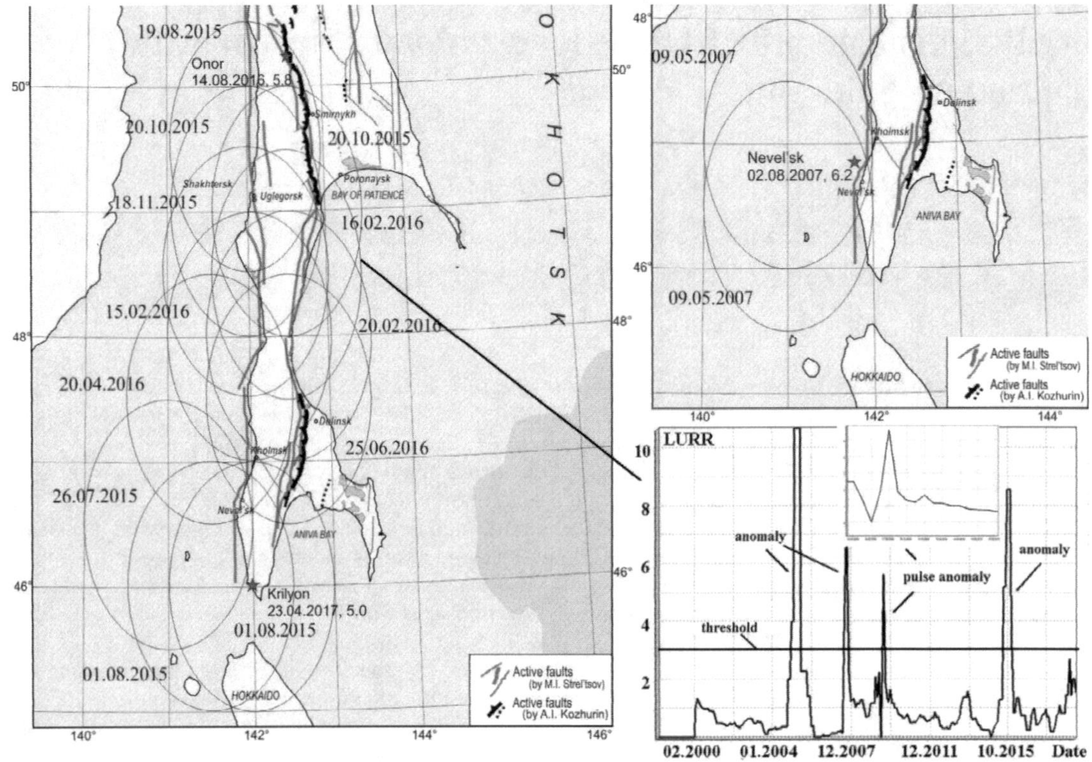

Figure 1. Examples of forecast zones (large and small) with a LURR parameter graph for an arbitrary calculation area.

Analysis of the data on the alarm zones for the three-year alarm period showed 17 earthquakes out of 19 with $M \geq 5$ to be retrospectively predicted. Also, 4 out of 15 periods turned out to be false, that is, for three years, no any earthquake with $M \geq 5$ occurred in these 4 zones. Thus, the result of this work turned out to be worthy of the forecast topic – 75 % of the forecasts were successful for 89 % of earthquakes.

Despite the good results, it should be noted that the main problem for the qualitative statistical evaluation of the results was and is the uncertainty of the forecast parameter – the alarm period. It is selected arbitrarily, but it is fixed for all forecast areas. In the work [1], a period of three years was taken as a basic level. At the same time, calculations showed that the waiting period (the time from the alarm announcement to the earthquake) was in most cases small, but alarm periods of three years or more [1] lead to their intersection in space and time. Note that the number of false alarms and missed goals depends on this value (the duration of the alarm).

2. Methods and results

In this paper, we will try to evaluate the effectiveness of forecasts using the LURR method, taking into account all the above aspects. To begin with, we will do this at the simplest level for predictive topics in the terminology of capturing, missing a target and false alarms, but we will get these parameters depending on the duration of the alarm period. Recall that the alarm period is announced for the alarm area for a certain fixed period of time, while the waiting period for an earthquake may be significantly less. For alarm periods from 6 to 36 months (in increments of 3 months), we will determine the number of completed forecasts. The minimum alarm period of six months is justified, because with a sliding window of 360 days, the period for which a forecast can be made cannot be less than half of it (the window), that is, 180 days (6 months).

The analysis results of the forecasts efficiency are shown on the graph (figure 2). We will add the cases of alarm periods overlapping for several forecast zones to the indicators of capturing and missing a target and false alarms.

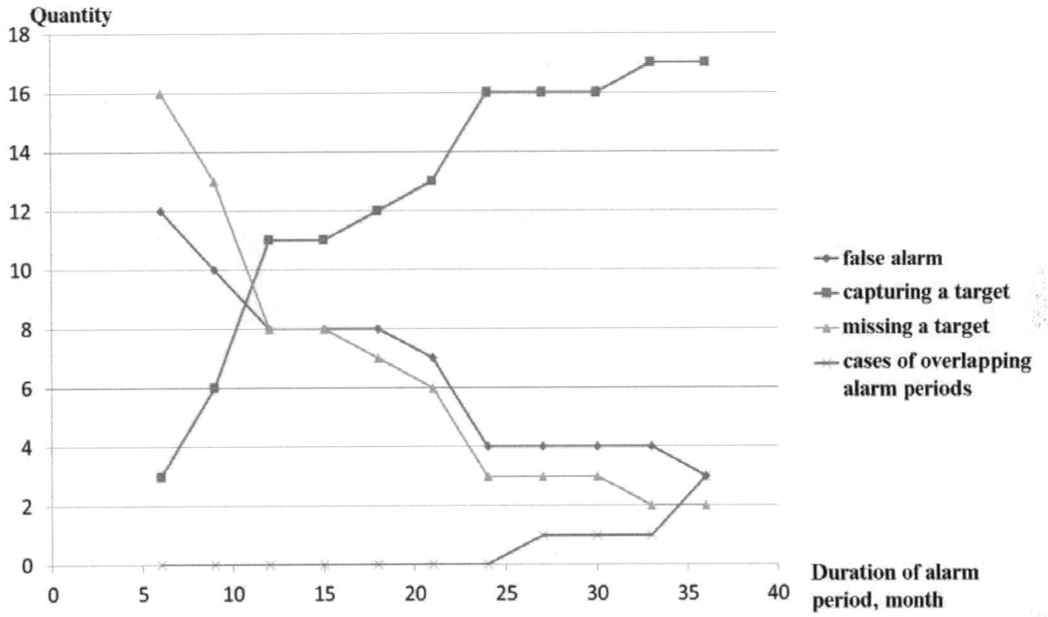

Figure 2. Forecast parameters.

Starting from the period of 27 months, the first cases of overlap between the alarm periods with each other appear and this progresses in the future. Thus, the maximum efficiency can be achieved within a period of 24 months, while avoiding overlaps. Such estimates are quite acceptable, but the announced forecasts have different initial conditions also for the area where anomalies are noticed. It turns out that the quality of the forecast is determined not only by the time of implementation, but also by the area of the alarm zone. For example, the M 7 Uglegorsk earthquake – one of the strongest events in the central region of Sakhalin – has occurred 6 months after the anomaly, and the anomaly was noted in only one calculated area. This is one of the best retrospective results. In order to take into account both of these parameters, for each alarm zone, we will determine the proportion of the area that is involved in the forecast from the total area where anomalies are being searched (36 calculated areas).

$$K_e = \left(\frac{N}{N_0}\right) \times \left[\frac{S_{et} \times n}{\sum_{i=1}^{n} S_{si}}\right] \times \left[\frac{N_z \times S_0}{N_l \times \sum_{i-1}^{k} S_{lk} \times X^2}\right], X = \frac{\left(\sum_{i=1}^{n} S_i\right) \times T_{al}}{S_0 \times T_0}, \quad (1)$$

Where N – the number of earthquakes in the alarm zones during the considered alarm period T_{al}; N_0 – the number of objects (earthquakes) of the retrospective forecast; S_{et} – the area of the reference forecast zone (the minimum possible equal area of one ellipse); n – the number of alarm zones (in our case there are 15 of them); S_{si} – the area of the i-th alarm zone where the target was captured during this alarm period T_{al}; N_z – the number of alarm zones where a target was captured during this T_{al} alarm period; N_l - the number of alarm zones where there are no earthquakes with $M \geq 5$ during this alarm period T_{al} (false alarms); S_0 – the total area of all 36 calculated areas (taking into account the zones overlap, the real area is considered); T_0 – total observation period (1997-2019); T_{al} - alarm period

S_{lk} – the area of the k-th alarm zone, where there are no earthquakes with $M \geq 5$ during this alarm period T_{al}; X is a coefficient that indicates the degree of anxiety of the studied territory for the entire observation period, which determines the probability of an arbitrary earthquake accidentally hitting the alarm zones for a given alarm period.

Figure 3 shows two types of calculation (K_e), the difference of which is in the calculation of the anxiety coefficient X. According to the first type, the area of the alarm zone is multiplied not by the declared alarm period T_{al} (except for false alarms), but by the time spent before the implementation. According to the second type, the area of all zones is multiplied by the alarm period T_{al}.

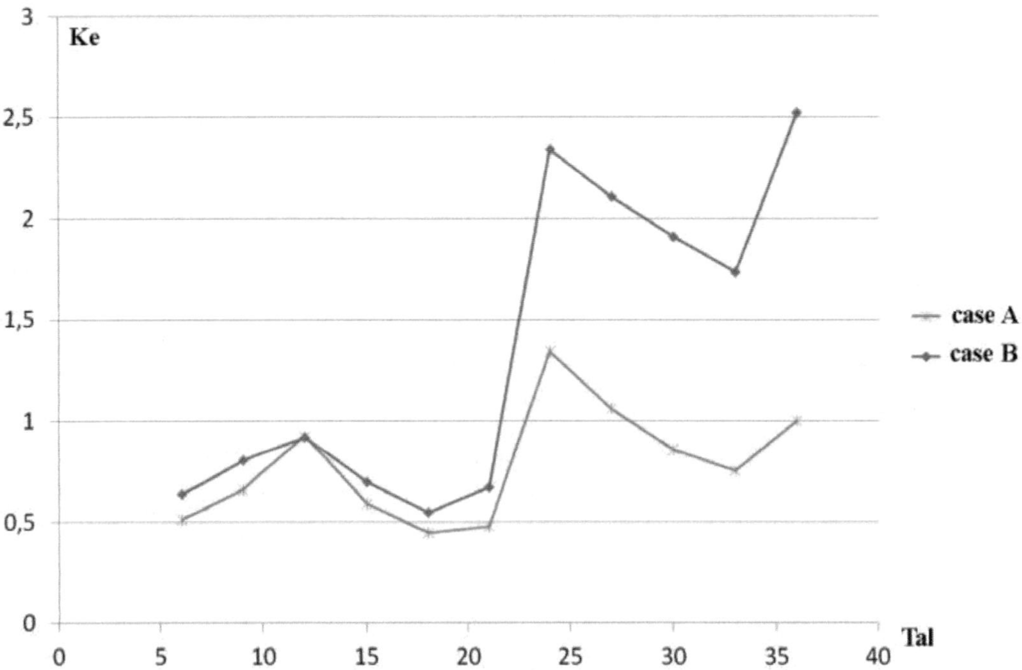

Figure 3. The forecast effectiveness depending on alarm period.

Note that the values of X, and, accordingly, the values of K_e are noticeably distorted when the alarm zones massively overlap during large alarm periods. Thus, the acceptable coefficient values can only be obtained before the alarm period of 24 months. The same period turns out to be the most favorable for obtaining the best predictive statistics (figure 2). We use this result for the next check of the randomness (non-randomness) of the presented retrospective forecasts. We used a random number generator (the generator is available on the website https://randomus.ru) to get 10 sets of 19 earthquakes in the period from 1997 to 2019. Each earthquake has a date (month and year) and its location generated. Despite the fact that an earthquake could occur anywhere in Sakhalin, we will complicate our case, giving the generator a choice of 36 zones where anomalies are guaranteed to be present (table 1). This is despite the fact that a total of 119 zones were considered, but not all of them turned out to have enough earthquakes to perform the calculation using the LURR method [1].

Table 1. Calculated areas for the retrospective forecast of earthquakes M = 3.3-5 of Sakhalin Island (1997–2019).

	Latitude (°N)	Longitude (°E)	The number of earthquakes in the region	
			in total	M = 3.3–5
1	46.0	141.5	2091	458
2	46.0	142.0	2324	520
3	46.0	142.5	1814	382
4	46.5	141.0	1588	335
5	46.5	141.5	2469	474
6	46.5	142.0	3082	569
7	46.5	142.5	3401	1683
8	46.5	143.0	640	327
9	47.0	141.0	1433	322
10	47.0	141.5	2396	420
11	47.0	142.0	3258	539
12	47.0	142.5	3241	532
13	47.5	141.5	1843	378
14	47.5	142.0	2727	486
15	47.5	142.5	2123	310
16	48.0	142.0	1378	347
17	48.0	142.5	1479	377
18	48.5	142.0	996	337
19	48.5	142.5	994	340
20	49.0	142.0	1006	370
21	49.0	142.5	1022	365
22	49.5	142.0	864	325
23	49.5	142.5	835	332
24	52.0	142.0	706	334
25	52.0	142.5	982	500
26	52.0	143.0	893	448
27	52.5	142.0	922	484
28	52.5	142.5	1170	617
29	52.5	143.0	1094	578
30	52.5	143.5	732	381
31	53.0	142.0	989	521
32	53.0	142.5	1029	535
33	53.0	143.0	933	481
34	53.0	143.5	769	419
35	53.5	142.5	880	462
36	53.5	143.0	808	433

Thus, if an earthquake occurs in one of the 36 zones within 2 years after the appearance of the anomaly, it is associated with a successful forecast (S), if not, then with a not successful forecast respectively (UnS). The final result for 10 sets is presented in Table 2.

Table 2. Model earthquakes obtained by a random number generator (the first sample).

Set	1	2	3	4	5	6	7	8	9	10
S/UnS	10/9	9/10	11/8	11/8	12/7	8/11	7/12	12/7	10/9	7/12

3. Conclusion

The result of modeling with random samples for Sakhalin is indicative. The ratio of successful forecasts to unsuccessful ones is close to 1 on average (the randomness of the generated samples is obvious). The most successful implementations of forecasts were 2 samples out of 10 with a 63% implementation percentage (against real sample with 89%). With such results, it already seems pointless to consider the number of false alarms, since this can only further spoil the already bad results. Such an experiment can partially prove that the predictions made using the LURR method are not random. At the same time, the proposed assessment of the LURR method efficiency remains at the qualitative level. However, the results obtained for Sakhalin Island allow us to consider it empirically reliable, especially in view of the practical implementation of two forecasts proposed in real time (for the Onor and Krilyon earthquakes). As a result of the work performed, a method for evaluating the effectiveness of the previous studies is proposed, which allows determining the optimal value of the alarm period and using this in real-time forecasting.

Acknowledgments
The data used in the work were obtained with large-scale research facilities «Seismic infrasound array for monitoring Arctic cryolitozone and continuous seismic monitoring of the Russian Federation, neighbouring territories and the world» (http://www.gsras.ru/unu/).

References

[1] Zakupin A S and Boginskaya N V 2021 Mid-term earthquake prediction using the LURR method on Sakhalin Island: a summary of retrospective studies for 1997–2019 and new approaches *Geosystems of Transition Zones* **5(1)** 27–45
[2] Yin X C, Zhang L P, Zhang H H, Yin C, Wang Y, Zhang Y, Peng K, Wang H, Song Z, Yu H and Zhuang J 2006 LURR's twenty years and its perspective *Pure Appl. Geophys.* **163** 2317–2341
[3] Zakupin A S, Levin Yu N, Boginskaya N V and Zherdeva O A 2018 Development of medium-term prediction methods: a case study of the August 14, 2016 Onor (M = 5.8) earthquake on Sakhalin *Russ. Geol. Geophys.* **59(11)** 1526–1532
[4] Zakupin A S, Bogomolov L M and Boginskaya N V 2020 Using the load/unload response ratio and self-developing processes methods of analyzing seismic sequences to predict earthquakes in Sakhalin *Izv. Atmos. Ocean. Phys.* **56(7)** 693–705
[5] Zakupin A S and Boginskaya N V 2020 Mid-term assessments of the seismic hazard on Sakhalin Island by the LURR method: new results *Geosystems of Transition Zones* **4(2)** 160–177
[6] Rebetsky Yu L Concerning the theory of LURR based deterministic earthquake prediction *Geosystems of Transition Zones* 2021 **5(3)** 192–222

Geodynamic risks of developing rockburst-hazardous deposits in the Far Eastern region

Yu V Fedotova*, P A Anikin and M I Potapchuk

Mining Institute of the FEB RAS, Khabarovsk Federal Research Center, Khabarovsk, 680000, Russia

E-mail: *fjulia@mail.ru

Abstract. Geodynamic activity of the Earth's crust depends on the parameters of the movement of lithospheric plates and stress fields, both inherited in aseismic areas and modified in seismic active ones. Geomechanical processes occur in rock mass under the influence of various natural (endogenous and exogenous) and man-made (anthropogenic) factors. The degree of influence of these factors on the change in the energy saturation of blocks of various ranks will depend on various factors. In the geological environment of natural and man-made systems, there are dynamic phenomena of various energy levels from acoustic noise (microseismic) to rockburst and mining-induced earthquakes, that is, there is always a geodynamic risk in the geological environment of natural and man-made systems, especially in mining ones. The conditions of the geodynamic risks occurrence at specific deposits are considered on the example of rockburst hazardous deposits in the Far Eastern region of the Russian Federation. The influence of natural seismic activity on the realization of mining-induced seismicity is shown.

1. Introduction

Geodynamic activity of the Earth's crust depends on the parameters of the movement of lithospheric plates and stress fields, both inherited in aseismic areas and modified in seismically active areas. For rock massifs represented by the most solid strong and brittle rocks capable of accumulating significant stresses and unloading in a dynamic form, the problem of technogenic seismicity in the development of deposits is the most acute and relevant not only for a number of regions of Russia, but also for many mining provinces of the world. Especially in seismically active regions [1–3].

Geomechanical processes occur in rock mass under the influence of various natural (endogenous and exogenous) and man-made (anthropogenic) factors. From the point of view of the development of ideas of the change mechanisms in the energy saturation of blocks of various ranks, the process of cracking in a hierarchical block rock mass and its reflection in stress and strain fields during large-scale mining operations is of the greatest interest for research. As can see, various factors can be the trigger of the processes of deformations and displacements in a hierarchical block rock – natural, man-made, and most often, their various aggregates. The degree of influence of these factors on the change in the energy saturation of blocks of various ranks will depend on various factors [1–8].

Let's consider schematically the influencing factors, and how they affect the change in the process of cracking in the rock mass. The relationship diagram (figure 1) represents the main factors and causes affecting this process.

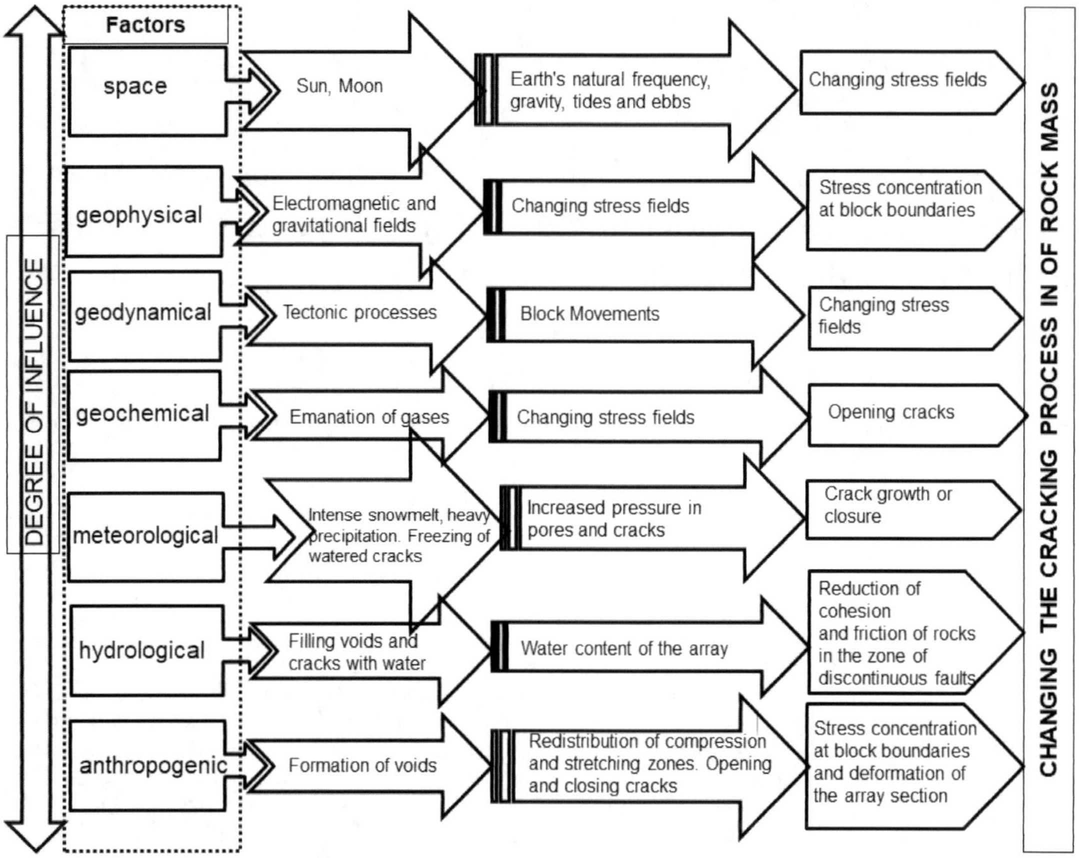

Figure 1. Factors affecting the processes of fracturing in rock massifs of mined deposits.

Since the process of cracking is the result of the factors manifestation reflection (or their combination) in the rock mass, the degree of qualitative (strong, medium, weak) influence of individual factors is determined depending on the analyzed scale level.

A large-scale impact of mining operations, anthropogenic factors influence the processes of cracking to a greater extent, although the role of others is no less significant under certain conditions. This is due to the fact that the natural and technical system is a complex hierarchical block environment, and its destruction is realized in the weakest zones (discontinuous disturbances filled with permeable and water-absorbing rocks and being the boundaries between the blocks). In the zone of the mining operations impact, where significant volumes of rock mass are extracted and moved and significant developed spaces are formed, firstly, the rate of the rocks deformation increases compared to natural, and secondly, the influence of dynamic effect from mass explosions is imposed on the stability of the mined deposits various sections. With the participation of natural (in addition to anthropogenic) factors in the process of cracking, there is an ambiguous reaction of the rock mass to their impact (either an increase or a decrease in cracking) [2].

In the geological environment of natural and man-made systems, there are dynamic phenomena of various energy levels from acoustic noise (microseismics) to rockbursts and earthquakes caused by mining operations, that is, there is always a geodynamic risk in the geological environment of natural and man-made systems, especially in mining ones [1–8].

Geodynamic risks in mining operations are determined by the level of two main factors: natural and man-made seismicity of massifs (the number of recorded rockbursts and mining-induced earthquakes) within the mine fields of the mined deposits.

IOP Conf. Series: Earth and Environmental Science **946** (2021) 012011 doi:10.1088/1755-1315/946/1/012011

2. Results and discussions

In the Far Eastern region, a number of ore deposits are being developed, prone and dangerous for rockbursts (figure 2). The most difficult conditions are characterized by the Antey deposit (Zabaikalsky Krai), which is being worked out by the Priargunsky Production Mining and Chemical Association (PJSC PPMCA), the Yuzhnoye and Nikolaevskoye deposits developed by MMC Dalpolymetal (Primorsky Krai). According to the results of studies carried out in previous years, it was found that the deposits of the Far Eastern region, prone and dangerous for mountain impacts, are located within the tectonically active Amur Plate, characterized by high structural heterogeneity, tectonic fragmentation and the presence of areas of increased stress [9].

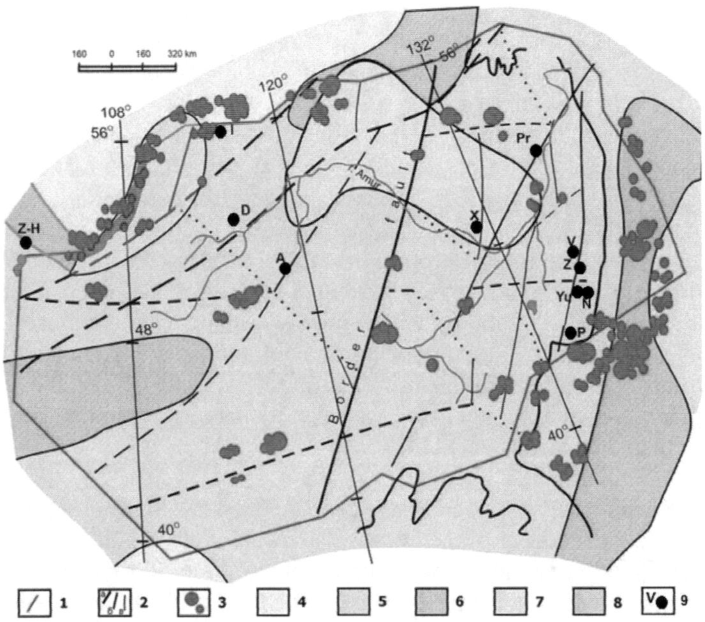

1 – faults bounding the Amur geoblock; 2 – faults bounding the inner blocks of the lithospheric plate: a – Border deep fault dividing the Amur Plate into blocks; I order; b, c – other, mainly interblock, neotectonic faults within the Amur Plate; 3 – earthquakes registered by the stations network of the FRC GS RAS from January 2019 to May 2021 with M ≥ 4; 4 – areas of tectonically neutral stress state; 5 – compression areas; 6 – strike-slip areas; 7 – areas of compression with a strike-slip; 8 – extension areas; 9 – ore deposits dangerous and prone to rockburst: Z-H – Zun-Kholbinskoye; I – Irokin-dinskoe; D – Darasun; A – Antey; X – Khinganskoe; Pr – Perevalnoye; V – Vostok-2; Yu – Southern; Z – Forgotten; N – Nikolaevskoe; P – Partisanskoe.

Figure 2. Geodynamic position of rockburst hazardous deposits in modern structures of the Amur geoblock.

The Amur lithospheric plate (or geoblock) is a tectonic element of the Central Asian Orogenic Belt, which is sandwiched between the North Asian and Sino-Korean craton [10]. According to the interpretation of seismic materials, the modern horizontal movement of the Amur Plate occurs under external force from the North Asian Craton (Siberian Platform). On the other hand, it is obvious that the stress-strain state of the eastern part of the Amur Plate is influenced by geodynamic conditions associated with the subduction of the Pacific Plate under the continent, and from the northwest – by the zone of active formation of the Baikal rift system [11]. It should be noted that most of the deposits developed within the Amur Plate are characterized by a complex mining and geological structure, which is due to their frequent localization near the interface of different structural floors of the geological section. At these sites, there is an increased tectonic disturbance of mine fields, which is combined with heterogeneity in the physical and mechanical properties of the host medium, as well as the complex morphology of ore bodies and ore-modified rocks. All deposits are post-magmatic and are divided into vein, stockwork and scarn.

In addition, it should be borne in mind that according to previously obtained results [10], in historical time intervals (months, years, decades, etc.) within the Amur craton, selective geodynamic activation of faults or their parts was observed, which is associated with the time sequence and direction of a certain seismic process. In this regard, block-forming faults are activated in recent times not along their entire length, but in separate sections. Based on the results of field measurements at specific deposits, the spread of the expected values of the modern compression of the upper part of the Earth's crust of the Amur Plate can range from 20 to 100 MPa and maybe more. Moreover, the position of the known impact–prone deposits of the Amur Plate, both in relation to its boundaries and in relation to the distinguished neotectonic elements, differs significantly and this circumstance cannot be ignored in their structurally scalable geomechanical assessment [10, 11].

The current natural stress-strain state of the subsurface in the inner regions of the Amur Plate is determined by geodynamic events occurring on its borders. Currently, the main compression forces come from the North Asian Craton and from the subduction zone of the Pacific Plate under the Asian continent. This is expressed by the presence of highly active seismic belts: from the north and northwest – Baikal-Stanovoy, from the east – Sakhalin. The inner region of the Amur Plate is characterized by relatively low seismicity [10].

The spatial distribution of earthquakes registered by the network of stations of the Federal Research Center "Unired Geophysical Survey of the Russian Academy of Sciences" (FRC GS RAS) from January 2019 to May 2021 with a magnitude greater than 4 [13] and the geodynamic position of rockburst hazardous deposits in the modern structures of the Amur geoblock are shown in figure 2. As can be seen from the data in the figure, in recent years, geodynamic processes have continued to intensify within the northwestern boundary of the Amur Plate near Lake Baikal and the northeastern one near Sakhalin Island and the eastern one in the Sea of Japan in the area of influence of the Pacific subduction zone. At the same time, the activation of the northern boundary of the plate has intensified in the area of influence of the Baikal continental rift zone, which is actively developing at the present time, as well as in the area of the Border fault passing almost along the center of the plate. The results of geomechanical studies at such deposits as Antey (Zabaikalsky Krai), Nikolaevskoye, Yuzhnoye (Primorsky Krai) and a number of others indicate that their impact hazard is largely determined by the high level of tectonic stresses acting in the massifs [11, 12, 14, 15].

The excess of horizontal compressive stresses over vertical ones is determined by their geodynamic position within the tectonically active Amur Plate, characterized by high structural heterogeneity, tectonic fragmentation and the presence of the increased stress areas [16, 17].

An earlier analysis of the rockburst hazard at the underground mines of the Eastern Primorye and Transbaikalia revealed a complication of the mining situation and an increase in the geodynamic risk during mining operations due to the growth of the developed spaces and the depth of development. There is an activation of geodynamic processes occurring in the form of restructuring and self-organization of a block rock mass in a natural and man-made stress field. These processes are accompanied by displacements and movements along tectonic disturbances of various scale levels, the release of significant elastic energy and the manifestation of mining-induced seismicity [10–12, 14, 15].

It should be noted that the probability of a rockburst or a mining-induced earthquake is due not only to the geodynamic situation of the deposit areas shown in figure 2. Firstly, the carried out studies of the physical and mechanical properties of the exploration wells core from these fields have shown the presence of strong and brittle rocks capable of collapsing in a dynamic form. Secondly, visual inspections of underground workings showed the presence of peeling, shooting and dynamic formation of rock blocks. Thirdly, seismic events of various energies (up to 10E6 J) were registered by monitoring systems at these deposits. At the same time, the depth at which the probability of the implementation of rockbursts and mining-induced earthquakes is very high at different deposits is individual. This circumstance is due not only to the depth, but also to the technology used for developing a particular deposit.

From the point of view of geodynamic risks at the studied objects, the probability of the implementation of rockbursts and mining-induced earthquakes is currently possible at ore deposits: Irokindinskoye, Antey, Perevalnoye, Yuzhnoye, Nikolaevskoye and Partizanskoye.

Previously performed geodetic studies have established the specifics of fluctuations of different-scale blocks of rocks under the influence of natural and mining-induced seismicity, and the analysis of seismoacoustic data has shown the nature of the development of cracking in certain areas, preceding the destruction of the massif sections in dynamic form [10–12]. In the course of the studies carried out on rockburst hazardous deposits, areas with varying degrees of probability of mining-induced earthquakes during seismic activity of the blocks containing these deposits were identified [14]. The influence of natural seismic activity on the realization of man-made seismicity is estimated [15].

During the development of deposits, such dangerous geodynamic phenomena as slabbing, shooting, rockburst, mining-induced earthquake are recorded. It should be noted that the geodynamic phenomenon may not lead to the destruction of mine workings, but the presence of such events clearly indicates the danger to mining operations. But, from the above list, the latter phenomenon is a powerful dynamic event with catastrophic consequences for the mines.

It is quite obvious that the consequences of the geodynamic phenomenon for the mine will be determined not only by the energy of the event, but also by the location of the source. If the source is sufficiently remote from mining operations, there will be push, sound and shaking by which the energy of event will be identified; if the source is located near mining operations, a mining-induced earthquake in its generally accepted formulation will take place. That is, there is an influence of various factors that need to be taken into account when assessing risk:

- The further away the zone of geodynamic activity is from the zone of influence of the cleaning front, the less likely the destruction of the workings.
- The further away the zone of geodynamic activity is from the zone of influence of a discontinuous disturbance, the less likely the destruction of the workings located in this zone.
- The further away the mining contour is from the zone of geodynamic activity, the less likely its destruction in the case of dynamic manifestations of rock pressure.

At the same time, preventive measures are being carried out at the enterprises developing the deposits under study to bring the workings into a state that ensures a reduction in stresses in the rock mass and the safety of mining operations. From this point of view, the probability of catastrophic dynamic phenomena in certain areas of the developed deposits is very low.

In the future, developing the direction of work on the assessment of geodynamic risk, neural networks are supposed to be used to make decisions on managing the state of the deposit being worked out. This is due not only to the variety (qualitative and quantitative) of the analyzed parameters used in the assessment of geodynamic risk, but also to the differing degree of their influence on the development of the geodynamic situation both individually and with various combinations of the parameters under consideration.

3. Conclusion

The conditions of geodynamic risks occurrence at specific deposits are considered on the example of high-impact deposits in the Far Eastern region of the Russian Federation. It is shown that dynamic phenomena of various energy levels from acoustic noise (microseismics) to rockburst and man-made earthquakes always take place in the geological environment of natural and man-made systems, that is, geodynamic risk is always present in the geological environment of natural and man-made systems, especially mining. The analysis of geomechanical processes at rockburst-hazardous deposits in the Far Eastern region of the Russian Federation showed a high degree of risk of the rock mass sections destruction in dynamic form for the Yuzhnoye and Nikolaevskoye deposits.

Acknowledgments

The research was carried out with the financial support of RFBR grant No. 20-55-53028 GFEN_a.

References

[1] Sashurin A D 2012 Geomechanics in mining: basic and applied research *Eurasian Mining* **1** pp 17–19

[2] Kozyrev A A, Panin V I and Fedotova Yu V 2012 Safety problems in mining operations in highly stressed rock massifs *Notes of the Mining Institute of St. Petersburg* **198** pp 150–156
[3] Ware K, Westion J F and Dickinson J A 2012 Investigation of a Canyon System off Cote D'Ivoire to Determine Present Day Risk of Slope Instability for a Field Development *SUT Conference Paper* (UK) pp 5443–508
[4] Kuzmin Yu O 1999 *Modern geodynamics and assessment of geodynamic risk in subsurface use* (M.: AES) 220 p
[5] Terbrugge P J, Wesseloo J, Venter J and Steffen O K H 2006 A risk consequence approach to open pit slope design *J. of the South African Inst. of Mining and Metallurgy* **106** 7 pp 503–511
[6] Bykov A A and Porfiriev B N 2006 On risk analysis, concepts and risk classifications *Problems of risk analysis* **3** 4 pp 319–337
[7] Adams D J 2005 *Proc. 1st Int. Seminar of Strategic vs. Tactical Approaches in Mining (Sandton South Africa) September* Rockburst research: a strategy for safety and profitability *J. of the South African Inst. of Mining and Metallurgy* pp 269–275
[8] Stacey T R, Terbrugge P J and Wesseloo J 2007 Risk as a Rock Engineering Design Criterion *Challenges in Deep and High Stress Mining* ed Y Potvin, J Hadjigeorgiou and T R Stacey (Perth: Australian Centre for Geomechanics)
[9] Seminskii K Zh 2008 Hierarchy in the zone-block lithospheric structure of central and eastern Asia *Russ. Geol. Geoph.* **49** 10 pp 771–779
[10] Rasskazov I, Saksin B, Usikov V and Sidlyar 2018 A The assessment of the impact of natural and anthropogenic factors on the current stress-strain state of rock massif of burst-hazardous ore deposits in the East of Russia *E3S Web of Conferences* **56**
[11] Prosekin B A and Ilyin E A 2010 Mining pressure monitoring system at the Antey deposit *St. Petersburg Notes of the Mining Institute* **188** pp 95–98
[12] Potapchuk M I, Kursakin G A and Sidlyar AV 2014 Improvement of safety of development of bump hazardous vein deposits of Eastern Primorye *Eurasian mining* **1** 21 pp 18–22
[13] Geophysical Survey of RAS [*Electronic materials*] http://www.ceme.gsras.ru
[14] Iuliia Fedotova 2020 *VIII Int. Sci. Conf. Problems of Complex Development of Georesources* Investigation of the deformations character of the Rudny fault shores during the development of the Yuzhnoye Deposit, Primorye Kraj *E3S Web of Conferences* **192**
[15] Rasskazov I Yu, Fedotova Yu V, Sidlyar A V and Potapchuk M I 2020 Analysis of manifestations of technogenic seismicity in the impact-prone rock mass of the Nikolaevsky deposit *Mining information and analytical bulletin* **11** pp 46–56
[16] Sankov V A 2014 Modern geodynamics of intracontinental regions: instrumental and geological-geomorphological estimates of movements and deformations of the earth's crust of Central Asia *Geod. and tectonoph.* **5** 1 pp 159–182
[17] Imaeva L P et al 2017 Geodynamic activity of the newest structures and fields of tectonic stresses of Northeast Asia *Geod. and tectonoph.* **8** 4 pp737–768

The experimental results of the geophysical pulse voltage generator

S A Gulyakov* and N S Stovbun

Institute of Marine Geology and Geophysics Far Eastern Branch of Russian Academy of Science, Yuzhno-Sakhalinsk, 693022, Russia

E-mail: *gulyakov_97@mail.ru

Abstract. The paper is devoted to the development process of the geophysical pulse voltage generator. The peculiarity of the generator lies in the non-specific purpose of this type of construction. Its main function is the controlled effect on the active faults of seismically dangerous zones. The results of the field experiment with the constructed device from 2018 to 2020 are presented. During the experiments, estimations and theoretical analysis we have obtained the seismic noise recorded by means of molecular electronic instruments and determined the periods of the electrical pulses. Operational parameters of the generator were identified on the basis of experiment results. The software and hardware elements of the system were upgraded and, as a result, a new version of the geophysical pulse voltage generator was developed.

1. Introduction

The paper studies the electroseismic effects in order to control deformation processes in geologic environment. A new type of external effects on earthquake source was found in the end of the last century in Russia, this is a simulated and natural electromagnetic effect on earth's formations and crust in deformed condition. The experiments on stimulation of seismic events in Pamir and northern Tian Shan using a magnetohydrodynamic generator (MHD generator) are widely known [1]. However, the amount of electrical sounding on earth's crust was reduced due to the high cost of MHD. Consequently, the Research Station of the Joint Institute for High Temperatures developed and constructed the thyristor assemblies EIS-100 (1983), EIS-300 (1984), EIS-630 (1986), ERGU-600-2 (1987) with their maximum currents reaching 100, 300, 630 and 1400 A respectively [1]. Considering the relevance of the experiments and the fact that equipment was made by special request, the Institute of Marine Geology and Geophysics (IMGG FEB RAS) started to develop its own construction for simulated electrical effect on seismic activity – the geophysical pulse voltage generator.

2. The first version of the pulse generator

The geophysical pulse voltage generator is a device that consists of two elements: a power source and a high-voltage commutator. The high-voltage commutator is a separate device and its main function is to stop/start the timed electric charge for affecting seismic activity. The core element of the commutator is an electric transistor switch. The transistors are used as an electrical switch is conditioned with the fact, that these electronic devices can be simply switched with low voltage pulses, that allows forming the pulses with the same on-off time. Insulated-gate bipolar transistor (IGBT) was selected since it is a fast-acting semiconductor device (switching time up to 200 ns) with its power channel resistance low enough. It is capable of carrying a current up to 310 A at a voltage of 4 kV. The block diagram of the commutator is shown in figure 1.

A user adjusts the amount and on-off length of the pulses by means of the input device, which is constructed out of tact switches. A microcontroller (MC) receives and processes all the settings that are entered. The MC uses special software to switch on or off the IGBT according to the settings with a gate driver chip, which is intended to make galvanic isolation between analog and digital parts in order to make the scheme optimized and stable. The construction of the first version of the voltage commutator is shown in figure 2.

Figure 1. The block diagram of the commutator used in generating of electric pulses.

Figure 2. The first version of the voltage commutator.

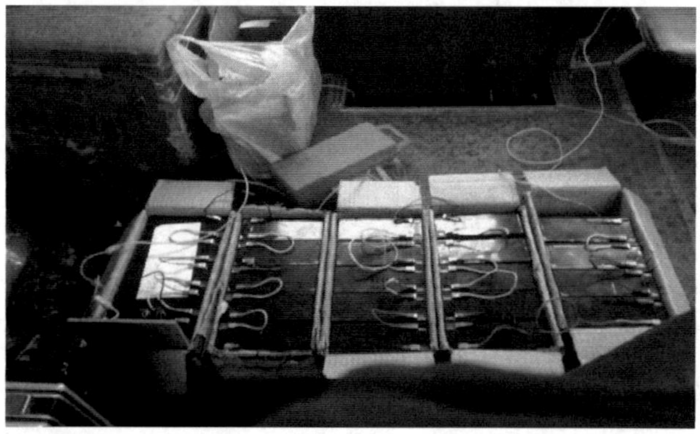

Figure 3. A set of batteries used as a power source in the experiments.

A set of batteries (up to 30 rechargeable cells) connected in series was used as a power source (figure 3) of 300-400 V, that is not practical, but for the first experiment it could provide continuous current of 5 A or less, which met the requirements.

3. The first experiments

The first experiment was conducted in autumn 2018 on the proving ground of the complex observations «Karymshina» of the Institute of Cosmophysical Research and Radio Waves Propagation (IKIR FEB RAS, The Kamchatka Peninsula). Geophysical equipment of the complex observation station continually measures and records the parameters of the solar activity, magnetosphere, ionosphere, natural magnetic field of the earth and acoustic earth emission. Before the experiment, we used a molecular electronic hydrophone along with listed geophysical sensors in order to study its sensitivity to measuring the earth acoustic emission when the electric pulses are started. Instead of vertical earth electrodes, five-meter buried pipes of dry well were used with the distance of 35 meters between them. The experiment lasted for one hour, one hundred pulses were started with the on-time of five second and the off-time of thirty seconds. The maximum current determined was 0.9 A (figure 4). After the experiment the residual potential difference between the two electrodes was measured (Figure 5) [2].

During the experiment, the geophysical complex did not register the simulated electric pulses and a conclusion was made, that the current was not high enough to affect the seismic activity. It resulted in the reconstruction of the pulse voltage generator – the commutator parameters, its software and hardware parts were changed.

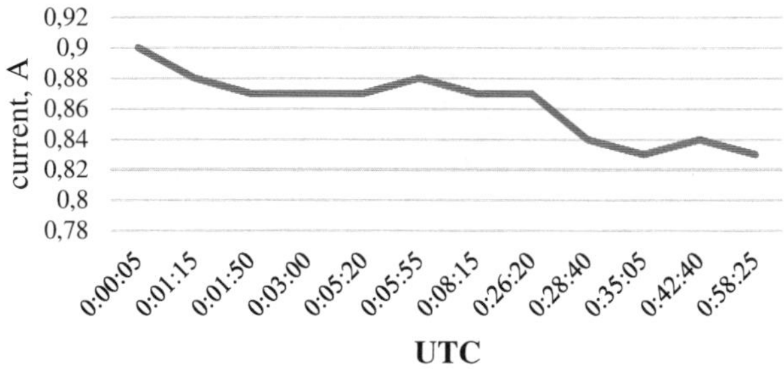

Figure 4. The graph of the change in the current during the experiment.

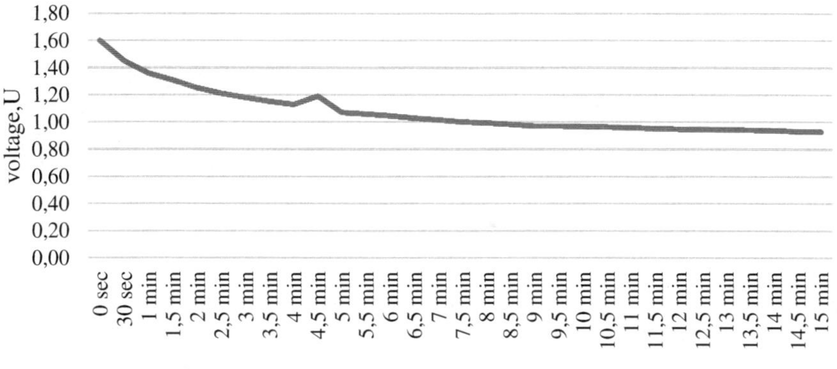

Figure 5. The dynamics of the change in the potential difference after the disconnecting the voltage source from the electrodes.

Further experiments were conducted in summer 2019 in geophysical observatory «Mihnevo» of the Institute of Geosphere Dynamics (IDG RAS, Moscow). The observatory studies the properties of the interrelated disturbance of inner and outer geosphere. During the experiment the peak current was up to 3.2 A. The result was achieved through increasing the amount of vertical earth electrodes to four and making the dipole length to 180 meters. However, the output data have not been published in open access so, there is no way to analyze the experiment results. Consequently, IMGG FEB RAS has purchased its own special wave recorder for molecular-electronic sensors in order to continue the experiments on Sakhalin Island [3]. In addition, there were disadvantages in using the set of batteries as a power source for the electric pulses: high cost and heavy weight, no air transportation, permanent maintenance of the construction since a drop of electromotive force (EMF) occurs, that negatively affects an experiment. The idea of using the set of batteries as a power source was rejected in favor of an AC-DC converter (figure 8) which was developed and constructed directly in the research equipment-sharing center of the Institute of Marine Geology and Geophysics. With small weight it is capable of converting the 220 V alternating voltage of 50 Hz into stabilized direct voltage of 360 V with maximum current of 4.2 A.

4. The experiment on the geophysical proving ground of IMGG FEB RAS

In 2020 the experiments have been continued on the geophysical proving ground of IMGG FEB RAS located in village of «Petropavlovskoe». A molecular electronic hydrophone MTAS-30 and a seismometer CME-6111 were used during the experiments in order to register the seismic changes [4-6]. We used a multi-electrodes system located on closely-spaced pits [7] to register the geoelectric processes in the near-surface part on the tectonosphere-atmosphere interface. A diesel generator as a primary source of the alternating voltage was used, and a developed AC-DC converter was connected to it in order to get the required direct voltage for the experiment. Four galvanized pipes of 2.5 meters long were used as earth electrodes with the dipole of 300 meters long. As a result, the peak current reached 5.5 A.

IOP Conf. Series: Earth and Environmental Science **946** (2021) 012012 doi:10.1088/1755-1315/946/1/012012

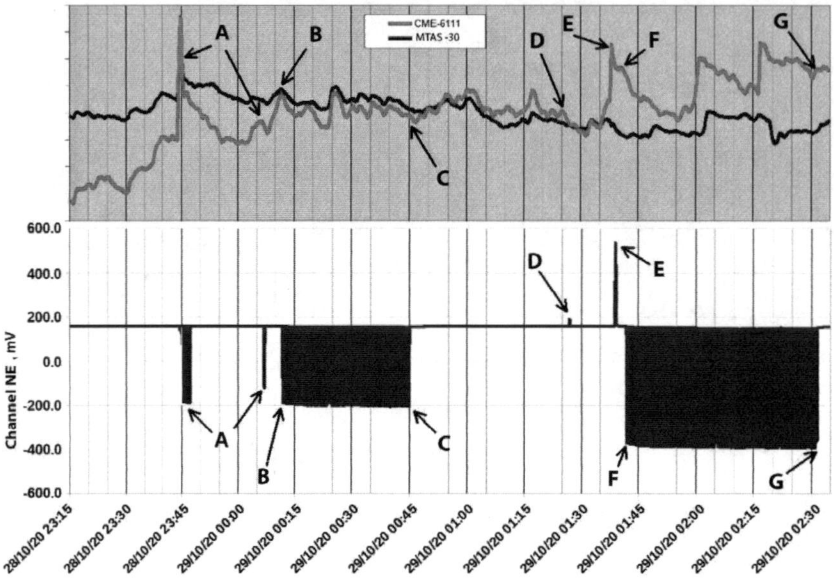

Figure 6. The envelope of the seismic noise signals of molecular electronic devices (top) and the change in the potential difference on the measurement channels between the electrodes of the north-eastern (NE) pit of the geoelectric measurement system (bottom) during the experiment (UTC time).

The figure 6 shows that every phase of the experiment with the pulse generations and the resulting current in the outer layer of the earth's crust matches the response recorded by molecular electronic sensors on the level of seismic noise. A-stage – two pulses started for the equipment checking. B-stage – the first pulse starts (polarity: electrode 1 (–), electrode 2 (+)), the pulse length – five seconds, the pause length – fifteen seconds, number of the pulses is one hundred, the current is 5.2 A at the start of the pulse and 4.95 A at the end. C-stage – the first series of the pulses ended. D-stage – planning to start the second series of the pulses (polarity: electrode 1 (+), electrode 2 (–)), pulse is not more than one second (failure to start the series of the pulses). E-stage – a test pulse to start the second series (polarity: electrode 1 (+), electrode 2 (–)), the pulse length is not more than 5 seconds (the generator overload occurred). F-stage – the second series of pulses (polarity: electrode 1 (–), electrode 2 (+)), the pulse length – ten seconds, the pause length – twenty seconds, the number of pulses is one hundred, the current is from 5.2 A to 5.4 A. G-stage – the end of the second series.

You can observe the obtained data more thoroughly in the work [8].

5. Developing a new version of the pulse voltage generator

Throughout the field experiments, the work of the first version of the voltage commutator was analyzed, the parameters and the modes were reconsidered, that was used to develop in parallel a new version of the device. The design of the new voltage commutator is shown in figure 7. It has an interface which consists of a matrix keyboard with an LCD display which is used to flexibly set the parameters and modes for an experiment, the power supply requires only one battery (the first version required three batteries to power the device), a graphite screening along with software and hardware solutions led to a better noise-immunity. The IGBT used in the new version is capable of carrying out a current up to 40 A with collector voltage of 1.5 kV.

The new version has three modes to choose: first to set the amount of the pulses, their on and off time, second to operate with the time of the experiment instead of the pulses counting – a user sets the time during which the device is working and the on and off time of the pulses, the software calculates the amount of the pulses to generate within the time range and starts the process, third mode is meant for test pulses before the whole experiment in order to check the parameters of the equipment.

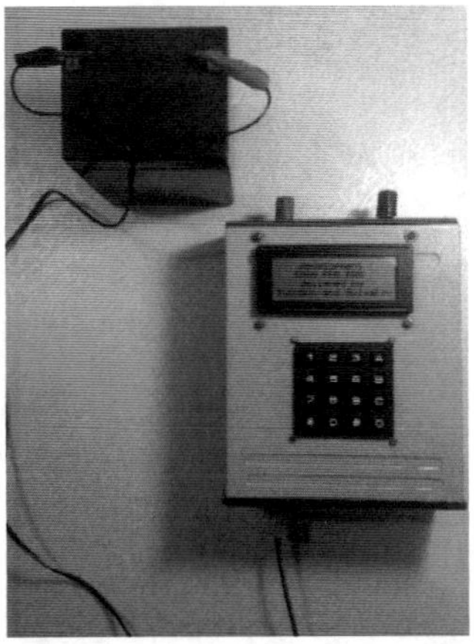

Figure 7. The design of the new version of the voltage commutator.

Figure 8. The design of the developed AC-DC converter.

6. Conclusion

As a result of the field experiments, observation, laboratory simulation and theoretical analysis which were carried out over the last four years, we developed and constructed the equipment of a new type for electrical sounding. The results of the experiments showed the general realization of the simulated electromagnetic effect on the active tectonic structure in real conditions.

The next phase of the experiment is the development of a network of seismological stations in Central Sakhalin Fault in order to monitor the seismic and trigger effects. In order to improve the quality of the values recorded by the geophysical equipment the current going through the dipoles must be increased significantly [9]. At present, the switching element of the geophysical pulse voltage generator is capable of carrying out up to 40 A while the maximum current, which the power source can provide, is four times lower. This drawback can be solved by getting more powerful source or developing a set of voltage capacitors.

References

[1] Volyhin A M, Bragin A M and Zubovich A V et al 1993 *Reflection of Geodynamical Processes in Geophysical Fields* (Moscow: Science)

[2] Kolesnikov V P 2007 *The basis Interpetation of Electrical Sounding* (Moscow: Scientific World)

[3] Kamenev P A, Kostylev D V, Boginskaya N F and Zakupin S A 2019 Geophysical surveys in the southern part of the Central Sakhalin Fault based on new integrated network *Geosystems of transition zones* **3** 4 pp 390–402

[4] Kostylev D V, Bogomolov L M and Boginskaya N F 2019 *Proc. Conf. on Geodynamical Processes and Natural Hazards* (Yuzhno-Sahalinsk: IMGG FEB RAS) p 35

[5] D V Kostylev et al 2019 *IOP Conf. Ser.: Earth Environ. Sci.* 324 012009

[6] Kostylev D V, Boginskaya N F 2020 Seismoacoustic observations using molecular-electronic hydrophones on Sakhalin and the South Kuril Islands (Kunashir Island) *Geosystems of transition zones* **4** 4 pp 486-499

[7] Lybushin A A, Bobrovskiy V C and Shopin S A 2016 Experience of complexation of global geophysical observations *Geodynamics & tectonophysics* **7** pp 1–21

[8] Dudchenko I P, Kostylev D V, Gulyakov S A and Stovbun N S 2021 Geophysical pulse voltage generator for seismic and electric exploration of the subsurface *Geosystems of transition zones* **5** pp 46–54

[9] Bogomolov L M 2013 Search for new approaches to explain mechanisms of interrelations between seismicity and electromagnetic effects *Vestnik of Far Eastern Branch of Russian Academy of Sci.* 3 pp 12–18

Geomechanical parameters of sedimentary rocks of Southern Sakhalin

P A Kamenev[1,*], L M Bogomolov[1], O M Usoltseva[2], P A Tsoi[2,3] and V N Semenov[2]

[1]Institute of Marine Geology and Geophysics Far Eastern Branch of the Russian Academy of Sciences, Yuzhno-Sakhalinsk, 693022, Russia
[2]Chinakal Institute of Mining of the Siberian Branch of the Russian Academy of Sciences, 54, Krasny ave., Novosibirsk 630091, Russia
[3]Novosibirsk State Technical University, 20, K. Marksa ave., Novosibirsk 630073, Russia

E-mail: *p.kamenev@imgg.ru

Abstract. The samples of terrigenous rocks of the Sakhalin were selected for geomechanical studies: the rocks of the Nevelskaya and Kholmskaya Formations from the Petropavlovsk quarry; ones of the Kholmskaya Formation on the west coast of the south of Sakhalin, rocks of the Junonskaya Formation on the east coast. The Kholmskaya Formation N_1hl is represented by siltstones and interlayers of fine-grained sandstone. The Nevelskaya Formation N_1nv is represented by siltstones with inclusions of tufogenic material. The Junonskaya Formation T_2-Jun is represented by variegated greyish-green jaspers. Their static and dynamic parameters (strength limits, static Young's modulus and Poisson's ratio, cohesion and angle of internal friction) were determined. The obtained results are characterized by a significant spread of values, which is likely explained by the significant fracturing of the initial samples and the effect of anisotropy.

1. Introduction

The development and exploitation of hydrocarbon fields, in conditions of high tectonic activity, in most cases are accompanied by the study of the geomechanical properties of the drilled rocks [1, 2]. For the selection and optimization of the bit, it is extremely important to know the values of the uniaxial compression strength of rocks. The Poisson's ratio values, the angle of internal friction, and cohesion are necessary to calculate the optimal density of the drilling mud. Moreover, all these parameters must be known before the construction of the well. Thus, the importance of the mechanical properties of rocks is obvious, but a researcher or an engineer addressing such problems faces significant problems. There are no regional geomechanical databases, it is impossible to find a drill sample for the laboratory tests, the drilling core sampling is expensive. As a result, there are very few regional works in this direction. The recent work [3] involving similar studies of the Daginskaya Formation may confirm this state of the art. Actually, core material studied in [3] was too limited. In such situations, sampling rocks is done from outcrops located near the site of interest or from a locally outcropping stratigraphic unit. Another approach to the study of the mechanical properties of rocks is the use of empirical relations obtained based on the logging data or directly from the data of modern acoustic logging [4]. This work aims to continue studying the mechanical properties of sedimentary rocks of the south of Sakhalin by laboratory methods. Taking into account the development of modern laboratory equipment, the needs of oil and gas production, the development of the construction industry and infrastructure, it is reasonable to continue these works in order to clarify the geomechanical parameters of sedimentary rocks of southern Sakhalin. This work can be considered as a continuation of [4, 5].

2. Main results

At the initial stage of the study, we selected rock samples of the Nevelskaya and Kholmskaya Formations from the Petropavlovsk quarry located a few dozen kilometers from Yuzhno-Sakhalinsk. The next stage of the study was sampling on the east and west coast of the south of Sakhalin (figure 1).

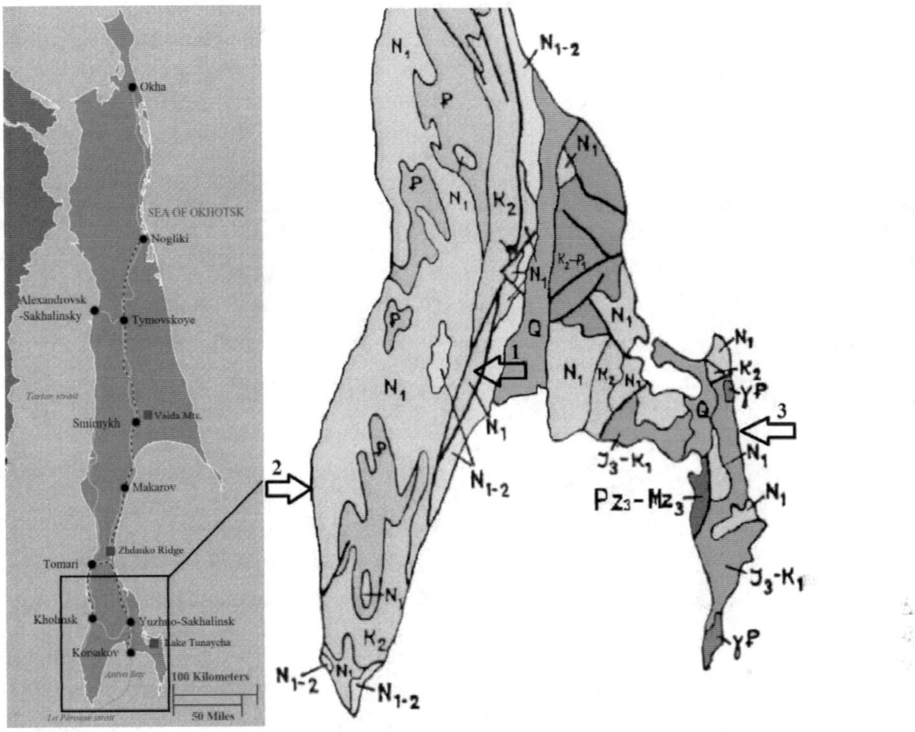

Figure 1. The map of Sakhalin Island with indication of sampling locations.

The samples on the east coast were collected in the Korsakovsky district at Cape Ptichiy. The series is represented by variegated greyish-green jaspers, which are fractured. The cracks in samples are healed mostly with calcite or hardened siliceous-clay rocks of the Junonskaya T_2-Jjun Formation. The rock samples on the west coast were taken in the Nevelsky district, 3 kilometers north of the village of Shebunino. The series is represented by strong, from light grey to grey, sometimes bleached with a yellowish tinge by siltstones of the Kholmskaya Formation N_1hl with layers of fine-grained sandstone. Strong, from light grey to grey sandstones of the Kholmskaya Formation N_1hl, from fine-grained to coarse-grained, with inclusions of tuff material. The Kholmskaya and Nevelskaya Formations deposited mainly on the southwestern coast of Sakhalin and the adjacent shelf. The northern border of the formations can be traced along the southeastern shore of Sakhalin, in the southern part of the Poronaisky district. The deposits of the Kholmskaya Formation N_1hl nearby the Petropavlovsk quarry are divided into three bundles [6]. The lower one is represented by sandstones, tuffs and tuffstones; tuffaceous siltstones, siltstones and claystones are less common. The middle pack includes overlapping tuffs, sandstones, siltstones, claystones, tuffaceous siltstones. The upper pack is mainly tuffaceous siltstones and siltstone with rare interlayers of tuffstones, tuffs, sandstones and tuffites.

Sandstones are mainly volcanics, glauconite is often found in them. Siltstones are from dark grey to black in color, often fractured, with abundant inclusions of claystones, sandstones and tuffs. The Nevelskaya Formation N_1nv is divided into lower and upper sub-formations. The lower sub-formation is represented by a flysch interlayer of siltstones, tuffstones, sandy tuffites, claystones. The Verkhnenevelskaya Sub-Formation is composed of finer-grained rocks containing in large quantities the remains of diatoms, up to the transition of rocks into tufodiatomites and diatomites. The sandstones of the Nevelskaya Formation are mainly volcanics or represented by tuffstones. Geomechanical parameters (uniaxial compression strength (UCS), static Young's modulus and Poisson's ratio, cohesion and internal friction angles) were determined using an INSTRON 8802 testing machine, a radial confining pressure chamber, strain gauges for measuring axial and radial deformations by INSTRON. The methodology for rock samples testing is described in [7]. Moreover, the propagation velocities of elastic compressional and shear waves were measured using an ultrasonic control and measuring device Pundit Lab. Laboratory density studies were carried out by weighing samples of standard sizes (cylinders 60 by 30 mm and 30 by 30 mm) shown in figure 2. The measurements of density and elastic wave velocities allowed determination of dynamic values of Young's moduli, E_{din}, and Poisson's ratios, ν_{din} [4, 5].

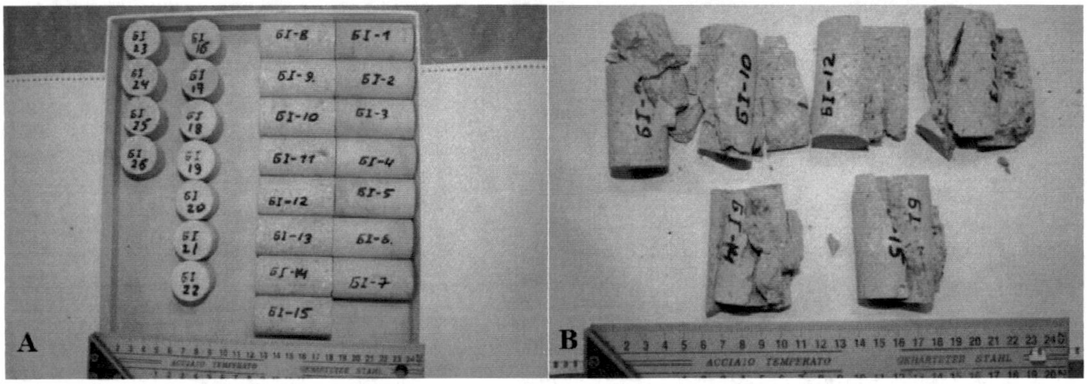

Figure 2. Ready test cores (A) and cores after static tests (B).

The deformation and strength properties of core samples were studied using the following experiments: uniaxial compression – 13 tests, extension – 20 tests, triaxial compression – 11 tests. During the triaxial compression test, simultaneous loading with axial force and radial pressure was carried out up to a radial confining pressure value of 5 MPa, then with constant radial confining pressure – loading with axial force until destruction (so-called Karman test) in the figure 3. The values of the dynamic (E_{din}) and static (E_{st}) Young's modulus, Poisson's ratio (ν_{din}) (ν_{st}), uniaxial compressive strength (UCS) were obtained as a result of testing samples of siltstones of the Kholmskaya N_1hl and Nevelskaya N_1nv Formations. The values of cohesion C_0 and the angle of internal friction φ were determined as well (Tables 1, 2, 3).

Dynamic tests were performed only for rocks selected in the Petropavlovsk quarry. The tensile strength varies in a wide range from 8.6 to 193 MPa for the same type of rocks of the Kholmskaya and Nevelskaya Formations. In the rock of the Junonskaya Formation, it varies from 3.4 to 63.2 MPa. The static shear modulus is from 2.12 to 25 GPa and 3.5 to 5.9 GPa, respectively.

The static Poisson's ratio is from 0.10 to 0.31, in the case of samples of the Junonskaya Formation – 0.17. The cohesion is from 7 to 13.5 MPa and 7.7 to 7.9 MPa in the Junonskaya Formation. The angle of internal friction is 28.3–58.2 degrees and 47.6–50.5 degrees, respectively. Thus, quite heterogeneous characteristics have been obtained. Such a spread is due to several reasons at once. There are various types of static tests: uniaxial compression, tension and triaxial compression, which results are involved for the comparison. The following reasons significantly affect the mechanical properties: a natural fracturing (latent in the case of core sample), a probable anisotropy of properties associated with stratification. It should also be noted that rather large variations in the values of static and dynamic properties are available and given in Table 1.

Table 1. Test results of samples of siltstones of the Kholmskaya and Nevelskaya Formations from the Petropavlovsk quarry.

Sample No	Test type	UCS, MPa	E_{st}, GPa	ν_{st}	E_{din}, GPa	ν_{din}	ρ, g/cm^3	C_0, MPa	φ, deg
1–1	Triaxial compression 5 MPa	60.9	4.61	0.11	8.98	0.26	1.99	13.1	29.7
1–2	Triaxial compression 5 MPa	91.2	9.59	0.17	–	–			
2	Uniaxial compression	18	2.12	0.15	8.27	0.24	1.97	–	–
3–1	Uniaxial compression	73.5	10.02	–	35.27	0.27	2.50	–	–
3–2	Uniaxial compression	43.6	2.09	0.12	–	–			
4	Triaxial compression 5 MPa	193	25.04	0.12	33.86	0.23	2.52	21.5	53.9
5	Triaxial compression 5 MPa	156	19.84	0.13	24.35	0.27	2.30	21	49
6	Uniaxial compression	89.4	12.95	0.1	22.27	0.26	2.27	26.1	28.3
7	Uniaxial compression	54.8	11.48	0.17	20.33	0.19	2.24	7	58.2
8	Uniaxial compression	69.6	9.05	0.1	19.10	0.24	2.19	–	–

Table 2. Test results of siltstones samples of the Kholmskaya Formation from the western coast.

Sample No	Test type	UCS, MPa	E_{st}, GPa	ν_{st}	C_0, MPa	φ, deg
2–1	Uniaxial compression	31.6	4.043	0.137	11.4* 13.5**	41.5* 50.2*
2–2		53.4	6.617	0.146		
2–3		51	6.613	0.313		
avg		**45.3**	**5.758**	**0.199**		
2–4	Triaxial compression 5 MPa	102.9	7.798	0.129		
2–5		81.7	8.895	0.164		
2–6		83.1	9.386	0.137		
avg		**89.2**	**8.693**	**0.143**		
2–7	Extension	10.2	–	–		
2–8		11.1	–	–		
2–9		9.6	–	–		
2–10		8.6	–	–		
avg		**9.9**	–	–		
2–11	Uniaxial compression	28.2	4.257	0.165	6.7* 8.3**	43.7* 54.3**
2–12		32.6	4.446	0.152		
2–13		23.1	2.439	0.175		
avg		**28.0**	**3.714**	**0.164**		
2–14	Triaxial compression 5 MPa	75.2	6.056	0.144		
2–15		81.1	7.176	0.142		
2–16		74.2	8.136	0.169		
avg		**76.8**	**7.123**	**0.152**		
2–17	Extension	5.8	–	–		
2–18		5.1	–	–		
2–19		6.2	–	–		
2–20		5.3	–	–		
avg		**5.6**	–	–		

Notes. *– the envelope is constructed according to the extension and uniaxial compression circles; **– the envelope is constructed according to the extension and triaxial compression circles.

Table 3. Test results of samples of jasper of the Yunonskaya Formation from the eastern coast.

Sample No	Test type	UCS, MPa	E_{st}, GPa	ν_{st}	C_0, MPa	φ, deg
	Uniaxial compression					
3–1		34.9	3.552	0.177		
					7.7*	47.6*
					7.9**	50.5**
3–2	Triaxial compression 5 MPa	63.2	5.957	–		
3–3		4.1	–	–		
3–4	Extension	3.4	–	–		
3–5		11.4	–	–		
3–6		4.1	–	–		
avg		**5.8**	–	–		

Notes. *– the envelope is constructed according to the extension and uniaxial compression circles; **– the envelope is constructed according to the extension and triaxial compression circles.

One can see from the Table 1 that the results of static tests give large dispersion of the values of geomechanical parameters of tested siltstones from the Petropavlovsk quarry. The dispersion static moduli E_{st} for siltstone samples from the western coast is not so large: there is nearly two – fold difference between maximum and minimum value. The similar difference in E_{st} values is in the case of Table 3 (samples from the eastern coast). Meanwhile previous data on E_{st} of Sakhalin siltstones samples gave even higher dispersion (maximum and minimum values were different by order of magnitude or more [8, 9]). So, the mean values of geomechanical parameters in Tables 1–3 seem informative.

3. Discussion and Conclusion

Three series of sedimentary rocks were selected in the western, central, and eastern parts of southern Sakhalin. Studies of the mechanical properties of cores manufactured from the selected rocks have been carried out. Static and dynamic characteristics (strengths, static and dynamic Young's modulus and Poisson's ratio, cohesion, and internal friction angle) have been obtained by methods, which are conventional for high-strength rock. This allowed obtaining the refined data on geomechanical parameters of some sedimentary (but consolidated) rocks. The tests revealed a considerable dispersion of static Young's modulus, cohesion, and internal friction angle. Such dispersion is caused likely by the internal (hidden) fracturing of the tested rocks samples. From methodical viewpoint this dispersion is to signify that «short» sampling of 3 items is not enough to triaxial compression tests of sedimentary rocks like siltstones. Samplings of 5–7 items are necessary to get more confident evaluations. So, presented mean values of geomechanical parameters may be preliminary estimates. Additional experiments are planned to involve tests of similar samples of sediment rocks under various steps of lateral pressure. Also, the anisotropy effect is to be investigated experimentally.

In spite of dispersion the obtained averaged characteristics may be applicable for geomechanical modeling aimed to the construction of wells and their hydraulic fracturing, to slope stability assessments.

Acknowledgements

The authors are grateful to Alexander Shemenda for valuable comments and discussion of the primary results of the work.

The tests have been carried out using the equipment of the Shared Use Center for Geomechanical, Geophysical, and Geodynamic Measurements at the Siberian Branch of the Russian Academy of Sciences.

IOP Conf. Series: Earth and Environmental Science **946** (2021) 012013 doi:10.1088/1755-1315/946/1/012013

References

[1] Zoback M D 2007 *Reservoir Geomechanics* (Cambridge: Cambridge University Press) p 505
[2] Dubinya N V 2019 An Overview of Wellbore Methods of Investigating Stress State of the Upper Layers of the Earth's Crust *Izv. Phys. Solid Earth* **55** pp 311–326
[3] Zhukov V S 2020 Estimating the strength and elasticity of rocks in the Dagi formation on the Sakhalin shelf *MIAB. Mining Inf. Anal. Bull.* **4** pp 44–57
[4] Kamenev P A and Bogomolov L M 2017 Depth distribution of internal friction coefficient and cohesion in sedimentary rocks of Sakhalin island *Geophysical Research* **18** pp 5–19
[5] Kamenev P A, Usoltseva O M, Tsoi P A, Semenov V N and Sivolap B B 2017 Laboratory research of geomechanical parameters of sedimentary rocks massifs in the South Sakhalin. *Geosystems of Transition Zones* **1** pp 30–36
[6] Tyutrin I I and Dunichev V M 1985 *Tectonics and petroleum potential of the northwestern part of the Pacific belt* (Moscow: Nedra) p 174
[7] Usoltseva O M, Eremenko A A, Shaposhnik Yu N, Tsoi P A and Semenov V N 2019 *IOP Conf. Series: Earth and Environmental Science* Laboratory studies of geomechanical properties of deep-level rocks **262** 012081
[8] Skorikova M F 1970 *Elastic properties of rocks of the southern part of Sakhalin and their use in the interpretation of geophysical observations* (Moscow: Nauka) p 176
[9] Skorikova M F 1965 Physical properties of Southern Sakhalin rocks *Exploratory geophysics* **7** pp 30–39

IOP Conf. Series: Earth and Environmental Science **946** (2021) 012014 doi:10.1088/1755-1315/946/1/012014

Results of reinterpretation of marine geochemical survey on the Sakhalin shelf

Yu V Kostrov* **and P A Kamenev**

Institute of Marine Geology and Geophysics Far Eastern Branch of Russian Academy of Science, Yuzhno-Sakhalinsk, 693022, Russia

E-mail: *kos-geo@yandex.ru

Abstract. Geochemical survey of hydrocarbons (HC) all over the world is a reliable tool of the complex of geological exploration, which allows to localize hydrocarbon saturation in structures exposed by seismic exploration, as well as to identify non-structural deposits. In 2011, a marine geochemical survey of the sorbed gases of bottom sediments was carried out on the shelf section of the northwest of Sakhalin Island. Based on the results of geochemical studies, 12 maps of the distribution of hydrocarbon and non-hydrocarbon gases in the work area and 7 maps of the distribution of metals in bottom sediments were constructed. Promising areas were distinguished by anomalies with the maximum content of parameters. The research area is characterized by a complex structure, located within the Baikal synclinal zone of the North Sakhalin oil and gas basin, which is part of the rift system of the Cenozoic sedimentary basins the Sea of Okhotsk. In 2019, the authors began to re-process and reinterpret the data in order to clarify the results. The work was based on modern theoretical foundations and methodological approaches of oil and gas prospecting geochemistry. The interpretation of the results was carried out on the basis of the model of interpretation of geochemical anomalies developed by the authors. Maps of anomalies were constructed according to 11 geochemical criteria and two geological and geochemical sections. The complex interpretation of geological and geochemical data was carried out taking into account the results of seismic exploration and drilling in a single project. According to the results of the complex interpretation, 6 promising sites were identified, which are ranked according to the degree of prospects.

1. Introduction

The efficiency of forecasting, prospecting and exploration of hydrocarbon accumulations (HC) can be significantly increased as a result of the use of geochemical prospecting technologies for oil and gas fields. These methods and technologies are becoming important in poorly studied and hard-to-reach regions where the resolution of traditional geological and geophysical studies for oil and gas is not high enough. Their complex use can lead to a significant optimization of the entire process for oil and gas, including in conditions of non-structural (non-anticline and complex) traps [1, 2].

To date, the theoretical foundations of geochemical methods have been sufficiently developed, including the theory of geochemical fields; models of the formation anomalous geochemical fields over hydrocarbon accumulations in various geological conditions. Optimal complexes of search indicators are used, separate forecast for individual stages and stages of exploration for oil and gas; integration of geochemical prospecting methods for oil and gas fields with other methods, including geophysical, at various stages of exploration for oil and gas. New methods (with vibration, the use of

IOP Conf. Series: Earth and Environmental Science **946** (2021) 012014 doi:10.1088/1755-1315/946/1/012014

geochemical surveys with seismic exploration, remote types of research, etc.) and modifications of search operations are scientifically substantiated [3, 4].

On Sakhalin Island, geochemical work was carried out in the northeast of the island in the period 1945–1979. Prospecting geochemical works in the period 1945–1951 identified promising areas where the following deposits were later discovered: Gilyako-Abunan (1950), Erri (1953), Odoptu-susha (1955), Tungor (1958), Nelma (1964), Northern Okha (1967). In the period 1976–1981, geochemical oil and gas prospecting works were carried out on Sakhalin by methods of surface gas and fluorescence evaluation. Among the recent regional works, one can single out the article [5].

2. Data and main results

In 2011, a marine geochemical survey of the sorbed gases of bottom sediments was carried out on the Astrakhan section of the shelf of the northwest of Sakhalin Island (figure 1A).

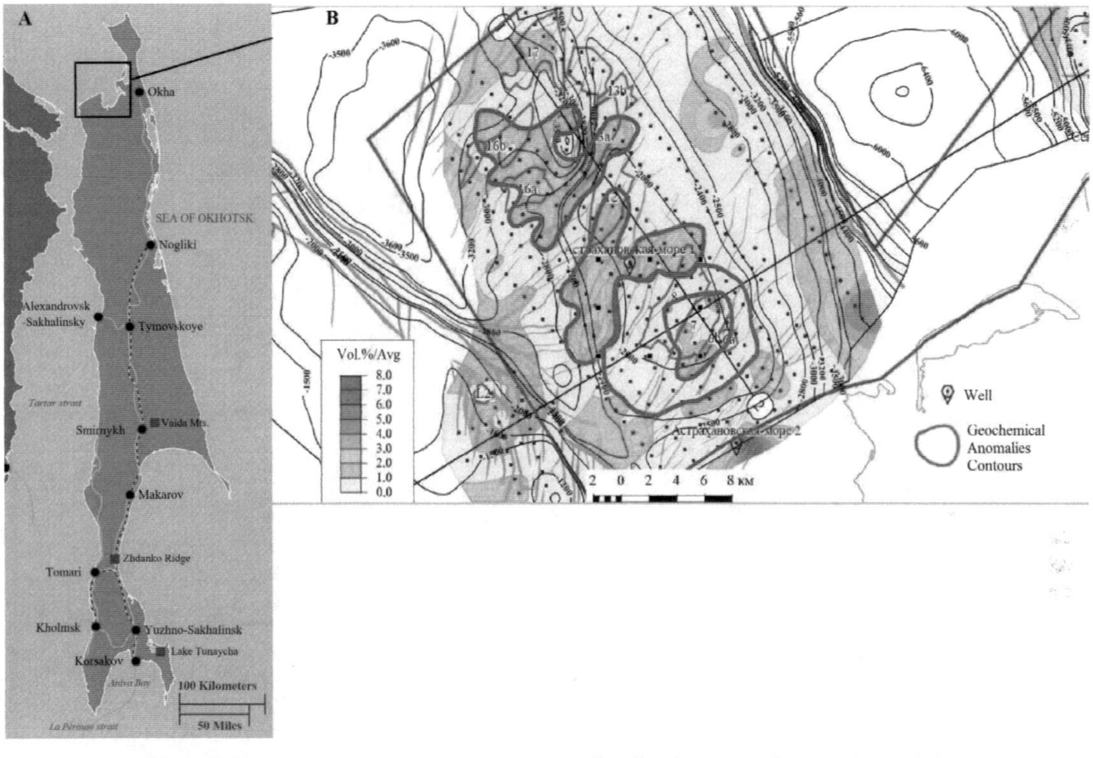

Figure 1. Location map (A) and contrast distribution over the HC C_2-C_4 (B).

Geochemical studies included sampling of bottom sediments in the volume of 410 samples, 20 samples of sediments on land, sampling of near bottom water in the volume of 25 samples. Laboratory chemical-analytical studies included: degassing of samples on the ship, chromatographic analysis for free gases (Ne, H_2, CO_2, N_2+O_2, C_1-C_6,), (figure 1B); chromatographic analysis for light hydrocarbons (vaporous C_7-C_9, including aromatic); chromatographic analysis for heavy hydrocarbons (C_{10}-C_{24}); trace element analysis for 28 elements; isotopic analysis of gases δ13C; total organic carbon (TOC); biomarker study. A total of 430 gas-geochemical samples and 860 lithochemical samples were analyzed. Based on the results of geochemical studies, 12 maps of the distribution of hydrocarbon and non-hydrocarbon gases in the work area and 7 maps of the distribution of metals in bottom sediments were drawn. The samples are represented by sands with inclusion of clay and silt. The prospect areas were identified by anomalies with the maximum value of geochemical parameters. At the same time, the permeability of faults was not taken into account, since when compared with seismic survey materials, the location of the selected prospect areas was questionable. In this and similar cases, geochemical methods are good and relatively cheap alternative to other research methods (geophysical, drilling, etc.).

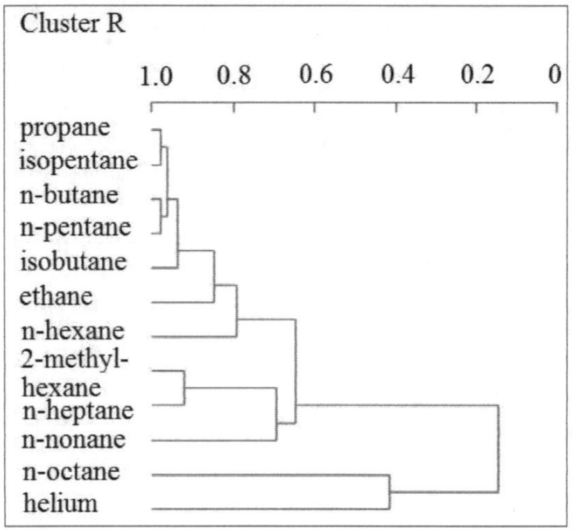

In 2019, we started to re-process and reinterpret geochemical data in order to clarify the results.

As the primary processing of an array of analytical data, R-type cluster analysis was applied (figure 2), based on a multidimensional statistical procedure for calculating Euclidean distances (similarity/ dissimilarity of properties). The interpretation was carried out on the basis of a model developed by the author in the fields of Western Siberia and modified in accordance with the conditions of a marine geochemical survey on the Sakhalin shelf [6].

Figure 2. Distribution of alkane hydrocarbons and helium in bottom sediments.

The main factors affecting the generation of the anomaly are: the formation pressure, the depth of deposit, the thickness structure above the deposit (mainly the quality of the caprock), fault tectonics and the structural position of the deposit. On the basis of the model described above, a model of interpretation of the Astrakhan site has been developed, taking into account the geological structure and conducted laboratory study and criteria have been pre-selected for various areas of the forecast.

1) The main criteria for assessing the permeability of discontinuous faults are helium, hydrogen, HC gases - C_1-C_4, heavy HC C_7+, bitumen HC, heavy metals (V, Ni, Co).
2) The main criteria for the presence of hydrocarbon deposits are: low-contrast annular or areal anomalies C_2-C_4, $C_{5\text{-}6}$, helium, hydrogen, carbon dioxide, the sum of the limiting hydrocarbons. Additional criteria for the presence of a deposit are: the content of heavy metals, the V/Ni ratio, the types of bituminoids, the bituminosity coefficient, Fe_2+/Fe_3+.
3) The main criteria for assessing the phase saturation of hydrocarbon deposits are: the dryness coefficient of gases, the maturity coefficient of gases, the epigenetic coefficient of hydrocarbon gases, the type of HC gases, the value of butane and pentane coefficients (iC_4/nC_4; iC_5/nC_5).

Based on the results of processing geochemical data and adapted interpretation model, 11 most informative indicators were selected. Anomaly maps are constructed in the Isoline GIS (geoinformation system) based on the selected indicators. Also, a seismic project has been uploaded to the Isoline GIS, including the surfaces of discontinuous faults and structural maps. Geological and geochemical sections have been constructed in two directions (submeridional and sublatitudinal). Based on the materials obtained, taking into account the drilling results (logging, core, well testing results), a comprehensive interpretation was carried out.

The geochemical field according to the sum of the limiting HC C_2-C_4 is medium-contrast, medium-rugged (figure 1B and 3). A low-contrast annular anomaly is observed in the near-water part of the Astrakhan structure, and a low-contrast areal anomaly is observed in the boundaries of blocks 7–8.

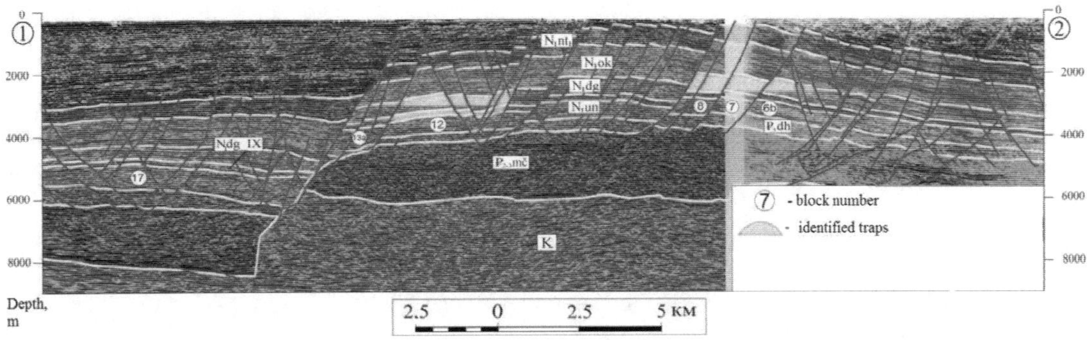

Figure 3. Geological and geochemical section through the Astrakhan structure.

Based on the results of the complex interpretation, a map of total anomalies was created, on which the contours of anomalies identified by all major geochemical parameters were placed. According to the number of matching contours and the accuracy of the coincidence, several sites with varying degrees of reliability and, accordingly, prospects were identified and a perspective production map was created (figure 4).

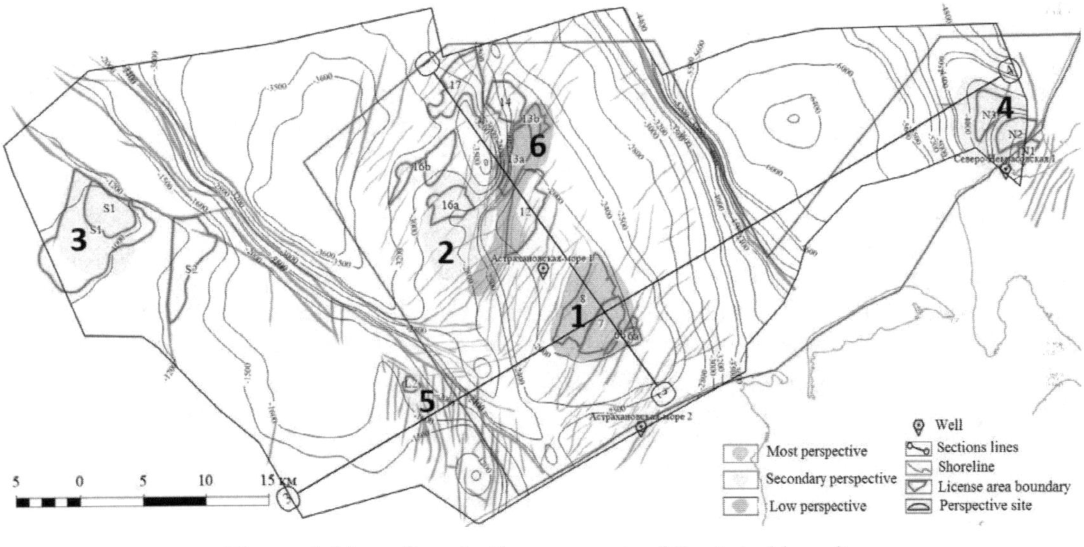

Figure 4. Map of production prospects of the Astrakhan site.

The most promising section 1 is allocated in the vaulted part of the southern block of the Astrakhan structure. This area is distinguished by all geochemical indicators, the contours of all anomalies practically coincide with each other and with the contours of traps No. 7, 8 (figure 3), which are highlighted on all major stratigraphic horizons. The probable phase saturation of the prospective site 1 is oil. Four more sites have been identified with a slightly lower degree of confidence.

Based on the results of the work, recommendations were developed for subsequent geochemical work on the shelf:

- it is necessary to conduct experimental work to assess the properties of bottom sediments in order to develop an optimal sampling technique;
- when conducting offshore geochemical searches for oil and gas, it is recommended to select control sites (HC fields, proven deposits) on the shelf areas with similar conditions;

- when conducting geochemical searches for oil and gas on land, samples should be taken from a depth of at least 1 m.

Based on the conducted research, we have proposed an optimal algorithm for conducting geochemical surveys at sea.

Sampling:

- sampling of bottom sediments. The work can be carried out at sea depth from 1 to 1500–2000 m. Sampling depth is 2–3 m from the sea bottom surface;
- degassing of samples immediately after sampling;
- express analysis on a mobile chromatograph.

Analytical work:

- analytical work in a stationary "town" geochemical laboratory;
- construction of maps and graphs of the distribution of geochemical parameters and coefficients (the selection of parameters for each area is individual);
- delineation of anomalous zones and interpretation of the received data.

Interpretation of results:

- geochemical zoning by phase saturation and conditions of preservation of hydrocarbon deposits.
- comparison of the results with information about the geological features of the area;
- identification of promising areas for drilling.

3. Conclusion

A comprehensive interpretation of geological and geochemical information carried out using modern software products, involving all available geological and geophysical information, allows us to obtain the most reliable results on structural and non-structural deposits, the degree of conductivity/closure of faults and the phase composition of the forecast saturation. During the work, 12 maps of the distribution of hydrocarbon and non-hydrocarbon gases in the area of study and 7 maps of the distribution of metals in bottom sediments were constructed. According to the results of the conducted research, 6 promising sites have been identified, which are ranked according to the degree of prospects, which in turn will significantly reduce the risks during further exploration.

The most informative geochemical indicators on this area were:

- helium and the sum of heavy hydrocarbons C_7+ for assessing the permeability of structural discontinuities;
- low-contrast annular or area anomalies C_2-C_4, C_5-$_6$, the sum of alkanes and cycloalkanes – as criteria for the presence of hydrocarbon deposit.

Acknowledgments

We express our great gratitude to the staff JSC "Rosneft" Kharakhinov V.A. and Levin A.V. for fruitful discussion of the study results.

References

[1] Koryukin G L 2009 *Geochemical criteria for the search for oil and gas deposits in the conditions of the Gulf of Ob* (Moscow: Anniversary collection of LLC Gazflot) p 34–48

[2] Kurchikov A R, Belonosov A Yu and Timshanov R I 2011 New approaches in conducting geochemical oil prospecting studies *Gornie Vedomosti* **6** pp 104–110

[3] Timshanov R I, Belonosov A Yu and Sheshukov S A 2018 Geochemical surveys in different exploration and prospecting stages *Oil and gas geology* **3** pp 103–109
[4] *Handbook of Oil and Gas Geochemistry* 1998 (St. Petersburg: JSC "Nedra Publishing House) p 576
[5] Shakirov R B, Venikova A L, Sokolova N L, Obzhirov A I, Veselov O V, Maltceva E V, Kuziv F V and Leksin V K 2021 Peculiarities of anomalous gas-geochemical fields in the East Deryugin graben of the Sea of Okhotsk *Geosystems of Transition Zones* **5** pp 229–239
[6] Kostrov Yu V 2011 Oil and gas geochemical research of highly explored areas using geoinformatical technologies *Geoinformatika* **4** pp 18–23

Dirty thunderstorms caused by volcano explosive eruptions in Kamchatka by the data of electromagnetic radiation

E I Malkin[1*], N V Cherneva[1], P P Firstov[1,2], G I Druzhin[1] and D V Sannikov[1]

[1]Institute of Cosmophysical Research and Radio Wave Propagation (IKIR) FEB RAS, Paratunka, Kamchatskiy kray, Russia
[2]Kamchatka Branch of Geophysical Survey RAS, Petropavlovsk-Kamchatksiy, Russia

E-mail: *malkin@ikir.ru

Abstract. During volcano eruptions, so called dirty thunderstorms are the sources of electromagnetic radiation. They are caused by ash-gas clouds formed during explosive eruptions. Thunderstorm activity in an ash-gas cloud during volcano eruption is monitored by radio equipment. The VLF direction finder, located at Paratunka, monitors thunderstorm activity in the region of Kamchatka Peninsula including dirty thunderstorms accompanying explosive eruptions of Shiveluch and Bezymyanniy volcanoes. In the paper, we analyze records of electromagnetic radiation associated with dirty thunderstorms occurring during volcano eruptions from 2017 to 2020. During that period 24 eruptions of Shiveluch volcano and 5 eruptions of Bezymyanniy volcano occurred. Seventeen and three of them, respectively, caused dirty thunderstorms. Two-stage scenario of development is typical for all the dirty thunderstorms. The first stage lasts for 5–7 minutes and accompanies eruptive column development. However, if the eruption begins according to a smooth scenario, the first stage may be weak. The second stage lasts for 20–80 minutes and is associated with eruptive cloud formation and propagation. The intensity of this dirty thunderstorm stage depends on eruption power as well as on the interaction of an eruptive cloud during its propagation with the clouds of meteorological origin. Based on the obtained data, that is indicated by the increase of cloud-to-cloud stroke number.

1. Introduction

Volcano explosive eruptions are accompanied by complicated geophysical processes occurring during injection of a great amount of burning ash and volcanic gases composed of 90 % of water steam. The presence of dissolved water in the magma up to 7 % from its composition results in its fragmentation (destruction) under geostatic pressure release and formation of a large amount of dispersed material (pyroclastics) which forms eruptive clouds together with volcanic gases.

Volcanic ash creates danger for population and urban lands in the far-field zone. It is also dangerous for airplane engines. Eruptive cloud propagation monitoring by different methods allows one to notify about ash hazardous situation. That makes it possible to take preventive actions to decrease the consequences. One of the methods to trace eruptive cloud motion is observation of electromagnetic pulses (atmospherics) accompanying lightning strokes that was described in a number of papers [1–3].

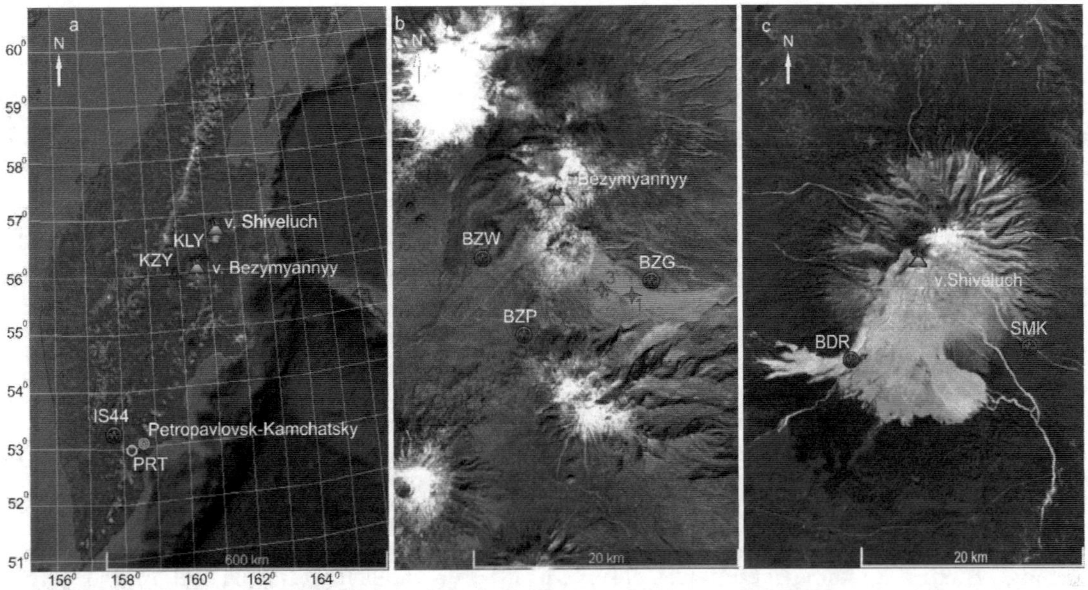

Figure 1. Location of Shiveluch and Bezymyanniy volcanoes and observation sites in the far-field zone on Kamchatka Peninsula (a); location of radiotelemetric seismic stations (RTSS) (BZW, BZP, BZG) near Bezymyanniy volcano and location of lightning strokes recorded by the WWLLN during the eruption on December 20, 2017 (b); location of RTSS (BDR, SMK) near Shiveluch volcano and location of lightning strokes recorded by the WWLLN during the eruption on March 9, 2019 (c) [15].

The Shiveluch volcano in Kamchatka is one of the most active. Its latest catastrophic eruption occurred on November 12, 1964. After a 16-year pause, the extrusive dome has been growing within a newly formed crater from August 1980 up to the present time with short pauses. That is accompanied by explosive eruptions [6]. During strong explosions, eruptive cloud rises up to the tropopause height (10–12 km) and propagates for hundreds of kilometers. In this case, significant lightning activity is observed [7]. In some cases, during the eruptions, pyroclastic flows were formed that is a mixture of hot gas (up to 800 °C), ash and stones moving with the velocity up to 700 km/h along the volcano slopes (figure 2).

Another the most active volcano of the world, Bezymyanniy volcano, is located in the central part of Kluchevskaya Group of Volcanoes on Kamchatka Peninsula (figure 1a). After the catastrophic eruption on March 30, 1956, an extrusive dome was constantly rising up in the newly formed crater (1.5×2.8 km, 700 m deep) [Gorshkov, Bogoyavlenskaya 1965] up to the end of 2012 with some periods of activation in the form of separate explosive eruptions and relatively calm periods. Manifestations of its volcanic activity are various: formation of extrusive domes and blocks; explosions of different intensity including directed ones; eruptions of pyroclastic flows and squeezing of small viscous lava flows. From the end of 2012 up to December 2016, the volcano was in relatively calm and in December 2016, the volcanic activity began to grow.

In the paper, we present observation results for the atmospherics accompanying separate volcano explosions for the period of 2017–2019.

2. Existing methods

A direction finder of very low frequencies (VLF) was developed at the Institute of Cosmophysical Research and Radio Wave Propagation (IKIR) FEB RAS to record electromagnetic radiation form lightning strokes. The software-hardware complex for recording atmospherics is described in the paper [4]. The VLF direction finder, located at Karymshina site, that is 390 km from Shiveluch volcano, recorded 17 dirty thunderstorms accompanying separate explosions of Shiveluch volcano and 3 dirty thunderstorms accompanying Bezymyanniy volcano eruption.

The World Wide Lightning Location Network (WWLLN, http://wwlln.com) allows us to record atmospheric coordinates with the accuracy up to several microseconds [8]. WWLLN records atmospherics from volcanic lightning and gives good results for the monitoring of eruptive cloud propagation in case of optimal location of the network sites [10]. Unfortunately, due to the sparse network of the sites on Kamchatka Peninsula, WWLLN records only powerful lightning strokes which are not more than 10 % from the atmospherics recorded in the same region by the VLF direction finder [5]. Thus, in this paper we use the VLF direction finder as the main source of data.

Data on atmosphere stratification were taken from the open sources (https://ruc.noaa.gov).

3. Data and analysis

Based on the data of Kamchatka Branch of the Geophysical Survey RAS [14] a catalogue of the strongest events on Shiveluch (table 1) and Bezymyanniy (table 2) volcanoes was made for the period of 2017–2019. Then, eruptive cloud height was estimated for the same period. During the first stage, azimuthal distribution of atmospherics in the sector of 0°–80° was considered for the selected events (figure 2a). It is clear from the figure that during the Shiveluch volcano eruption on May 11, 2017, lightning activity was observed for half an hour from the azimuth of 25.6°±10°. During the second stage, graphs for atmospheric counting rate dynamics (***v***) from the direction to the volcano were plotted (figure 2b).

Table 1. Parameters of six the most vivid events recorded during Shiveluch volcano explosions [8].

date	time	first phase time	maximum counting rate	pulse number	second phase time	maximum counting rate	pulse number	WWLLN
11.05.17	**18:24:00**	**4**	**16**	**47**	**28**	**14**	**183**	**4**
23.07.17	17:43:00	4	3	10	26	18	205	0
09.03.19	05:49:00	---	---	---	24	12	107	8
26.09.19	20:57:00	---	---	---	9	5	24	0
01.10.19	23:38:00	4	11	21	---	---	---	0
21.10.19	12:17:00	5	18	45	---	---	---	0

For a more detailed analysis we selected the events on May 11, 2017 (figure 2b), on July 23, 2017 and on October 21, 2019 (figure 3a, b) since they occurred under relatively similar height distribution of airflow velocities.

The eruption on May 11, 2017 is the most vivid event, since it began with an explosion followed by ash-gas mixture outflow. Eruptive cloud rose to the tropopause height (10 km). There are two phases in the counting rate dynamics (***v***) for this eruption. Phase I lasted for ~ 4 min and is characterized by a sudden increase followed by sudden decrease. More gradual increase of ***v*** and significant duration for ~ 30 min is characteristic for the next phase II (figure 2b).

Dynamics ***v*** with two phases was observed for some explosions on Alaska volcanoes, Augustine in 2006 and Redoubt in 2009 [13]. Phase I takes place during the first stage of the eruption on the areas of thermal formation and rising (gas pressure and automodel) that determines its short duration. Phase II characterizes eruptive cloud electrification on the section of floating when it becomes blurred and carried by the wind. In this case, eruptive cloud contacts with the clouds of meteorological origin that causes water steam crystallization and additional electrification causing powerful lightning activity [3].

A similar picture was observed on October 21, 2019, except for the duration of the whole eruption process. In the first case, the duration of the explosive earthquake was ~ 46 min, whereas in the second case the duration of the process was about 10 min. An analogous situation with short time of eruption (~10 min) is observed in all the cases containing only the first stage of volcanic thunderstorm. Thus, we can make a conclusion that presence of phase II of the volcanic thunderstorm on May 11, 2017, and its absence in the eruption on October 21, 2019, is determined by the amount of the erupted material in the form of gas-ash cloud.

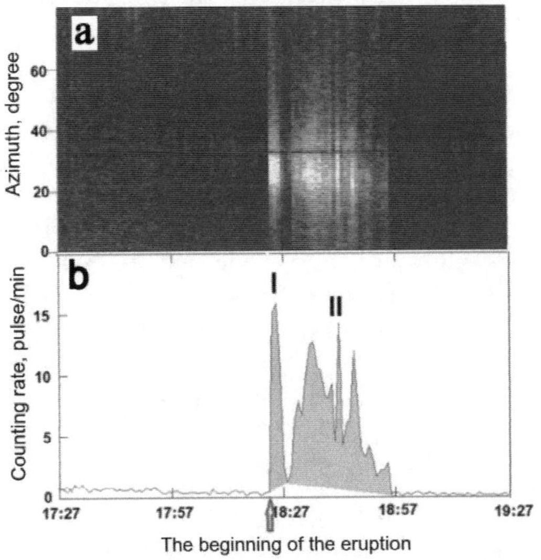

Figure 2. Azimuthal distribution of atmo-spherics in the sector of 0 °–80 ° according to the color gradient from blue (min) to yellow (max) (a), the atmospheric counting rate from the direction to Shiveluch volcano (b) during the eruption on May 11, 2017.

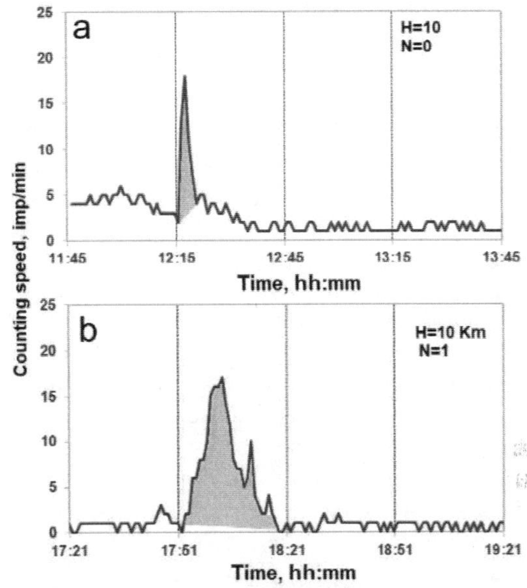

Figure 3. Atmospheric counting rate during Shiveluch volcano eruptions on October 21, 2019 (a) and on July 23, 2017 (b).

The eruption on July 23, 2017, (figure 3b) is opposite in dynamics and is characterized by gradual increase within ~10 min and gradual decrease within ~47 min of explosive earthquake intensity. There is no phase I in the dynamics of volcanic thunderstorm in this case. That, in its turn, confirms the assumption on the dependence of charge separation intensity on the velocity of ash-gas stream outflowing from the vent.

Table 2. Parameters of three events recorded during Bezymyanniy volcano explosions.

date	time	first phase time	maximum counting rate	pulse number	second phase time	maximum counting rate	pulse number	WWLLN
20.12.17	**18:24:00**	**3**	**4**	**9**	**38**	**54**	**790**	**4**
20.01.19	17:43:00	---	---	---	28	6	63	0
15.03.19	17:11:00	---	---	---	58	83	1138	13

In December 2016, Bezymyanniy volcano became active again. On March 9 and June 16, 2017, two explosive eruptions occurred and the third, the most powerful one, began on December 20, 2017, at 03:39. Intensive flowing of pyroclastic streams began at 03:45. When the streams were in the stagnation area, secondary eruptive clouds were generated. This process promoted the formation of a multilayer eruptive cloud. The eruptive cloud height was 15 km, the area, above which it propagated, was ~78 000 km^2 (figure 2a). According to the satellite data, the pyroclastic flow drift length was 6 km and that of mud flows was up to 18 km. Based on the modeling estimates of eruptive cloud motion, the erupted tephra mass on the land was ~ 3·107 t (~0.023 km^3) [9]. Three eruptions of Bezymyanniy volcano described in table 2 were accompanied by dirty thunderstorms. Their peculiarity is the absence of phase I. That confirms the fact that all eruptions accompanied by volcanic thunderstorms began according to the "smooth" scenario (without explosion).

4. Conclusion

During the past years, great attention is paid to electromagnetic radiation from dirty thunderstorms accompanying explosive eruptions. That is associated with the increase of population density on the Earth and ash hazard for the population and for the aviation, in the first instance. In the paper, we compare lightning activity dynamics during explosions. We distinguished the features for the development of two stages of dirty thunderstorms. They are associated with eruptive column formation during the first stage and eruptive column development during the second one. Data analysis allowed us to understand clearly the mechanism of Bezymyanniy volcano eruption occurred on December 20, 2017, and that of Shiveluch volcano eruption occurred on May 11, 2017. Two phases are distinguished in the dynamics $\boldsymbol{\nu}$ depending on eruption character and intensity. Phase I reflects the character of the eruptions onset. For the eruptions, which begin with magma fragmentation according to the explosion scenario, peak phase I is observed. It lasts up to 6 minutes. For the explosions, which begin according to the "smooth" scenario, only phase II is observed. It characterizes lightning activity in a developed eruptive cloud. This phase intensity depends not only on eruption mechanisms but on meteorological conditions in the region as well. That explains the fact that, on the whole, there is no relation between the dynamics $\boldsymbol{\nu}$, eruptive column height and the atmospherics number according to WWLLN data. Monitoring of electromagnetic pulses from dirty thunderstorm strokes expands the informative value of the eruption process and the characteristics of eruptive column propagation depending on meteorological state. Undoubtedly, during current development of technologies and computational methods, VLF radiation monitoring may be included into the system of notification on ash hazard for airplanes having multiple routes along the eastern coast of Kamchatka.

Acknowledgements

The work was carried out within the framework of State Task Subject (2021–2023) «Physical Processes in the System of Near Space and Geospheres under Solar and Lithospheric Impacts» (AAAA-A21-121011290003-0). The yound aurthors of the paper pay tribute of respect and appreciation to Firstov P.P., a great Russian researcher, whose scientific advising promoted the analysis and understanding of the obtained results.

References

[1] Behnke S A and McNutt S R 2014 *Bull. Volcanol.* **76** pp 1–12
[2] Behnke S A, Thomas R J, Edens H E, Krehbiel P R and Rison W 2014 *J. Geophys. Res. Atmos.* **119** pp 833–859
[3] Behnke S A, Thomas R J, McNutt S R, Schneider D J, Krehbiel P R, Rison W and Edens H E 2013 *J.V.G.R.* **259** pp 214–234
[4] Druzhin G I, Pukhov V M, Sannikov D V and Malkin E I 2019 *Vestnik KRAUNTs. Fiziko-matematicheskie nauki – Bulletin KRASEC. Physical & Mathematical Sciences* **27** pp 95–104
[5] Dowden R L, Brundell J B and Rodger C J 2002 *J. Atm and Solar-Terr. Phisics* **64** pp 817–830
[6] Shevtsov B M, Firstov P P, Cherneva N V, Holzworth R H and Akbashev R R 2016 *Nat. Hazards Earth Syst. Sci.* **16** pp 871–874
[7] Firstov P P, Akbashev R R, Holzworth R, Cherneva N V and Shevtsov B M 2017 *Izvestiya, Atmospheric and Oceanic Physics* **53** pp 24–31
[8] Firstov P P, Malkin E I, Akbashev R R, Druzhin G I, Cherneva N V, Holzworth R H, Uvarov V N and Stasiy I E 2020 *Atmosphere* **11** p 634

[9] Girina O A, Loupian E A, Melnikov D V, Manevich A G, Sorokin A A, Kramareva L S, Uvarov I A and Kashnitskiy A V 2018 *Sovremennye problemy distantsionnogo zondirovaniya Zemli iz kosmosa* **15** pp 88–99
[10] Hutchins M L, Holzworth R H, Rodger C J and Brundell J B 2012 *J.A.O. Tech.* **29** 1102–10
[11] McNutt S R and Williams E R 2010 *Bull Volcanol* **72** pp 1153–67
[12] Malkin E, Druzhin G, Firstov P, Cherneva N, Uvarov V, Sannikov D and Stasiy I E 2019 *E3S Web of Conferences* **127** 02021
[13] Smith C, Said R, Eaton A V and Holzworth R H 2018 *25th International Lighting Detection Conference March 12–15 Ft. Lauderdate Florida USA* pp 1–7
[14] Monitoring of volcanic activity (*Electronic materials*) http://www.emsd.ru/~ssl/monitoring/main.htm
[15] Google.maps service (*Electronic materials*) https://www.google.ru/maps
[16] World Wide Lightning Location Network (*Electronic materials*) http://wwlln.com

Probability assessment of possible ice encounters during the petroleum hydrocarbons transportation in the Sea of Okhotsk

V A Romanyuk[1*], **V M Pishchal'nik**[2] **and I V Nikulina**[2]

[1] RN-SakhalinNIPImorneft LLC, Yuzhno-Sakhalinsk, Russia
[2] Sakhalin State University, Yuzhno-Sakhalinsk, Russia

E-mail: *varomanyuk2020@gmail.com

Abstract. The increase in freight traffic along the Northern Sea Route necessitates the study of the ice regime of the freezing seas of Russia, including the Far Eastern ones (Bering Sea, the Sea of Okhotsk and the Sea of Japan). This paper, on the basis of the analysis of the Sea of Okhotsk's ice cover and the location of the edges of the Sea of Okhotsk's ice massif, evaluates the probability of an encounter with ice for the ships in transit during cargo transportation on the Northern Sea Route through the Far Eastern seas along two basic routes from the southern tip of the Kamchatka Peninsula: 1) traverse Cape Lopatka–Sea of Okhotsk–La Perouse Strait; 2) traverse Cape Lopatka–Sangarsky Strait. It was shown that at the stage of ice cover maximum development in the Sea of Okhotsk (February-March), route No. 2 was the most adequate and the safest for ships of ice class Arc4 and below. The section of route No. 1 with a 70 % ice encounter probability was 421 km long in February and 382 km long in March. That section of the route was dominated by very cohesive first-year thin ice up to 70 cm thick with inclusions of first-year ice of average thickness (up to 120 cm).

1. Introduction

In view of the development of oil and gas resources of the continental shelf, as well as global tendency of increase in air temperature, which contributes to ice extent reduction of the freezing seas of Russia, much attention is currently paid to the development of the Northern Sea Route (NSR). In January 2021, Arctic tanker of Arc7 ice class "Christopher de Margerie", without icebreaker escort, successfully completed the passage from the port of Sabetta along the NSR in easterly direction. This event in practice showed the effectiveness of using this transport artery and the need for further development of its transit potential.

When transporting goods along the NSR to the countries of the Asia-Pacific region, after passing the Bering Strait, the vessel has two options on route to its destination: 1) through the Bering Sea, the Sea of Okhotsk and Sea of Japan (Far Eastern transport corridor) and 2) through the Bering Sea and the Pacific Ocean. It should be noted that in terms of the severity of ice conditions, the Sea Okhotsk and Bering Sea are comparable to the seas of Russian sector of the Arctic. The maximum development of the ice extent in the Far Eastern seas is observed in February-March. Probability assessment of the encounter with predominant first-year ice, performed on the basis of regular ten-day aviation observations of the state of the ice extent in 1956–1991, was presented in [1]. Modern pentad (once every five days) satellite observation data enabled not only a more accurate probability assessment of encounters with ice of various ages, but also with ice of different concentrations. The objective of this paper was to assess the probability of ice encounter for the ships proceeding from the southern tip of the Kamchatka Peninsula (Cape Lopatka) to the countries of the Asia-Pacific region through the Sea of Okhotsk and along the Pacific coast of the Kuril Islands at the phase of maximum development of the ice extent (February-March) for 1981–2010.

2. Material and Methods

The data of satellite microwave radiometers for variations in the ice extent area with pentad sampling rate from 1981 to 2010 (nominal spatial resolution up to 4 km) of the Japan Meteorological Agency [http://ds.data.jma.go.jp] were used as the initial data. The area of the ice extent was calculated using ICE software package [2] and original authoring software tools. Computational error of ice extent area does not exceed +/- 1–2 % [3]. Ice cover was calculated as the ratio of the area occupied by ice to the gross area of the sea, expressed as a percentage.

Statistical calculations were carried out for two navigation routes: 1) traverse Cape Lopatka–port of Korsakov–point 38°30'N, 134°00'E; 2) traverse Cape Lopatka–Sangarsky Strait–point 38°30'N, 134°00'E. Due to the ongoing active discussion about the creation of a multifunctional transport hub in the port of Korsakov, a call at the port of Korsakov, the largest port of Sakhalin region, was included in Route No. 1. The location of ice massif edge at the stage of its maximum development was refined based on the archive of daily data from the MASIE multispectral system for analysing daily satellite observations [http://nsidc.org/data/masie] for the period from 2006 to 2010 (nominal spatial resolution up to 1 km).

The probability of ice encounter (V) was calculated by the formula:

$$V = \frac{N}{n} * 100\ \%$$

where, N – number of ice encounters, n - number of observtions.

When verifying and analysing the calculated data, the following tools were used: the archive of visible range satellite images of the ice extent (NOAA, Meteor, Kosmos, TERRA, AQUA, NPP spacecrafts), radar images (Sentinel-1, 2), atlases of ice and the boundaries of ice propagation [4–5], as well as the results of the analysis of complex hydro-meteorological studies carried out in different years in the Sea of Okhotsk [6].

All terminology and concepts of ice phenomena correspond to the "International Symbols for Nautical Ice Charts and the Nomenclature of Sea Ice" [7].

3. Trends in sea ice cover in the Sea of Okhotsk for the period 1882–2020

Analysis of the ice regime elements of the Sea of Okhotsk paid particular interest to the long-term fluctuations in the area of ice extent, which allowed assessing the intensity of ice processes and relating them to global climate changes occurring in the current period. [3, 8] developed the secular series of interannual variability of ice cover values in the Sea of Okhotsk, based on ship, aviation and satellite ice extent observations data and air temperature observation data at coastal hydro-meteorological stations. The analysis of the secular series allowed us to conclude that in recent decades there was intensive reduction in the area of ice extent.

Figure 1 shows the interseasonal variability of sea ice cover in the Sea of Okhotsk for the period 1882-2020. From 1882 to 1979, the magnitude of negative trend in ice cover was 0.3 % / 10 years. The decrease in ice cover from 1980 to 2020 was one order of magnitude more intense - 4.9 % / 10 years. Spectral analysis of the series under study made it possible to identify periods with fluctuations of 6, 9, and 17–18 years. The latest period of ice cover decrease (from 1980 to the present) coincided with the current period of warming and so far continued for 42 years.

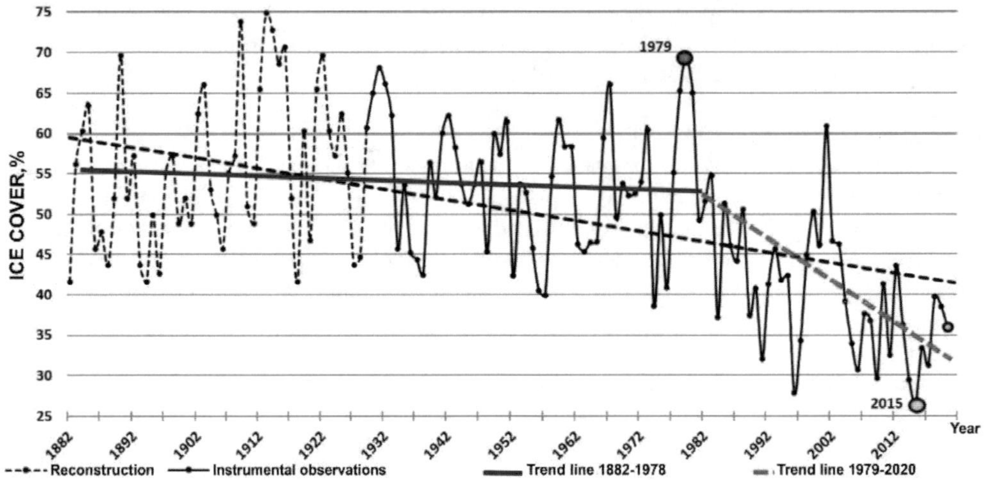

Figure 1. Interseasonal variability of sea ice cover in the Sea of Okhotsk for the period 1882–2020.

4. Probability assessment of possible ice encounter

4.1. Route No. 1: traverse Cape Lopatka–port of Korsakov–point 38°30'N / 134°00'E

Route No. 1 total length was 2 457 km (1 327 nautical miles). Analysis of the calculated data for the southern region of the Sea of Okhotsk allowed us to conclude that in mid-February the zone with a high probability of ice encounter (more than 60 %) was located along the eastern coast of Sakhalin at a distance of 120–250 km from the coast. South of 46°50'N latitude the width of the zone decreased to 60–100 km (figure 2). That ice mass was formed in the north-western region of the Sea of Okhotsk and drifted southward under the influence of winter monsoon [9]. In mid-February, the ice concentration was 9–10 points. The massif was dominated by thin first-year ice (up to 70 cm) with inclusions of first-year ice of average thickness (up to 120 cm).

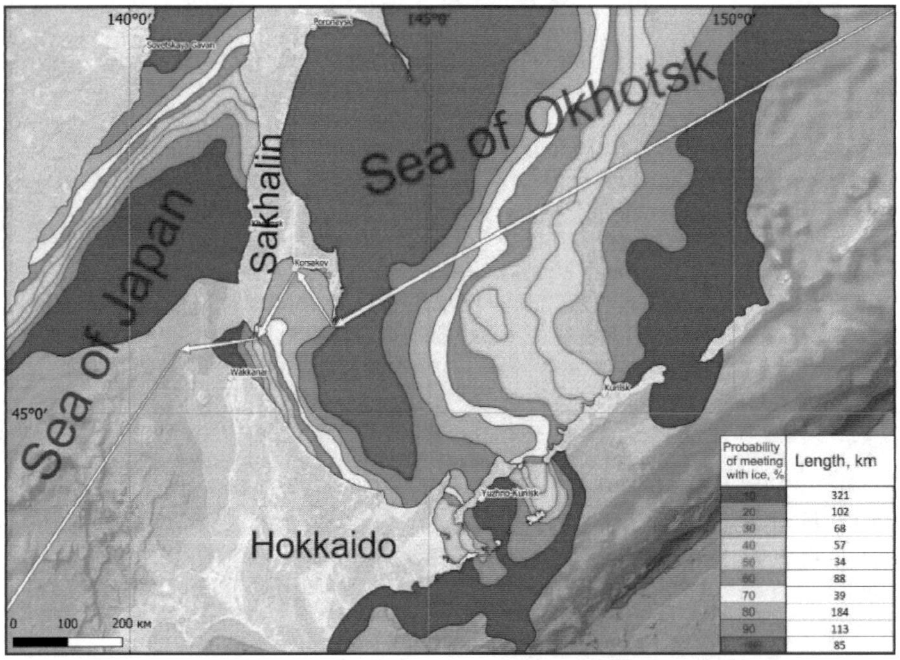

Figure 2. Probability of ice encounter on Route No. 1 on February 15.

On Route No. 1, on the 85 km segment, the probability of ice encounter was 100 %. On 336 km long section (from 144°25' to 143°30'E), the probability of ice encounter was 70–90 %. A characteristic feature of ice conditions in this area was the decrease in the concentration of ice extent to 4–8 points. For the rest of the route, the ice extent was represented by open ice with a probability of encounter from 10 to 60 %, which would not be a serious obstacle for ice class vessels. Thus, the overall length of the part of Route No. 1 with the probability of ice encounter of more than 1 % was 1 092 km. In the Aniva bay in mid-February, the probability of ice encounter was 70–80 %. The ice massif of the bay was represented by ice of local generation, the thickness of which did not exceed 15–25 cm.

Figure 3 shows the map of probability of ice encounter at the stage of the maximum development of ice extent in the Sea of Okhotsk (March 10). The border of the zone with high probability of ice encounter (more than 60 %) was located 50–125 km east of its position as of February 15. The overall length of the part of Route No. 1 with probability of ice encounter of more than 1 % was 1 240 km.

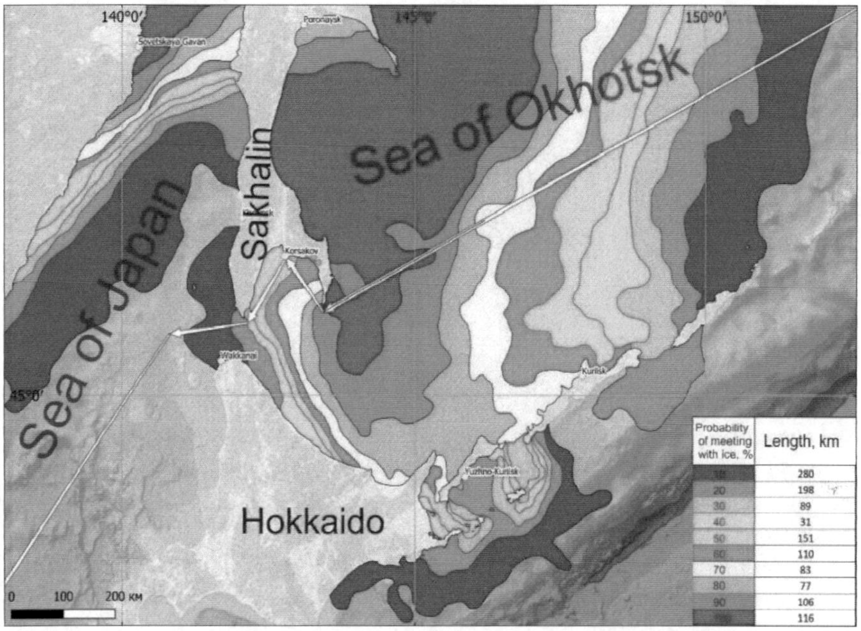

Figure 3. Probability of ice encounter on Route No. 1 on March 10.

4.2. Route No. 2: traverse Cape Lopatka–Sangarsky Strait–point 38°30'N / 134°00'E

The total length of Route No. 2 – 2 378 km (1 284 nautical miles).

The islands that form the western, southern and eastern boundaries of the Sea of Okhotsk are a natural barrier, forming a "trap" for ice drifting from the northern part of the sea. At a certain stage of filling the natural trap, ice begins to squeeze out through the Izmena Strait, Catherine Strait, Frisa Strait and the First Kuril Strait into the Pacific Ocean. The overall length of the part of Route No. 2 with the probability of ice encounter above 1 % on February 15 was 502 km (figure 4). The maximum value of the frequency of ice encounter when a vessel passes along Route No. 2 in mid-February is 40 % (area of Shikotan Island). The ice massif in these areas was represented mainly by one-year thin ice. On March 10, the overall length of the part of Route No. 2 with the probability of ice encounter of more than 1 % increased to 580 km (figure 5).

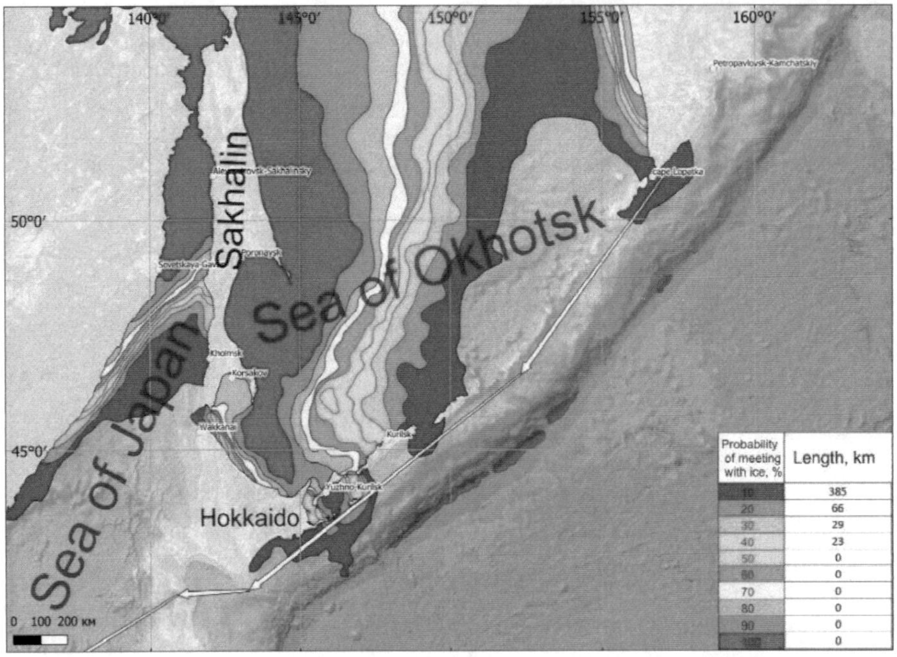

Figure 4. Probability of ice encounter on Route No. 2 in February.

As in mid-February, the greatest probability of ice encounter in the first decade of March was in the area of the Lesser Kuril Ridge (maximum probability was 50 %). Ice conditions during that time period were similar to those of mid-February. To exclude an ice encounter, the route could be laid further southeast, while its length would increase by ~ 70 to 100 km.

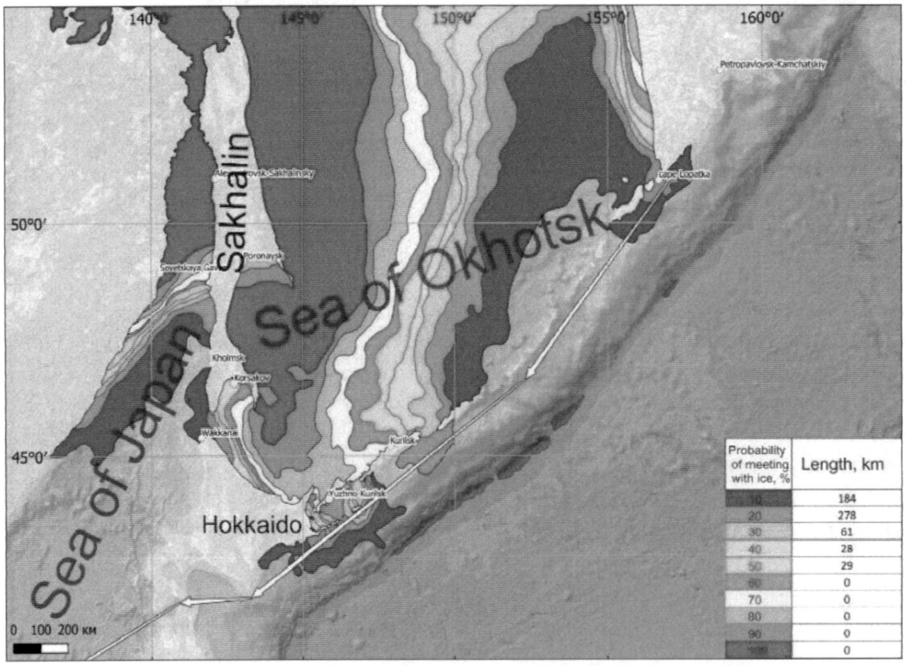

Figure 5. Probability of ice encounter on Route No. 2 on March 10.

5. Conclusion

1. The total length of Route No. 1 was 2 457 km, Route No. 2 – 2 378 km. When vessels take Route No. 1: traverse Cape Lopatka–port of Korsakov–point 38°30'N /134°00'E, the length of the section with the probability of ice encounter of more than 1 % was 1 092 km in mid-February and 1 240 km in the first decade of March. The length of the section of Route No. 1 with the frequency of ice encounters of 70–90 % in February and March was 336 and 265 km, respectively. The length of the route part with 100 % probability of ice encounter and the most difficult ice conditions (concentration of 9–10 points, ice thickness up to 120 cm and hummocking of 2–3 points) in March was 131 km, which was 31 km more than in mid-February.
2. On Route No. 2: traverse Cape Lopatka–Sangarsky Strait–point 38°30'N /134°00'E, the section with the highest probability (40–50 %) of ice encounter in mid-February and early March, was located to the southeast of the Lesser Kuril Ridge. The total length of that section was 200–250 km. Also, ice encounters were a possibility in the area of the First Kuril Strait and Catherine Strait. Those sections could be bypassed southeast, slightly increasing the length of the route.
3. In case of independent navigation of vessels of ice class Arc4 and below, from the point of view of the route length and ice encounter probability, the most favourable route was Route No. 2. With daily monitoring of ice conditions and adjustments of Route No. 2, the probability of ice encounter for ships could be completely eliminated.
4. For independent navigation of vessels during the period of maximum development of ice extent in the Sea of Okhotsk (February-March) along Route No. 1, the vessel should have a minimum ice class not lower than Arc5.

References

[1] Yakunin L P, Plotnikov V V and Petrov A G 2003 Ledovye usloviya *Proekt «Morya». Gidrologiya i gidrokhimiya morey* (SPb.: Gidrometeoizdat) **3** 1 p 348

[2] Pishchalnik V M, Bobkov A O, Minervin I G and Romanyuk V A 2015 *Svidetel'stvo o gosudarstvennoy registratsii programmy dlya EVM 2015660472* Programmnyy kompleks dlya rascheta ploshchadi ledyanogo pokrova v Okhotskom i Yaponskom moryakh po dannym DZZ PK «LED»

[3] Pishchal'nik V M, Romanyuk V A, Minervin I G and Batukhtina A S 2016 Analiz dinamiki anomaliy ledovitosti Okhotskogo morya v period s 1882 po 2015 g *Izv. TINRO* **185** p 228–239

[4] Tambovsky V S and Pishchal'nik V M 1993 The ATLAS of ICE of JAPAN and OKHOTSK SEAS (Yuzhno-Sakhalinsk) p 195

[5] Yakunin L P, Plotnikov V V and Petrov A G 2003 Ledovye usloviya (Ice conditions) *Gidrometeorologiya I gidrokhimiya morei* (SPb: Gidrometeoizdat) **8** 1 p 348

[6] Plotnikov V V 2002 *Izmenchivost' ledovykh usloviy dal'nevostochnykh morey Rossii i ikh prognoz* (Vladivostok: Dalnauka) p 172

[7] WMO Sea-Ice Nomenclature 1970 (Geneva)

[8] Romanyuk V and Kozeka F 2015 *Proc. 30th Intern. Symp. on Okhotsk Sea & Sea Ice. Mombetsu, Hokkaido, Japan* Recovery ice cover Okhotsk and Sea of Japan for the period from 1929 to 2014 pp 254–256

[9] Minervin I G, Romanyuk V A, Pishchal'nik V M, Truskov P A and Pokrashenko S A 2015 Zoning the ice cover of the Sea of Okhotsk and the Sea of Japan *Herald of the Russian Academy of Sciences* **85** 2 pp 132–139

On the relationship of power lines outages caused by thunderstorms with Forbush decreases of cosmic rays

L P Shadrina[1,*], **V I Kozlov**[2] **and Yu M Grigoriev**[3]

[1]Academy of sciences of RSYa, Yakutsk, Russia
[2]Yu G Shafer Institute of cosmophysical research and aeronomy of SB RAS, Yakutsk, Russia
[3]M K Ammosov North-Eastern federal university, Yakutsk, Russia

E-mail: *lushadr@mail.ru

Abstract. It is known that power lines outages often occur during thunderstorms. Here are the results of comparing of power lines outages in Yakutia from 2012 to 2018 with the database of Forbush-Storm events. This database contains information on geomagnetic storms and Forbush-decreases of cosmic rays from 1996 to 2018. There are 3 classes of the events: if these two ground-based manifestations of solar wind disturbances occur simultaneously (Forbush with Storm, F+S) or separately (Forbush without Storm, F-S and Storm without Forbush, S-F). For 7 years in the summer time, 73 power lines outages associated with thunderstorms were recorded. It is shown that in 56 cases these outages occurred simultaneously with (F-S) class, 16 – with (F+S) class, and only in 1 case lightning outages were not associated with Forbush-Storm events (-F-S). In 19 cases of (S-F) class, not a single lightning outage was recorded. This means that lightning outages on power lines are mainly associated with decreases in the cosmic rays intensity, and during geomagnetic storms, power transmission disruptions occur when storms are simultaneous with Forbush-decreases of cosmic rays. Apparently, this indicates the significance of the effect of cosmic rays on atmospheric electricity, and it is more significant than the effect of geomagnetic storms.

1. Introduction

It is usually assumed that outages on power lines during thunderstorms occur due to sudden jumps in the geomagnetic field in time (dB/dt) [1–2]. Sharp fluctuations of the geomagnetic field – geomagnetic storms and substorms – produce distributed electric fields, which leads to powerful geomagnetically induced currents (GIT) in technology conductors. In GIT power systems, they lead to dangerous effects: saturation of transformers, their overheating and destruction; generation of parasitic harmonics of alternating current; disruption of standard protective relays and, as a consequence, disconnection of power lines; fluctuations in active and reactive power in the power system and, as a consequence, to false operation or failure of automation and relay protection, massive power supply disruptions in load nodes [3]. The resulting currents reach values of tens of amperes. According to [4], the main cause of accidents on power lines are GIT generated in conductive materials during magnetic storms. In this case, geomagnetic storms, as a rule, should lead to thunderstorms and related outages on power lines. But this is not always so.

It was noted in [5] that the Dst index of the geomagnetic field, whose sharp decreases are a sign of geomagnetic storms, has a positive relationship with the number of thunderstorms, but there is evidence of an increase in the number of thunderstorms on days without the manifestation of magnetic storms. And according to [6], a negative correlation of thunderstorm activity with the parameters of solar activity and a positive correlation with the flow of galactic cosmic rays (GCR) was found. It was also found [7–10] that GCR can be one of the factors influencing the formation of clouds and associated cyclones, thunderstorm activity and other climatic phenomena, which are the main source of ionization in the stratosphere and troposphere.

In [11–13], similar mechanisms of the influence of GCR on cloud cover, on the transparency of the atmosphere, on the control of thunderstorm activity, and even on the initiation of lightning discharges was described. In this regard, it is important to understand what is still more significant for thunderstorms – geomagnetic storms or cosmic rays.

2. Data analysis

We compared an amount of power lines outages caused by thunderstorms (further thunderstorm outages) in Yakutia from 2012 to 2018 with the database of Forbush-Storm Classification (FSC) [14]. This database contains information on geomagnetic storms and Forbush-decreases in galactic cosmic rays (GCR) from 1996 to 2018. There are 3 classes of the events: if these two ground manifestations of solar wind disturbances occur simultaneously (Forbush with Storms, F+S) or separately (Forbush without Storms, F-S and Storms without Forbush, S-F).

Figure 1 shows a sheet from the database of Forbush-Storm Classification (FSC) [14]. Columns B–F contain information about geomagnetic storms (date, day in a year, an hour of storm beginning, main phase of the storm duration ΔT, geomagnetic index Dst amplitude ΔDst, in nT), columns G–I show information about Forbush-decreases (date, an hour of Forbush beginning, magnitude of CR decrease, in %), columns J–L show class of events: Forbush without Storms (F-S), Forbush with Storms (F+S), Storms without Forbush (S-F).

	A	B	C	D	E	F	G	H	I	J	K	L
1	Storm						Forbush			Class		
2	N	Date	Day	Begin	ΔT	ΔDst	Date	Time,UT	Magn,%	F-S	F+S	S-F
3							03.01.2017	3:00	1,3	1		
4	1	04.01.2017	3	6	17	39	04.01.2017	23:00	1,0		1	
5	2	05.01.2017	4	14	11	31						1
6	3	18.01.2017	17	5	11	55	17.01.2017	21:00	1,4		1	
7	4	21.01.2017	20	17	11	32						1
8	5	26.01.2017	25	13	7	37	26.01.2017	8:15	0,9		1	
9	6	31.01.2017	30	9	11	35	31.01.2017	3:00	1,2		1	
10	7	03.02.2017	33	2	8	30						1
11							15.02.2017	21:00	1,3	1		
12	8	17.02.2017	47	3	5	32						1
13	9	21.02.2017	51	22	12	32	22.02.2017	19:00	1,1		1	

Figure 1. An example of a sheet from the database of Forbush-Storm Classification (FSC) [11].

Data on thunderstorm outages in summer time of 2012 to 2018 in one of the regions of Yakutia were provided by Yakutskenergo. 73 thunderstorm outages were recorded for 7 years under investigation. There can be from one to several power lines outages during any thunderstorm, table 1 shows an example of comparing FSC data and thunderstorm outages in summer 2017. 1st column shows the date of the event, 2nd column – the FSC class, 3rd column – the number of thunderstorm outages during this and the next day, 4th column – the intensity of the geomagnetic storm (the Earth's magnetic field decrease in the main phase of the storm, in nanoTesla) and/or the Forbush-decrease of GCR intensity (in %). The last column contains a note, special data. For example, in the given example on July, 19 the event was attribute to class Forbush-decrease without a geomagnetic storm (F-S), but the storm began a day later, on July, 20, i.e. maybe this event can be attributed to the class (F+S).

Table 1. Example of data analysis from June, 01 to August, 31 of 2017.

Date	Class	Number of thunderstorm outages	Intensity of GMS (nT) and FD (%)	Note
June, 16	S-F	--	61 nT	
June, 22	F-S	2	3%	
June, 25	F-S	2	2%	
July, 02	F-S	7	7%	
July, 08	F+S	1	44 nT +1%	
July, 10	F-S	1	1%	
July, 16	F+S	8	115 nT +8%	
July, 19	F-S	1	1%	At July,20 GMS 43 nT
August, 03	F+S	2	67 nT +2%	
August, 14	F-S	3	3%	
August, 16	F-S	1	1%	
August, 22	S-F	--	54 nT	
August, 31	S-F	--	109 nT	

Table 1 shows that there were 10 days with 28 thunderstorm outages, from 1 to 8 per day, in 2017. 6 geomagnetic storms (GMS) and 10 Forbush-decreases (FD): 7 events (F-S), 3 events (F+S), and 3 events (S-F) were observed in this summer of 2017.

One can see that 3 dates – July, 8, July, 16 and August, 3, assigned to the class (F+S), are accompanied by thunderstorm outages. Moreover, during 2 from that 3 dates – on July, 8 and August, 3 – geomagnetic storms were moderate, Forbush-decreases were small, and there were only 1 and 2 power lines outages on these days. On July, 16 there was a large geomagnetic storm and a large decrease in the CR intensity – and 8 power lines outages.

On 7 dates – June, 22 and 25, July, 2, 10 and 19, August, 14 and 16 Forbush without storms (F-S) were observed, and on July, 2 there was an intense Forbush-decrease of GCR intensity, 7%. During all these days there were from 1 to 3 thunderstorm outages, and on July, 2, an intense Forbush-decrease were accompanied by 7 thunderstorm outages in a day.

Three dates were attributed to the FSC events looked as a geomagnetic storm without Forbush-decrease (S-F) – on June, 16, August, 22 and 31, two moderate storms and one large storm. None one of them were accompanied by power lines outages.

A similar analysis was carried out for the summer time of each year from 2012 to 2018.

3. Results and discussion

The summary data of the events analyzed for all 7 years are given in Table 2. For each FSC class, "yes" means the number of days when thunderstorm outages occurred during the Forbush-Storm event, "no" means that there is no such coincidence. For example, one can see that in 2012 there were 7 days with thunderstorm outages, 5 events took place during (F-S) class, 2 – during (F+B) class. There were 2 events of class (S-F), and both of them were not accompanied by power lines outages.

In common, Table 2 shows that from the 73 cases thunderstorm outages during summer 2012–2018 under the analyze, most of them (56) occurred during Forbush-decreases without geomagnetic storms, i.e. during class (F-S) registration. There are 16 cases associated with the Forbush-decreases with storms class (F+S), and only in 1 case of thunderstorm outage is not associated with Forbush-storm events (this class of events is designated (-F-S)). Moreover, in 19 events of geomagnetic storm registration without Forbush-decrease – events of class (S-F) FBC, not a single thunderstorm outage was recorded. This indicates a significantly more important role of the intensity of GCR in comparison with geomagnetic storms for the generation of thunderstorms and related power lines outages.

Table 2. The comparison of power lines outages caused by thunderstorms with the Forbush-Storm classification.

Years	F-S		F+S		S-F		-S-F		Number of thunderstorm outages
	Yes	No	Yes	No	Yes	No	Yes	No	
2012	5		2			2			7
2013	5		1			5			6
2014	15		2			1			17
2015	9		5			4			14
2016	4		2			2		1	7
2017	7		3			3			10
2018	11		1			2			12
Sum	**56**		**16**			**19**		**1**	**73**

Apparently, this indicates that the occurrence of thunderstorms and related outages on power lines depend more on the intensity of cosmic rays than on geomagnetic storms. This is consistent with the conclusions of [7–10]. One of the possible physical mechanisms of such an effect was described in [11–13]: "The different charges necessary for the emergence of a thunderstorm cloud are formed by cosmic rays and radioactive elements in the air and soil. At the stage of mature and disintegration of a thunderstorm cloud, the main generator of dissimilar charges necessary for the formation of a strong electric field in the cloud and lightning are the lightning itself, in the highly branched and ionized traces of which huge amounts of dissimilar ions are formed. Lightning discharges in clouds are initiated by broad atmospheric showers (SHAL) formed by cosmic rays of extra high energies ($\varepsilon > 10^{14}$ eV)". The relation of GCR with atmospheric phenomena is not limited by clouds, they control thunderstorm activity and initiate lightning discharges, affect atmospheric circulation, weather and climate, and in general, the processes of formation and development of baric systems of moderate and high latitudes [10].

Also, the fulfilled studies show the usefulness of using the Forbush-Storm classification of the events [14] to understand the impact of space weather on terrestrial processes.

4. Conclusions

This paper presents the results of a comparison of thunderstorm outages on power lines for the summer of 2012–2018 with Forbush-decreases of GCR and geomagnetic storms based on the Forbush-Storm classification of the events. The results are the following:

1. Most of the thunderstorm outages on power lines occur on days when there are Forbush-decreases of GCR (56 out of 73).
2. During geomagnetic storms thunderstorm outages on power lines occur if these geomagnetic storms occure simultaneously with Forbush-decreases of GCR (16 out of 73).

Taking into account the data presented here will help to assess the geoecological risks from the effect of interplanetary disturbances on ground systems.

Acknowledgements

This research was supported by grant from the Russian Foundation for Basic Research (RFBR) No 18-42-140002 p_a.

References

[1] Sokolova E Yu, Kozyreva O V, Pilipenko V A, Sakharov Ya A and Epishkin D V 2019 Variations of geomagnetic and telluric fields in the North-Western regions of Russia under space weather disturbances: Connection with the geoelectric structure and induced currents in power lines *Geophysical Processes and Biosphere* **18** pp 66–85

[2] Boteler D H, Pirjola R J and Nevanlinna H 1998 The Effects of Geomagnetic Disturbances on Electrical Systems at the Earth's Surface *Adv. Space Res.* **22** pp 17–27

[3] Kuznetsov V D 2012 Solar-Earth physics and its applications *Successes of Physical Sciences* (*SPS*) **182** pp 327–336

[4 Avakyan S V, Voronin N A and Dubarenko K A 2012 The influence of magnetic storms on the accident rate of electric power, automation and communication systems *SPbGPU Scientific-Technical Bulletin. Science and Education* **2** pp 253–265

[5] Gvozdarev A Yu and Sakharov Ya V 2007 Investigation of the relationship of the lightning activity of the Altai Mountains with the geomagnetic activity indices *Actual problems of geography* pp 88–90

[6] Dmitriev A N, Krechetova S Yu and Kocheeva N A 2011 Thunderstorms and forest fires from thunderstorms on the territory of the Altai Republic. *Gorno-Altaisk: RIO GAGU* pp 154 p

[7] Kuznetsov V V and Cherneva N V 2008 Investigation of forbush depressions and terminator effects in an atmospheric electric field at the Paratunka Observatory (Kamchatka) *Vestnik of KRAUNC. Earth Sciences* **1** pp 89–97

[8] Kozlov V I and Solovyev V S 2008 The change of Northern Asia cloudiness depending on cosmic rays intensity variations *Modern Problems of Remote Sensing of the Earth from Space* **5** pp 478 pp 483

[9] Kudryavtsev I V and Unger H 2011 Variations of atmosphere transparency under the galactic cosmic rays effect as a possible reason for their influence on the cloud formation *Geomagnetism and Aeronomy* **51** pp 668–676

[10] Veretenenko S V and Ogurtsov M G 2015 The nature of long-term correlations between cloudiness and the flux of galactic cosmic rays variations *Geomagnetism and Aeronomy* **55** pp 457–465

[11] Ermakov V I and Stozhkov Yu I 2004 Physics of thunderstorm clouds *PhIAN* (*Preprint*) 21 p

[12] Krymski G F 2002 Cosmic rays and near-Earth space 2002 *Solar-Earth physics* **2** pp 42–45

[13] Krymski G F 2018 Electrical charge accumulation on raindrops *NEFU* (*Bulletin*) **6** pp 84–85

[14] Shadrina L P 2020 Catalog of Forbush – Storm events classification Database (*Certificate* 2020621842)

Development of the dual-channel frequency meter for measurements with hydrostatic pressure sensor

N S Stovbun* **and S A Gulyakov**

Institute of Marine Geology and Geophysics Far Eastern Branch of Russian Academy of Science, Yuzhno-Sakhalinsk, 693022, Russia

E-mail: *nikolay19972016@gmail.com

Abstract. The paper describes the process and the results of development of the dual-channel frequency meter, which function is to measure the output frequency generated by bottom-mounted pressure sensors. The sensors are actively used to monitor the marine environment. AVR family microcontrollers were used as the computing core of the presented device. This solution allows to obtain far lower power consumption, which is especially important when operating with no industrial power supply system in the coastal zone. As a result, we can deploy a reliable monitoring equipment capable of long-term saving data and if necessary transmit it for further processing. The developed frequency-meter is able to continually record the ambient temperature, atmospheric pressure and dynamically varying output frequency, which depends on hydrostatic pressure (sea level). To obtain more accurate data, we implemented a frequency measure method called reciprocal counter with lower relative error not affected by value of the output frequency. A laboratory experiment has been conducted, which confirms the suitability of the developed frequency meter for field-oriented conditions.

1. Introduction

One of the scientific researches of the Institute of Marine Geology and Geophysics is a coastal wave monitoring for identifying dangerous marine phenomena. Long periodic wave registration is conducted with special wave recorders, which have sensitive elements constructed on the basis of crystal oscillators, which output vibrating frequency changes depending on hydrostatic pressure. The recorders are waterproof, which makes it possible to mount them on seafloor for continuous measurements. There are two monitoring methods: cable systems and autonomous wave registration. For the second method, we use the autonomous waves recorders (AWR) intended for long-term periodic registration of the absolute pressure of the sea. It is not connected to the coastal measuring construction and data is saved in non-volatile memory, which is read only after removing the AWR ashore [1]. The power supply is provided by internal battery enough for long periodic measurements.

The cable systems consist of the hydrostatic pressure sensor with a cable communication line connected to a coastal construction that measures the output values of the sensor. As a result, the system can record all the data in real time with further transmission of the information for analysis. In contrast to the AWR, the advantage of the cable system is no loss of the recorded data with losing the sensor, which may occur due to sea waves, ice and other reasons [2]. However, the cable system requires a communication line between the sensor and the coastal measuring construction and, considering the lack of the power supply system in coastlines, it must have a backup power source. In order to make the measuring process less complicated, we should decrease the amount of cables used for communication and powering the sensors. To simplify the problem, it is recommended to use the frequency output sensors, which can be connected with one coaxial load-bearing cable [3]. Meteorological condition must be taken into account with wave registration, which results in using meteorological station, that also requires its own power supply making it more difficult to install the measuring equipment. Based on the above, when installing a cable monitoring equipment, we have not only to record meteorological condition and the output frequency with required discreetness within certain range, but also make its power consumption low enough for long-term measuring with powering by either a small-capacity battery or alternative source of energy like a solar panel with an energy accumulator.

In order to install the monitoring equipment and carry out the experiment, the standard solution is a complex of a personal computer, PCI timer/counters and digital meteorological station [4]. Computer parts and software must be modified to meet required conditions [5]. This method is not optimized since a PC has an excessive computational capability compared to its high power consumption for study of the wave processes.

The developed dual-channel frequency meter is the measuring coastal low-power device, which function is to record the output frequency of the hydrostatic pressure sensors, atmospheric pressure and temperature with data accumulation in non-volatile memory. The paper describes the developing process of the device, its block diagram and the implemented method of frequency measuring.

2. Development of the dual-channel frequency meter

2.1. Block diagram

The measuring device presented in this article is based on microcontrollers (MC) as the «intellectual» core, that decreased the power consumption of the scheme from several amps to milliamps. Its electrical characteristic makes it possible to operate with its own power source in any coastal zone with no industrial power supply system. It has an SD card as the data storage, real time clock module for saving data with time series and a barometer module for meteorological parameters.

The block diagram of the frequency meter is shown in figure 1. When conducting an experiment with several sensors, the equipment must provide the simultaneous measuring and recording of the values [6]. Usage of the personal or industrial computer could be the universal solution since modern processors have high frequency rate and software or hardware multithreading. Usage of the microcontroller can lead to a data loss or an asynchronous measurement of the frequency which is not valid for the experiment. Considering the fact, that it takes time to record the data packet to the SD and an error may occur when two events must be handled, we have taken a decision to use three separate MCs connected to the common I2C bus.

For frequency measurement two separate slave ATmega8-16PU MCs are used. The slave MCs have hardware timer-counters which are able to count the voltage impulses and the time between them that can be used for period measurement. It is more advantageous than the usage of mass-produced timer-counter for personal PCs characteristics of which don't exceed the used ones.

A piezoresistive absolute pressure sensor LPS25HB is used for detecting the absolute pressure and the ambient temperature. It is a digital output MEMs barometer with communication through I2C bus. A real time clock module (RTC) is connected to the common bus in order to obtain time reference for the recorded values. It has its own power source (battery of CR2032 type), which makes it possible to count the time after the main power source shutoff. The RTC is used to make the frequency measurements simultaneous in order to obtain strongly identical discreetness for both sensors. The RTC has an integrated temperature-compensated crystal oscillator (accuracy of 2 ppm) which is used to configure the sample rate of the frequency meter. The frequency output of the oscillator can be selected by the I2C bus master. The synchronization signal is connected to the slave MCs, which perform the simultaneous frequency measurements with a sample rate chosen by a user for an experiment.

The central component of the scheme is the main ATmega32A-PU MC. It reads all the needed data from peripherals, forms a data packet with further transferring the information to the SD. According to the SD card specification, it can enter SPI bus mode, which makes it compatible with MCs of AVR family through SPI bus regardless of SD manufacturer or its classification. When using an SD in the PC it requires a special program driver to identify it as a memory space for file recordings. The MC uses its own data structuring on the SD without a file system. Data placed in the storage is a large thread of bytes with no information where it begins or ends. Since the MC does not use a file system for data recording it results in no files on the SD when plugging it in a PC. If an SD falls into a wrong user, one cannot read the information. We created our own software desktop application to read the data from the storage and convert it into usual files for further analysis. It was not one of the technical requirements but it can be an additional feature. The main purpose of rejecting file systems is to increase the speed communication and make it reliable for long-term accumulation. An SD card of 8GB capacity is enough for more than a year of accumulating, which meets the requirements of instrumental experiment.

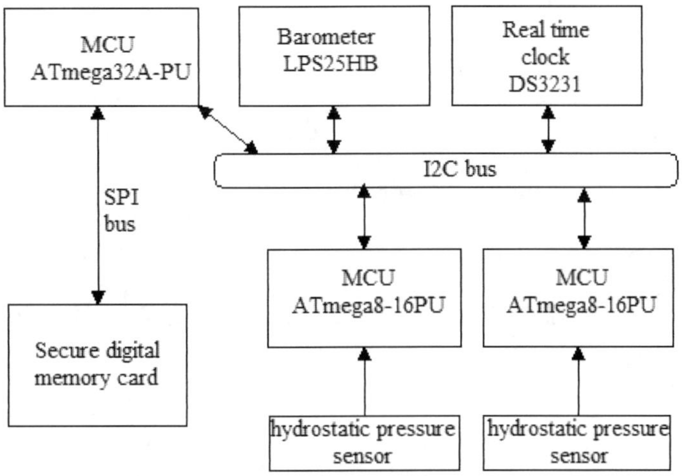

Figure 1. The block diagram of dual-channel frequency meter.

2.2 The method of frequency measurement

The frequency measuring method must cover the range and the resolution of the frequency generated by the hydrostatic sensors. The hydrostatic sensors produced by Russian company «SCTB ELPA» (Uglich, Yaroslavl region) were used to define the requirements to the frequency meter. The sensors have frequency range of 300–3000/42000–48000 Hz with error of ±0.06%.

The accuracy of the frequency measurement depends on the used frequency reference Fref (figure 2). Two crystal oscillators of 2-ppm accuracy have been used as the reference signal. There are many measurement methods, for instance, a gated counter where a timer-counter counts the frequency impulses (Fin) within a certain period. Counting the impulses per one second results in the Fin frequency of one-second discreteness. The disadvantage of the method is varying of the measurement resolution with the change in the input signal.

As it is shown in figure 2, the MC counts the input Fin signal within the Gate period. As we can see, using the gated counting leads to the MC missing the Tb and Te periods (figure 2). That is why special precaution must be taken, in particular, described in the source [7]. The resolution increases with the Fin frequency rising due to the Tb and Te periods decreasing. With low values of Fin frequency, it is recommended to measure the periods of the signal with further recalculation it into the actual frequency instead of impulse counting. Using this method makes it possible to consider the Tb and Te periods resulting in lower measurement error. In contrast to the first method, the measurement resolution decreases with the Fin frequency going up. The MC is able to switch to both programming methods with the Fin frequency going up or down and in consequence getting the maximum resolution not affected by Fin signal change. Implementation of this software solution is problematic when having dynamically changing Fin frequency from two hydrostatic sensors.

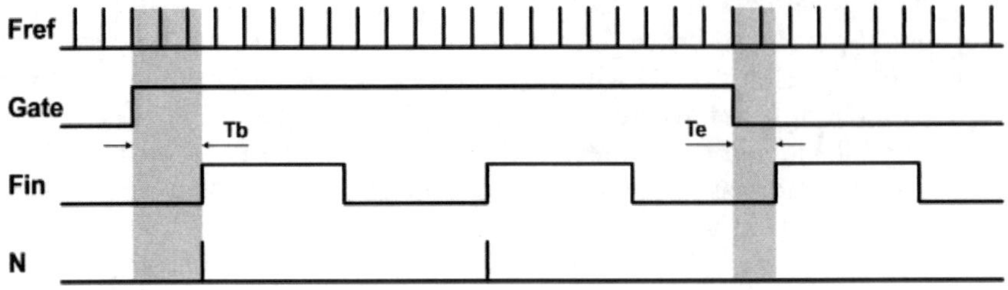

Figure 2. The diagram of the gated counter

The software reciprocal counter has been implemented in the dual-channel frequency meter and the final resolution remains stable regardless of the Fin frequency (figure 3) [8]. In this method the Gate period is not the time within a timer counts the Fref impulses, but it is only a sign to start the measurement, so the Fref signal is counted by the signal. The real measurement period starts at the first input signal and ends with the final one. So, if the Gate period stops, the counting keeps going until the last impulse is over. It results in the stable resolution not affected by the change in the Fref input. As we can see in the figure 3, the influence of the Tb and the Te periods is minimal. The software also considers the emergency situations when the Fref input unexpectedly disappears or it is beyond the range of the expected frequency.

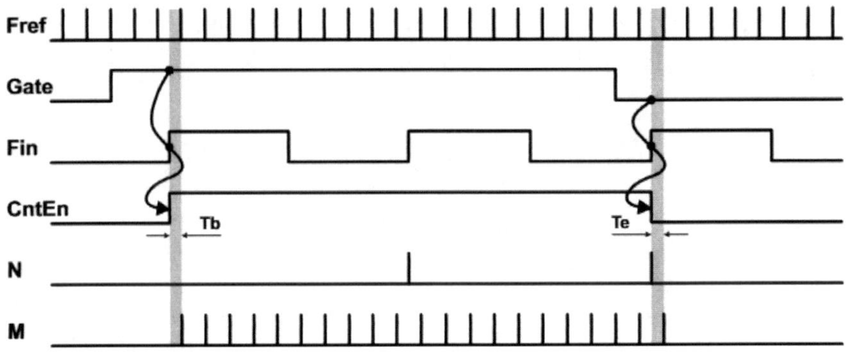

Figure 3. The diagram of the reciprocal counter

3. The results

The result is the developed low-power dual-channel frequency meter which synchronously measures two input frequencies generated by the hydrostatic pressure sensors, ambient temperature and absolute atmospheric pressure. It accumulates the data as long as the SD has enough memory size. The device can be used to observe the waves in coastal zones with no industrial power supply system. The design of the device is shown in figure 4. The scheme was mounted in the container case suitable for field-oriented conditions.

A laboratory testing and a functionality checking were carried out in order to set the parameters of the constructed device. A user can monitor the status of the scheme by the led indication mounted on the top of the case. A feeder on the front is used to connect the sensors.

The output data is text files in lines, one of which has the values for one measurement, for instance, for one second. The information can be easily imported in Microsoft Excel program or any other suitable one for operating with a large mass of data.

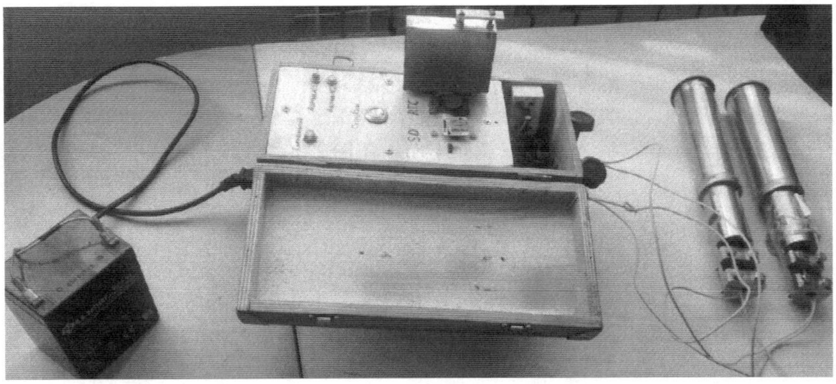

Figure 4. The design of the frequency meter with two connected sensors.

The completed instrumentation complex with the developed frequency meter and its mechanical and electrical characteristics is shown in figure 5. As we can see, there are two sensors placed on the sea bottom at a close distance from the shore where the frequency meter is set.

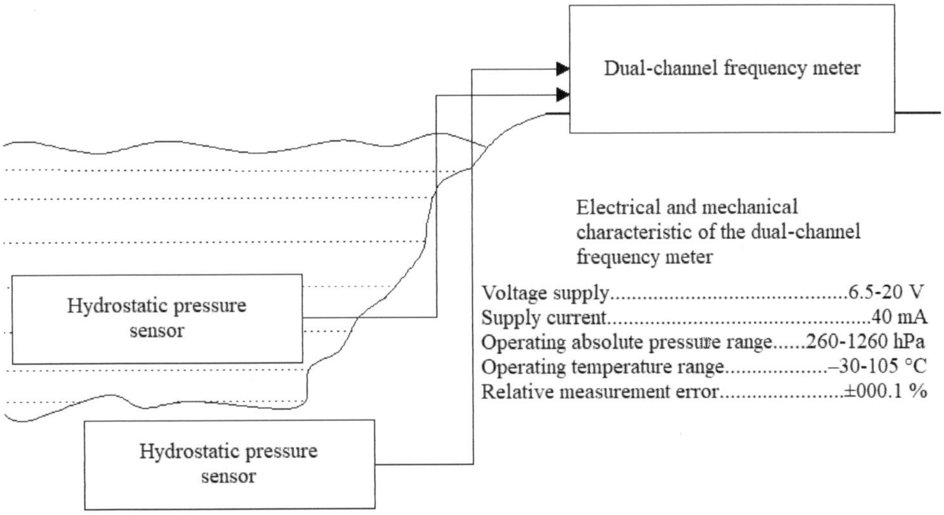

Figure 5. The instrumentation complex with dual-channel frequency meter.

4. Conclusion

We developed the dual-channel frequency meter for wave monitoring with its power consumption low compared to analogues. Usage of the microcontroller unit makes it possible to reject personal and industrial computers as a solution for instrumental observations. Due to the simple construction, it was possible to develop and construct the device directly in the research equipment sharing center of the Institute of Marine Geology and Geophysics. It is not only important to perform an experiment for estimating the waves characteristics but also to accumulate the experience of implementing the equipment similar to the one described in the paper.

References

[1] Kirillov K V 2018 The investigation of wave field using autonomous wave recorder ARV-K14 in the coastal area *Geosystems of transition zones* **2** pp 339–345

[2] Kovalev D P, Kovalev P D and Kirillov K V 2017 The investigation of dangerous marine pheonomena in the coastal zone based on the field observation results *Geosystems of transition zones* **1** pp 18–23

[3] Kovalev P D Kovalev D P 2010 *The methods of the Investigation of Dangerous Marine Phenomena in the Coastal Zone of the Oceans* (Vladivostok: Dalnauka) p 63
[4] P D Kovalev and D P Kovalev 2019 *IOP Conf. Ser.: Earth Environ. Sci.* 324 012016
[5] Kovalev D P and Ryshov A 2005 Hardware complex for observing hazardous marine phenomena *Technical issues of oceans development* **1** pp 197–200
[6] Kovalev D P and Kovalev P D 2018 *Investigation of the Generation Peculiarities of Infragravity and Edge Waves in the Coastal Zone Using the Data of the Field Experiments* (Yuzhno-Sahalinsk: IMGG FEB RAS) p 19
[7] Dudchenko I P, Gulyakov S A and Stovbun N S 2019 *Proc. Conf. on Geodynamical Processes and Natural Hazards* (Yuzhno-Sahalinsk: IMGG FEB RAS) p 32
[8] Lyutikov I V and Fomin A N 2016 *Metrology and Radio Measurements* (Krasnoyarsk: Siberial Federal Univercity) p 253

Comparison of forecast estimates of seabed subsidence of the Yuzhno-Kirinskoye field

V S Zhukov[1,*], **D K Kuzmin**[1], **Yu O Kuzmin**[1] **and I V Pleshkov**[2]

[1] Institute of Earth Physics, Russian Academy of Sciences, Moscow
[2] LLC "Rock Flow Dynamics", Moscow, Russian Federation

E-mail: *zhukov@ifz.ru

Abstract. The geodynamic consequences of the field development process include the subsidence of the earth's surface. Monitoring of deformations in offshore fields is difficult and it is necessary to evaluate them by various methods. It is important to investigate how much the calculated amplitude of the seabed subsidence is reduced by taking into account changes in the porosity and compressibility of the pore space with a decrease in reservoir pressure. The analysis of changes in the petrophysical parameters of the reservoirs of the Daginsky horizon during the projected development of the Yuzhno-Kirinskoye hydrocarbon field for depletion and a decrease in formation pressure by 10 MPa showed that the porosity decrease will be 0.038 absolute percent, the compressibility of the pore space will decrease by $0{,}08 \cdot 10^{-3}$ 1/MPa. With the help of the Petrel software, changes in the thickness of the productive layer from 80 to 120 cm were obtained, which can be taken as an estimate of the seabed subsidence in the area of the field. The application of the genetic model of the deformable formation by Kuzmin Yu showed that the maximum amplitude of the seabed subsidence to be 101 cm. Comparison of these estimates of the seabed subsidence indicates their proximity. Taking into account the dynamics of tectonophysical and petrophysical characteristics due to the long-term development of hydrocarbon deposits significantly changes the intensity of the deformation state of the rock mass and the earth's surface above the field.

1. Introduction

Long-term development of oil and gas fields is often carried out in conditions of reducing reservoir pressure. This often happens during the development of gas fields, since they, unlike oil fields, practically do not use reservoir pressure maintenance. As is known, the most well-known form of manifestation of negative geodynamic consequences accompanying the process of reducing reservoir pressure is extensive subsidence of the earth's surface [1–7]. For example, relative deformations can reach threshold values of 10^{-3}, which are designated as dangerous in regulatory documents [8, 9], with typical field sizes of 50–100 km and accumulated subsidence of several meters. It is extremely important to illustrate this by the example of offshore fields, since setting up full-scale monitoring of deformations, is impossible at such objects for natural reasons. In this situation, model assessments is both an element of the design of observation systems and a basic link in the interpretation of monitoring results, at the same time. In order to determine the optimal configuration of the observation system and its metrological characteristics, it is necessary to evaluate the level of possible deformations (the amplitude of displacements and their spatial distribution) by several different methods. It is important to investigate the question of how much such decreases in poroelastic parameters, especially the compressibility coefficient of the pore space, reduce the final, estimated amplitude of the subsidence of the earth's surface.

2. The results of experimental studies

The results of experimental studies using the method of modeling reservoir conditions [10] were analyzed for about 240 samples of sandstones that had an open porosity of 2.9–33.4 % in atmospheric conditions [11–13]. The results showed that the dependence of the average value of the porosity (m) on the effective pressure (P_{ef}) can be described with a high degree of confidence (0.99) (figure 1a) by the power dependence (1):

$$m = 22.604 \cdot P_{ef}^{-0.020} \quad (1)$$

An increase in the effective pressure (reduction of reservoir pressure) by 10 MPa (from 37 MPa to 47 MPa) during reservoir development will lead to a decrease in porosity from 20.967 % to 20.929 % or by 0.038 absolute percent of porosity.

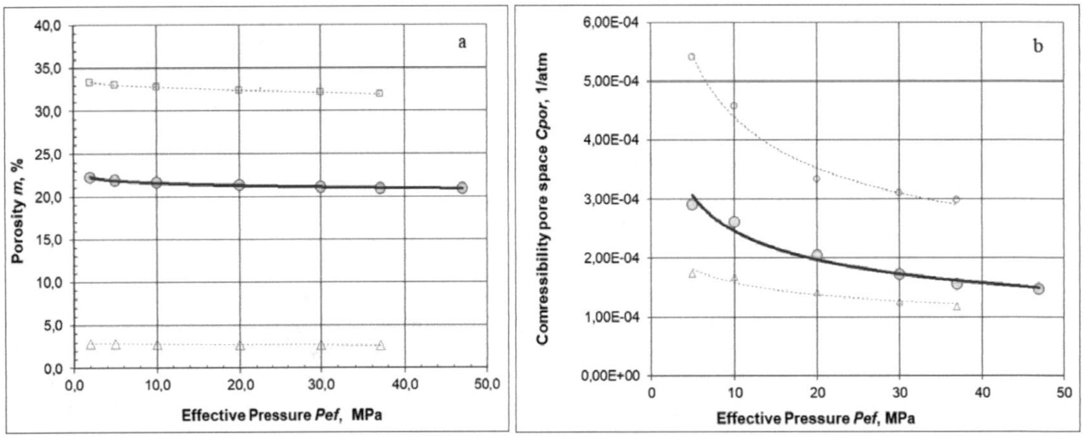

Figure 1. Changes in porosity (a) and compressibility (b) of the pore space with an increase in the effective pressure.

It is revealed that the dependence of the average value of the compressibility coefficient of the pore space (C_{por}) on the effective pressure (P_{ef}) (figure 1b) with a high degree of confidence (0.97) is described by the power equation (2):

$$C_{por} = 5.15 \cdot 10^{-4} \cdot P_{ef}^{-0.321} \quad (2)$$

An increase in the effective pressure during reservoir development from 37 to 47 MPa, i.e. by 10.0 MPa, will lead to a decrease in the average compressibility of the pore space from $1.56 \cdot 10^{-3}$ to $1.48 \cdot 10^{-3}$ 1/MPa or on $0.08 \cdot 10^{-3}$ 1/MPa, which means a decrease in the compressibility of the pore space by 5.3 % after reducing the reservoir pressure by 10 MPa relative to its magnitude of 37.0 MPa at the beginning of development.

3. The object of research

The Yuzhno-Kirinskoye oil and gas condensate field is located on the north-eastern shelf of Sakhalin Island at a distance of 35 km from the coast, the sea depth at the field varies in the range of 110-320 m. Drilling of eight exploration wells revealed the oil and gas content of slightly inclined layers of the Daginsky horizon of the Miocene, which is represented by sandstones, siltstones and clays with a thickness of up to 200 m [14–16]. The location of wells and blocks of the Yuzhno-Kirinskoye field is described in [14]. The values of the weighted average porosity for the wells of the field are on average 20–23 %. The main task of this work was to determine the nature and magnitude of changes in the pore space during the development of reservoirs of the Daginsky horizon and to estimate the magnitude of possible surface subsidence.

IOP Conf. Series: Earth and Environmental Science **946** (2021) 012019 doi:10.1088/1755-1315/946/1/012019

4. Methods and results of geological and mathematical modeling of the process of reducing reservoir pressure

In the hydrodynamic modeling programs PETREL, ECLIPSE, VIP Landmark and others, the problem of possible subsidence calculation is solved within the framework of the theory of elastic deformations, taking into account the distribution of reservoir and elastic properties of the reservoir in space. At the same time, in the first approximation, it is assumed that changes in porosity occur only in the reservoir due to a decrease in reservoir pressure. Considering the fact that the overlying layers are not subjected to dynamic influence, the displacement of the roof of the productive layer is usually assumed to be equal to the displacement of the seabed surface, taking into account the distribution in space of its filtration-capacitive and elastic properties and the value of the reduction in reservoir pressure.

The initial geological model before the start of development reflects the state of the productive reservoir at the time of starting the development, i.e. at an effective pressure of 37 MPa. A further increase in the effective pressure to 47 MPa accompanies the process of reducing the reservoir pressure by 10 MPa, and geological and mathematical modeling allows us to take into account changes in physical parameters, in particular the porosity of reservoirs, during the development of a field for depletion. The map of the effective thicknesses of the productive part of the reservoir was reconstructed, but already at a porosity value of 47 MPa. As a result, a map of the effective thicknesses of the productive reservoir was obtained taking into account the reduction of the volume of the pore space and the volume of the reservoir itself during development without maintaining the reservoir pressure.

Using the calculator built into the Petrel software, we obtained the difference in effective thicknesses (figure 2), which can be taken as the value of the change in the thickness of the productive layer. Assuming that their thickness and porosity do not change in the overlying strata, the areal distribution of the difference in effective thicknesses can be taken as the distribution of the value of the seabed subsidence in the field development zone.

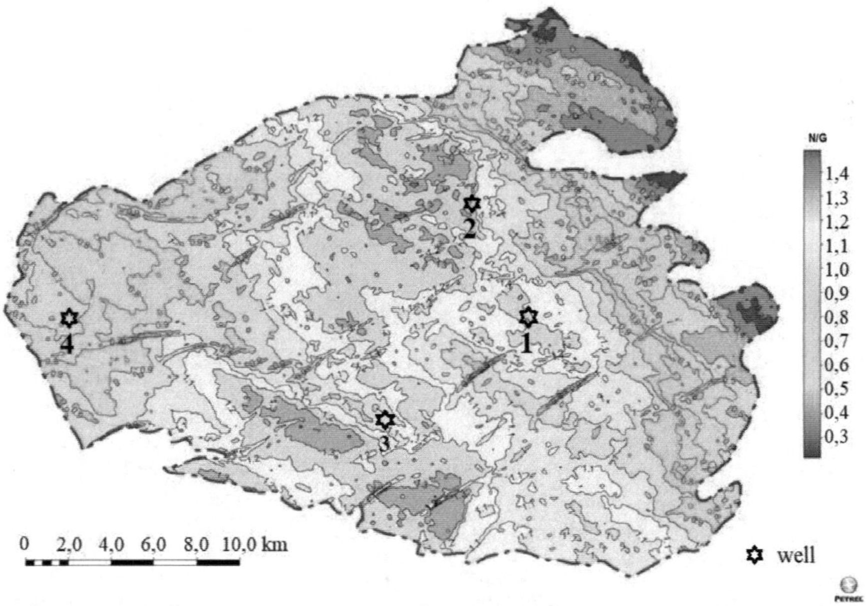

Figure 2. A map of changes in the effective reservoir thicknesses after a decrease in reservoir pressure by 10MPa (P_{ef} 47MPa).

Modeling with a genetic model of a deformable reservoir [17, 18] was successfully tested at a number of hydrocarbon deposits, including offshore and underground gas storage facilities [4, 6, 7, 19]. The priority in choosing this model is determined by a number of fundamental factors described in the works

[20, 21]. As is known, the formula for calculating, for example, the vertical displacements of the earth's surface (Uz) consist of the product of two factors:

$$Uz = F \cdot G \tag{3}$$

The physical factor F in (3) includes such parameters as porosity m, compressibility of the pore space C_{por}, as well as the change in reservoir pressure ΔP_{por}. The geometric factor G, in formula (3), takes into account the geometric dimensions of the layers taken on the basis of geological profiles along and across of the field [14, 15]. The productive layers were modeled by the prism with the following characteristics: width $2a$ (40 km), length $2b$ (80 km), depth of the upper d (2790 m) and lower D (2850 m) boundaries of the effective gas-saturated part of the formation and the average depth of the formation H (2820 m).

The average values of porosity and compressibility of the pore space at a reservoir pressure of 27 MPa at the beginning of development were used as initial values: m = 20.967 % and C_{por} = $1.56 \cdot 10^{-3}$ 1/MPa. To calculate the subsidence, a scenario was adopted that the reservoir pressure would decrease by P_{por} = 10 MPa. At the same time, the following values will be obtained: m = 20.929 %, C_{por} = $1.48 \cdot 10^{-3}$ 1/MPa. The vertical displacements were constructed taking into account changes in the porosity and compressibility of the pore space during the development of the field with a decrease in pore pressure to 17 MPa (figure 1a and 1b). For comparison and visual illustration of accounting the results of physical modeling of changes in the petrophysical parameters of reservoirs during development, displacement curves are shown, which have been obtained without taking into account changes in the porosity and compressibility of productive rocks (figure 3).

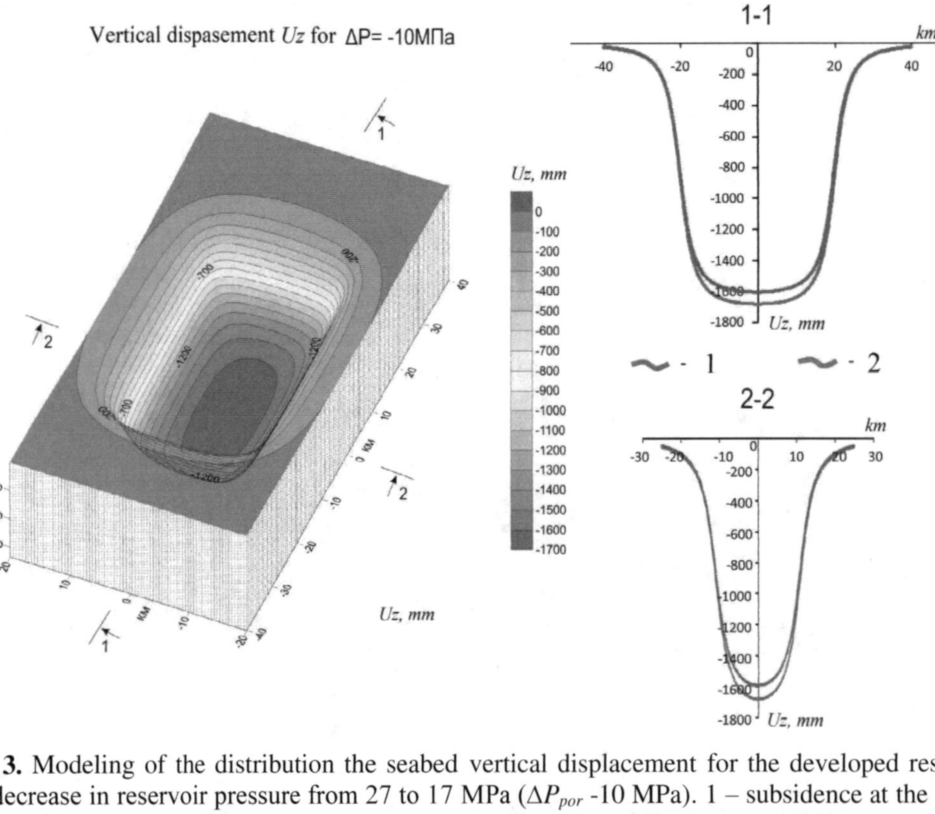

Figure 3. Modeling of the distribution the seabed vertical displacement for the developed reservoir with a decrease in reservoir pressure from 27 to 17 MPa (ΔP_{por} -10 MPa). 1 – subsidence at the values of porosity and compressibility at the time of the beginning of field development; 2 – subsidence taking into account changes in porosity and compressibility during decrease in reservoir pressure.

IOP Conf. Series: Earth and Environmental Science **946** (2021) 012019 doi:10.1088/1755-1315/946/1/012019

5. Discussion

The analysis of changes in petrophysical parameters of the oil and gas saturated reservoirs of the Daginsky horizon of Miocene age during the projected development of the Yuzhno-Kirinskoye hydrocarbon field for depletion and a decrease in fluid pressure in the reservoir by 10 MPa showed that the porosity coefficient will decrease by 0.038 absolute percent, the compressibility of the pore space will decrease by $0.08 \cdot 10^{-3}$ 1/MPa, which means a decrease in the compressibility of the pore space by 5.3 % relative to its value at the beginning of development.

The geological models constructed using the Petrel software reflect the geological structure and spatial distribution of the porosity of the productive reservoir both at the time of the start of development at an effective pressure of 37 MPa, and with a decrease in reservoir pressure by 10 MPa. They include maps of the distribution of effective thicknesses, maps of the distribution of the porosity of productive layers, which allowed, in the first approximation, to take into account changes in physical parameters, in particular the porosity of reservoirs, during the development of the field for depletion, accompanied by an increase in the effective pressure to 47 MPa. The difference in effective thicknesses at effective pressures of 37 and 47 MPa was taken as the value of the change in the thickness of the productive reservoir. Considering that the porosity and reservoir pressure of the fluids of the overlying layers do not change, their thickness also does not change, the areal distribution of the difference in effective thicknesses can be taken as the distribution of the magnitude of the seabed subsidence in the field area. Estimates of the seabed subsidence ranging from 0.2 to 1.34 meters were obtained. The most common subsidence values are in the range from 0.8 to 1.2 m. The maximum value of 1.34 m is found only in some local places.

The results of the application of the genetic model of the deformable formation have shown that when the reservoir pressure decreases by 10 MPa, the maximum amplitude of the seabed subsidence according to the data obtained at the beginning of development (P_{ef} = 37 MPa) will be 101 cm. This value is quite close to the value of 134 cm obtained by modeling using the Petrel software, but less due to taking into account the genesis of the anticlinal fold formed by tectonic forces. Applying the genetic model, we also tried to use data on the compressibility of the reservoir at P_{ef} = 47 MPa, which allowed us estimating already at a value of 95 cm. This allowed us to conclude that the difference in estimates, taking into account changes in compressibility during development and without it, is very significant. Thus, with a decrease in reservoir pressure by 10 MPa, it will be 6.0 cm, but the discrepancy by the end of development (a decrease in reservoir pressure by about 27–28 MPa) can be significantly larger, about 20–30 cm.

6. Conclusion

The comparison of the estimates of the seabed subsidence with a decrease in reservoir pressure by 10 MPa, performed using the Petrel software and the genetic model of the deformable formation, indicates their proximity. At the same time, taking into account the genesis of the anticlinal fold formed by tectonic forces makes it possible to slightly reduce the assessment of the seabed subsidence.

The conducted studies have convincingly demonstrated that taking into account the dynamics of tectonophysical and petrophysical characteristics due to the long-term development of hydrocarbon deposits significantly changes the intensity of the deformation state of the rock mass and the earth's surface above the deposit and, consequently, the idea of the level of geodynamic risk of oil and gas complex objects.

References

[1] Kuzmin Yu O 2002 *Fundamental basis of new technologies of oil and gas industry* Modern geodynamics of abnormal subsurface induced by development of oil and gas fields (Moscow, GEOS) pp 418–427

[2] Zhukov V S, Kuzmin Yu O 2003a Physical modeling recent geodynamic processes *Mining informational and analytical Bulletin* **5** pp 71–77

[3] Zhukov V S, Kuzmin Yu O 2003b *Gas resources of Russia in XXI century* Modeling of deformation and seismic processes in the development of hydrocarbon deposits (Moscow, VNIIGAZ) pp 456–469

[4] Zhardetsky A V, Zhukov V S, Moiseev P V, Kuzmin Y O 2003 Application of geological-mathematical modeling for monitoring of geodynamic processes in the operation of UGS *Karotazhnik* **102** pp 67–76

[5] Zhukov V S 2010 Assessment of changes in the physical properties of reservoirs caused by the development of oil and gas fields *Mining information and analytical bulletin* **6** pp 341–349

[6] Kuzmin Yu O, Descherevsky A V, Fattakhov E A, Kuzmin D K et al 2018 Inclinometric observations at the Y Korchagin field *Geophysical processes and biosphere* **53** 3 pp 31–41

[7] Kuzmin Yu O, Descherevsky A V, Fattakhov E A, Kuzmin D K, et al 2019 Analysis of the results of deformation observations by the inclinometer system at the V Filanovsky field *Geophysical processes and biosphere* **18** 4 pp 86–94

[8] Kuzmin Yu O 2019b Recent Geodynamics: from Crustal Movements to Monitoring Critical Objects *Physics of the Solid Earth* **55** 1 pp 65–86

[9] Kuzmin Yu O 2020 Actual issues of the use of geodetic measurements in geo-dynamic monitoring of oil and gas complex objects *Bulletin of SGUGIT* **25** 1 pp 43–54

[10] Zhukov V S, Lyugay D V 2016 *Determination of filtration-capacitance and elastic properties and electrical parameters of rock samples when modeling reservoir conditions* (Moscow: VNIIGAZ) 56 p

[11] Zhukov V S, Ivanov P Yu 2015 Changes in the physical properties of the reservoir as a result of an increase in the effective pressure during the development of the field (modeling on the example of the Yuzhno-Kirinskoe field) *Vesti gazovoi nauki* **24** 4 pp 144–148

[12] Zhukov V S, Churikov Yu M, Motorygin V V 2017 Changes in the structure of the pore space of the reservoirs of the Daginsky horizon when modeling reservoir conditions *Vesti gazovoi nauki* **31** 3 pp 238–246

[13] Zhukov V S, Kuzmin Yu O, Semenov E O 2018 Dynamics of physical properties of reservoirs in the development of oil and gas fields *Vesti gazovoi nauki* **12** 1 pp 82–99

[14] Dzyublo A D, Shnip O A, Altukhov E E, Tsherbakova A Iu 2016 Lithological-geophysical characteristics for the Daginsky horizon rocks of the Yuzhno-Kirinskoe field *Oil and gas Geology* **4** pp 39–46

[15] Rybalchenko V V, Gogonenkov G N, Parasyna V S 2014 Formation environments for the Yuzhno-Kirinskoe field on the Sakhalin shelf *Oil and gas Geology* **4** pp 42–52

[16] Aulova D Yu, Zhukov V S, Motorygin V V et al 2015 Assessment of the influence of clay content on the filtration and reservoir properties of the collector *Gas industry* **721** 4 pp 29–32

[17] Kuzmin Yu O 1999 *Recent geodynamics and evaluation of geodynamic risk in subsoil resources* (Moscow: Publ. "AEN") p 220

[18] Kuzmin Yu O 2010 Once again about the assessment of the subsidence of the bottom of the water area in the case of the development of the senomanskoe reservoir of one gas field *Mine Surveying Bulletin* **75** 1 pp 53–60

[19] Zhukov V S, Kuzmin Yu O, Poloudin G A 2002 Evaluation of the processes of subsidence of the earth's surface in the development of gas fields (for example, the North Stavropol field *Geology, Geophysics and development of oil and gas fields* **7** pp 54–57

[20] Abramyan G O, Kuzmin D K, Kuzmin Yu O 2018 Solution of inverse problems of recent geodynamics of subsoil in hydrocarbon fields and underground gas storage facilities *Mine Surveying Bulletin* **125** 4 pp 52–61

[21] Abramyan G O, Kuzmin D K 2019 Modeling of gradients of displacements of the Earth's surface in the developed oil and gas fields *Mine Surveying Bulletin* **132** 5 pp 56–62

Characteristics of maximal energy component of waves in Terpeniya Bay (Sakhalin Island)

A S Borisov* and K V Kirillov

Institute of Marine Geology and Geophysics Far Eastern Branch of Russian Academy of Science, Yuzhno-Sakhalinsk, 693022, Russia

E-mail: *a.borisov@imgg.ru

Abstract. Analysis of natural sea surface oscillations was performed in order to determine maximal spectral energy components in sea waves in the area of Terpeniya Bay. It has been found out that maximal components appear at periods of 5 and 8 seconds and reach energy of $3 \cdot 10^6$ cm^2·s during storms. During calm weather maximums can be observed in the range of swell waves, herewith maximums exist at periods of 12 and 15 seconds. Energy of these maximums doesn't go upper then $8 \cdot 10^5$ cm^2·s and $3 \cdot 10^5$ cm^2·s accordingly. Maximum components in the infragravity band of energy spectrum were estimated as well. Two steady energy peaks were determined at periods of 75 and 135 seconds, which are most probably relate to edge waves in considered water area.

1. Introduction

Significant wave heights, mean wave period and peak wave period are the basic parameters for description of sea waves due to its random nature. Knowing of these parameters is necessary for diverse spheres of human activity related to description of wave modes, shipping, coastal engineering [1, 2] and even to analysis of seismic oscillations [3]. Results of analysis of amplitude A_{max} and period T_p of maximum energy component in energy spectrum of waves in Terpeniya Bay are presented in this paper. Analysis was done on the base of observation of hydrostatic pressure performed from May 22 to October 6 in 2010. Record of hydrostatic pressure was captured at sample frequency 1Hz by pressure recorder mounted on the sea bottom at 17 m depth and 700 m distance from shoreline near Vzmor'e locality.

2. Experimental data and analysis

Relatively calm sea state with maximum wave height around 0.3 m as well as moderate and strong storms with maximum wave height up to 7 m were observed during all time of observation. Figure 1 is a record of sea level change. Estimations of T_p and A_{max} were performed by using power spectral density (PSD) of wave process. Sliding window with 100 minutes length and 95 % of window overlap were applied. Energy spectrum band from 2 to 600 s was analysed. Results are presented in the figure 2 as 2D distribution histogram. Most intensive and frequent were wind waves at the periods around 5 s. They prevailed upon others. Maximum components which appeared at the periods around 8 s reached the same energy, but were not so frequent.

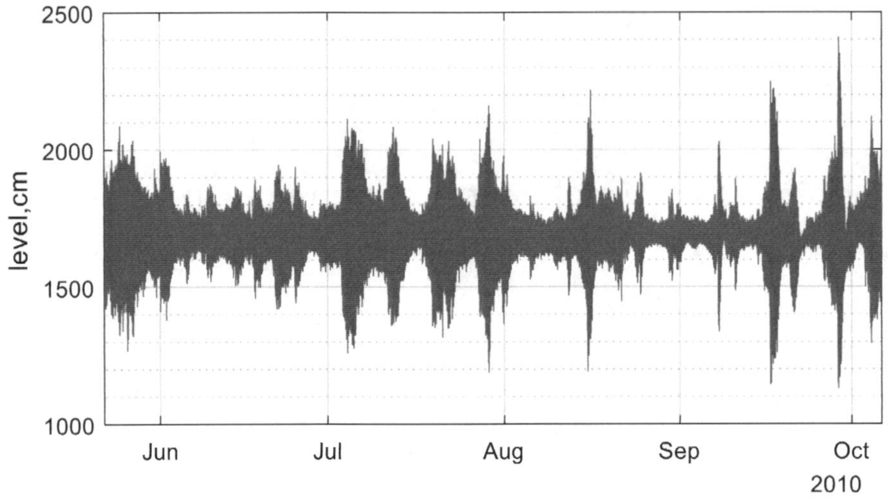

Figure 1. Water level changes during measurement time.

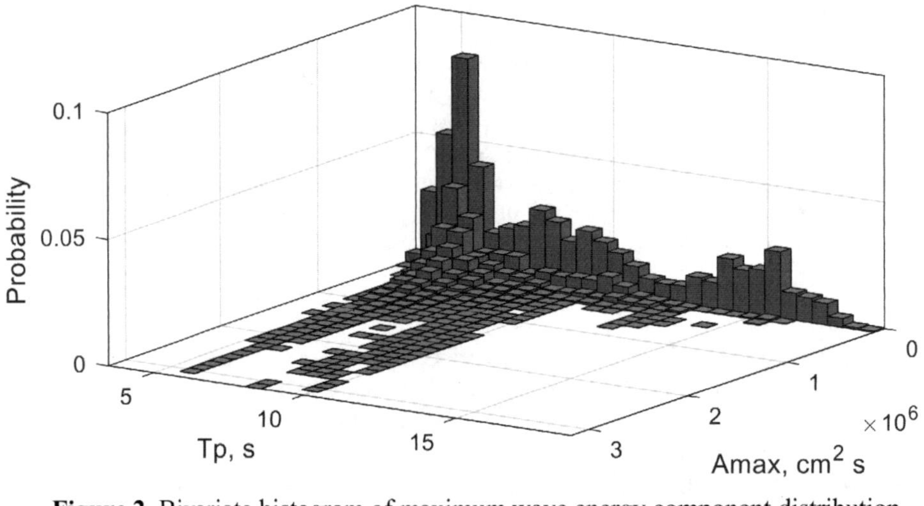

Figure 2. Bivariate histogram of maximum wave energy component distribution.

Two conditional maximums were detected in the range of swell wave periods. First one appeared in the vicinity of 12 s and another one – around 15 s. Spectral maximums at these periods were observed during calm weather. Herewith, components around 12 s had maximum amplitudes. But oscillations around 15 s were detected more frequent. It should be noted here, that energy spectrum in the range of swell waves has several steady peaks at periods of 13, 14, 15 and 17 s. This is visual in the figure 3. This fact is also mentioned in paper [4], where authors came across with separation of swell waves at close frequencies. Data for processing were taken by them from recorders mounted near Ostryi Cape in 2020.

Maximal energy of T_p component at wind wave periods reached value $A_{max} = 3 \cdot 10^6$ cm^2·s, and at swell wave periods maximal amplitude of energy component didn't exceed value $A_{max} = 8 \cdot 10^5$ cm^2·s for periods around 12 s and $A_{max} = 3 \cdot 10^5$ cm^2·s for periods around 15 s.

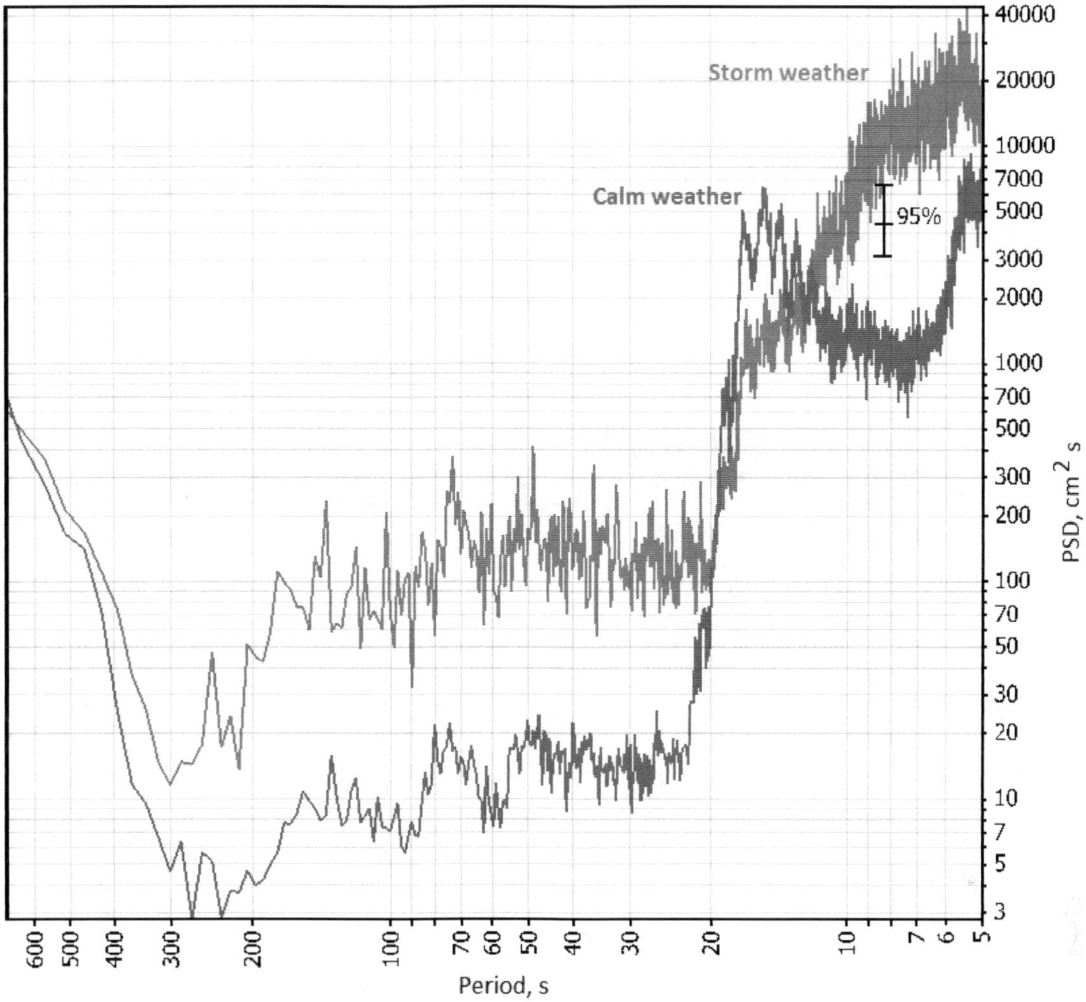

Figure 3. Power spectral density of wave during calm weather and during storm.

Investigation of parameters of infragravity (IG) waves has no less practical interest, especially in conditions of strong storms. According to research work [5], IG waves can appear as a result of nonlinear phenomena based on finiteness of amplitude of gravity sea waves. Frequency range of IG waves depends on frequencies of wind waves and occupies a range approximately from 0.005 to 0.05 Hz. Waves of this range differ from gravity waves with their effect on coastal constructions and sea medium. Degree of these differences depends on IG wave intensity relatively to wind waves. However, during IG waves propagation they can be considered as free waves having less steepness and therefore energy dissipation depending on it. That is why it is important to estimate relationships between IG and wind waves. Using only the characteristics of wind waves doesn't give an explanation of dynamic forces affecting transportation of sediments, deformation of bottom landform and seashore line, forming harmful process in port and etc.

Analysis of maximum energy components in frequency range of IG waves is considered below. One can see wave energy distribution versus time and period in the figure 4, where waveform and spectrogram of wave process during two weeks are presented. This chunk includes storm weather situation as well as situation with relatively calm water surface.

It is evident, that amplification of IG waves occurs during storms and simultaneously with amplification of wind waves. But sometimes IG energy increases during calm weather when swell waves come. Such a situation happened on May 31 (figure 4). Maximum amplification of IG waves observed at presence of wind and swell waves together. This is visible at the beginning of May 27 in figure 4.

Maximum IG wave components have peaks at periods of approximately 75 and 135 s. Energy of these components increases during storms or long-lasting presence of swell waves. One can see these peaks in detail in figure 3. Existence of oscillations at 75 and 135 s is probably caused by edge IG waves. In paper [5] IG waves at the periods of 95 and 170 s were detected. Using Stokes dispersion relationship authors of [5] found that these waves correspond to modes of edge waves.

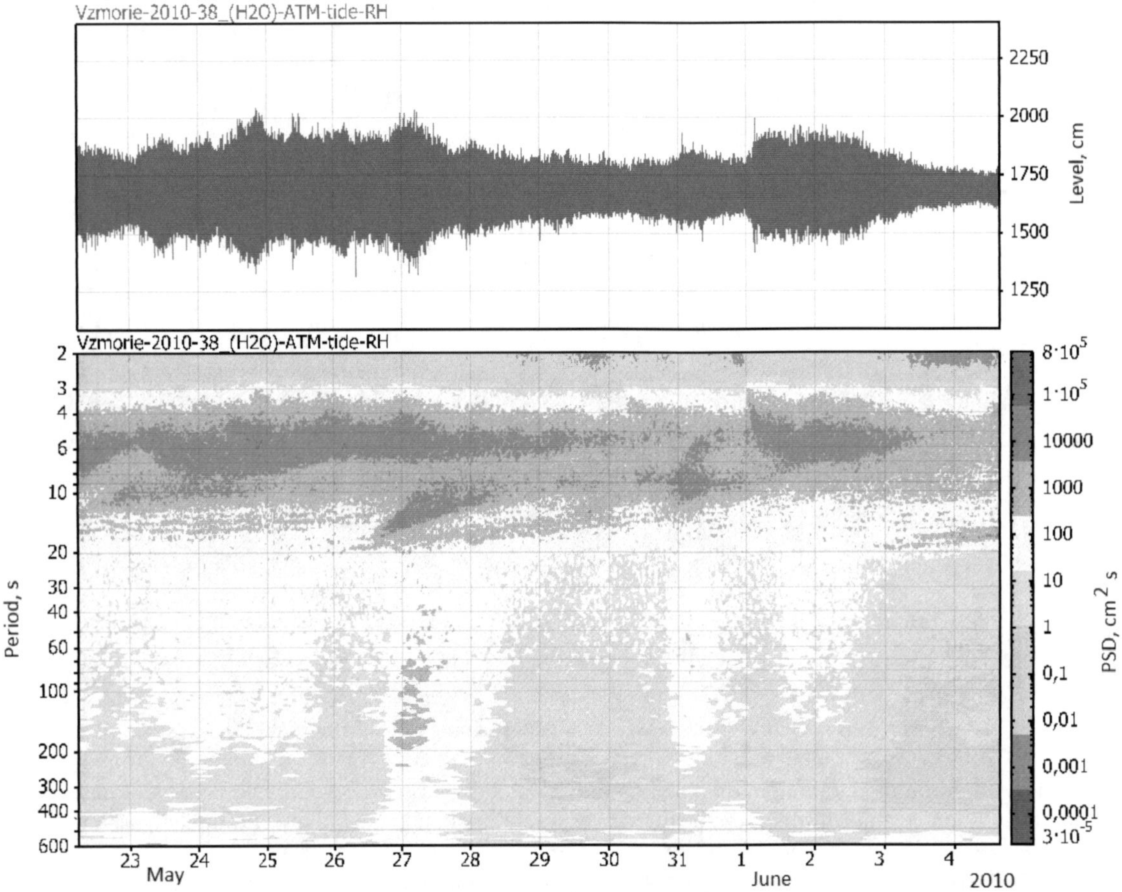

Figure 4. Waveform and spectrogram.

Estimation of interaction between IG and wind and swell waves was performed by the calculation of standard deviation in appropriate frequency ranges. Results are presented in figure 5. Analysis shows that amplification of IG waves occurs when total energy level in range from 2 to 22 s increases. But the best correlation is observed between IG and wind waves.

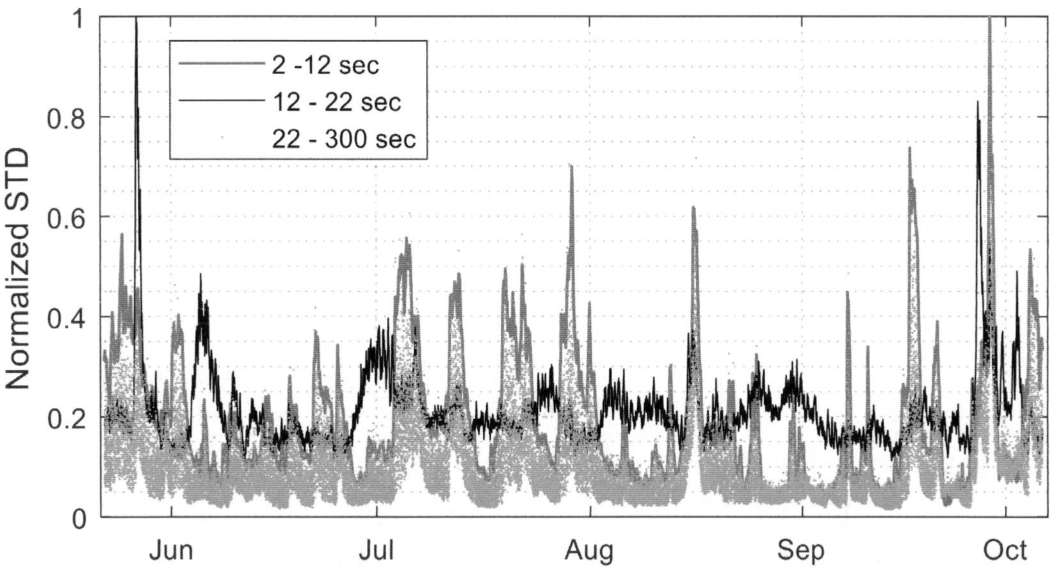

Figure 5. Standard deviation of wave process in three frequency ranges.

3. Conclusion

Periods of sea waves with maximum energy components were considered in the paper. Maximum energy in sea wave spectrum appears in wind wave band mostly at periods of 5 and 8 s and reaches value of $3 \cdot 10^6$ $cm^2 \cdot s$. During calm weather maximum components can appear in swell waves range around 12 or 15 s. Energy in this range don't get upper then $8 \cdot 10^5$ $cm^2 \cdot s$ and $3 \cdot 10^5$ $cm^2 \cdot s$ relatively. Two steady peaks were detected in IG wave range. One at 75 s and another one at 135 s. These two periods correspond to periods of edge waves in considered water area.

References

[1] Kovalev D P, Kovalev P D and Borisov A S 2020 *J. Marine Intellectual Technologies* **2** pp 108–117
[2] Shevchenko G V and Loskutov A V 2017 *J. Geosystems of Transition Zones* **1** pp 35–49
[3] Borisov A S , Kovalev D P, Kostylev D V and Levin Yu N 2019 *J. Geosystems of Transition Zones* **2** pp 201–208
[4] Kovalev P D, Shevchenko G V and Kovalev D P 2006 *J. Russian Meteorology and Hydrology* **9** pp 76–87
[5] Kovalev D P and Kovalev P D 2013 *J. Vestnik of the Far East Branch of the Russian Academy of Sciences* **3** pp 60–64

Using the paleotsunami data for the tsunami hazard assessment

V M Kaistrenko

Institute of Marine Geology and Geophysics Far Eastern Branch of Russian Academy of Science, Yuzhno-Sakhalinsk, 693022, Russia

E-mail: vict-k@mail.ru

Abstract. The article is focused on the development of statistical methods of the tsunami recurrence evaluation using paleotsunami data. The new key moment is the creation of a model to quantify the preservation potential of paleotsunami deposits. The article includes a brief overview of the results of studies of the variability and preservation of tsunami deposits. The model was tested on materials about paleotsunami on the coast in the Khalaktyrka area (a village within the city of Petropavlovsk-Kamchatsky), obtained earlier, for four time intervals set by the key-marker volcanic tephra layers in Kamchatka (Ksudach in 1907, Avachinsky in 1855 and 1779, Opala in 606). The maximum likelihood estimates of the number of tsunamigenic horizons for the indicated time intervals are given. The restrictions of the considered model are analyzed.

1. Introduction. Geological tsunami traces on the coasts

Tsunami is a dangerous natural phenomenon, at times attacking the Far East coast of Russia. Therefore, hazard and risk assessments are necessary for planning the development of tsunami-prone coasts and urban planning in the coastal zone [1]. However, the lack of data in the catalogs [2–3] of historical tsunamis does not allow to quantify the tsunami hazard with an acceptable accuracy of the order of 5–10 %. This requires 250–500 years of tsunami registration without gaps. The only alternative to this expectation is the use of paleotsunami data.

Research, started independently in different countries, has shown that strong tsunamis leave geological traces on the attacked coasts.

Figure 1 shows that tsunami deposits the most beach material in the "middle third" of the inundated coastal area. Near the shore, the material is weakly deposited from the high-speed flow, and in the zone of maximum run-up height, a little material remains in the flow.

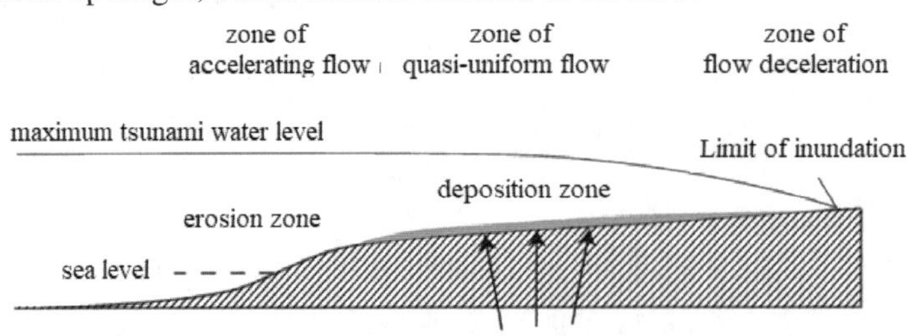

Figure 1. Schematic position of the zones of erosion and accumulation of material in the area affected by the direct tsunami flow [4].

Fresh deposits of tsunami can be traced almost to the boundary of inundation zone, but in the case of paleotsunami deposits, the situation is complicated by the incomplete preservation of thin layers.

The aim of this work is to create a method for obtaining the tsunami recurrence assessments using the data on paleotsunami deposits, taking into account their preservation potential.

2. Deposits of modern tsunamis and paleotsunami, their formation and preservation

When interpreting data on paleotsunami sediments on the Far East coast of Russia, the results of studying the features of sedimentation associated with modern large historical tsunamis of the 20th and early 21st centuries are of great importance [5–9].

Tsunami deposits on slopes are usually washed away by atmospheric precipitation over time, but in the conditions of coastal peatlands, the slopes are very small, and the beach material deposited by the tsunami is fixed with vegetation over the years. In the long-term process of fixation, tsunami deposits are deformed by plants growing from below and roots of a new vegetation cover, as well as by insects living in this environment (bioturbation). One of the urgent tasks in the study of tsunami deposits is the analysis of their preservation potential on the scale of geological time.

For several years, authors [10–12] traced the dynamics of deposits of the largest recent tsunamis in relation to their preservation potential. The most important factor was the initial thickness of the sediments. The probability of preservation of deposits less than 10 cm thick is quite small, but deposits with a thickness of more than 10 cm were well preserved. In a number of regions, anthropogenic impact turned out to be an important factor. The final conclusion drawn in the cited papers is following: the data on tsunami deposits require correction for variability and preservation.

3. Features of statistical accounting of data on paleotsunami on the example of the Khalaktyrka region, Kamchatka

For a long time, the main type of natural materials used by tsunami specialists were data on maximum tsunami run-up heights collected in catalogs [2–3]. The main feature of such material is the data on the heights that were *reached* by the tsunami waves. The presence of paleotsunami deposits found at a certain level above the ocean indicates that the tsunami has *exceeded* this level. What the maximum run-up height was in this case, is unknown.

The method of joint analysis of data on historical tsunamis and paleotsunamis is considered on the example of the coast in the Khalaktyrka region, Kamchatka (figure 2).

Figure 2 shows typical features of paleotsunami layers. As a rule, tsunami deposits at wave heights of less than 10 m are located in spots. Surface processes reduce the preservation of the layers, and as a result, each section contains its own set of deposits, with partial overlapping sets from adjacent sections. As can be seen from figure 2, volcanic tephra deposits can be traced along the profile much more clearly and reliably.

Accordingly, for the correlation and dating of the paleotsunami deposits, the tephrostratigraphy and tephrochronology method was used [13], based on the previously studied Holocene key-marker tephra layers in Kamchatka. This approach made it possible to determine the stratigraphic position, relative and absolute age of tsunamigenic layers in geological sections. In the study area on the coast of the Khalaktyrka beach, 13 tsunamigenic horizons (Ts1–Ts13) were identified [14]. Tsunami deposits are usually represented by thin (0.5 to 20 cm) layers of dark gray sea sands.

It is obvious that the formation of deposits of various tsunamis took place in different conditions. Therefore, the average probability of sediment preservation should be used in conditions that are close to homogeneous.

For this purpose, the entire time interval in which tsunamigenic layers were identified is divided into four intervals of shorter duration (k = 1, 2, 3, 4), within which the conditions for the formation of deposits are more uniform. For this, clear time boundaries were used, set by the deposits of key-marker tephra layers associated with the eruptions of the volcanoes such as Ksudach in 1907 (Ks-1907), Avachinsky in 1855 (Av-1) and in 1779 (Av-2), Opala in 606 (Op), Ksudach in 236 (Ks1) (figure 2). For the same purpose, a two compact groups of sections in the "middle" parts of profiles 1 and 2 were studied (figure 2). There are sections 302–306 (at a distance of 270–500 m from the shoreline) and 309–313 (at a distance of 250–440 m from the shoreline) [14].

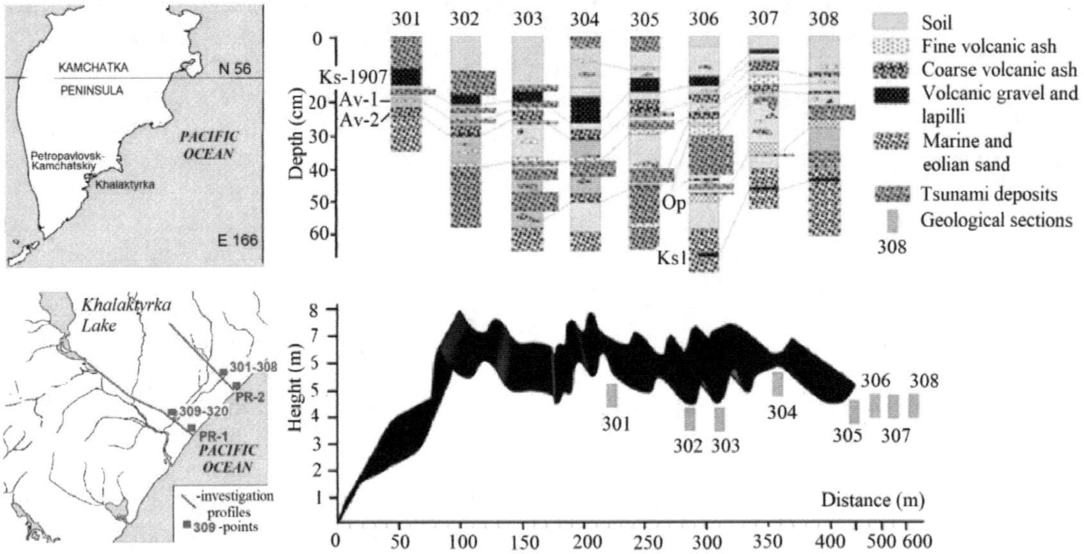

Figure 2. Paleotsunami study area on the Pacific coast of Kamchatka and the location of profiles and sections. Geomorphological profile 2 and section diagrams [14].

The data on the number of tsunamigenic layers in each of the studied sections that fall within the time interval between the horizons of key-marker volcanic ash are summarized in Table 1.

The positions of sections 302–306 and 309–313 are of different heights. However, the height of the beach ridge $h = 8$ m is associated with tsunami deposits in these sections, since the tsunami reached them, having overcome the high beach ridge (figure 2).

Assume that there were N paleotsunamis at some coastal location during the time period T, with the deposits of which the height h above the sea level is associated. It is known that the sequence of large tsunamis is close to Poissonian one [15]. Therefore, the probability of such an event is given by formula:

$$P_N = e^{-\varphi(h)\cdot T}\frac{[\varphi(h)\cdot T]^N}{N!}, \tag{1}$$

where the parameter φ (h), depending on the threshold height h, is the average frequency of tsunami manifestations, which is called the tsunami recurrence function (RF).

According to the definition, tsunami recurrence function (RF) is the average frequency of tsunami occurrence in a given place x, with maximum run-up height being equal to or more than the threshold height h

$$\varphi(h) \stackrel{\mathrm{def}}{=} \langle\frac{\overline{N(run\text{-}up\ height \geq h)}}{T}\rangle, \tag{2}$$

where N (run-up height $\geq h$) is the number of tsunamis with maximum height $\geq h$ occurring during the time period T.

Table 1. The number of tsunamigenic interlayers in the sections related to a considered time interval and an estimate of the probability of their preservation.

Number of section	Time interval (years)			
	Today – 1907, $k = 1$, $T_1 = 113$ years	1907–1855 $k = 2$, $T_2 = 52$ years	1855–1779 $k = 3$, $T_3 = 76$ years	1779–606 $k = 4$, $T_4 = 1173$ years
302	1	1	1	0
303	1	1	0	0
304	2	0	0	0
305	2	0	0	0
306	1	0	0	0
309	2	0	1	3
310	0	0	0	0
311	1	0	0	3
312	1	0	0	2
313	1	0	2	2
Tsunamigenic interlayers	Ts1–Ts2	Ts3	Ts4–Ts5	Ts6–Ts8
Maximum likelihood estimates of the tsunamigenic interlayers numbers, N_k	**2**	**1**	Uncertain ≥ 2	Uncertain ≥ 3
Probability of the tsunami deposits preservation, p_k	**0.6**	**0.2**	≤ 0.2	≤ 0.33

Moreover, the exponential approximation of the tsunami recurrence function is acceptable for large tsunami heights $h \geq 0.5$ m [16]:

$$\varphi(h) = f \cdot \exp\,[- h / H^*(x)]. \tag{3}$$

Parameter f is the asymptotic frequency of large tsunamis in the region, which generally slowly changes along the coast and can be considered to be a regional constant. Parameter H^* is the characteristic tsunami height for selected location x, which is proportional to the average coefficient $K(x)$ of the tsunami height transformation (amplification) from the open ocean to the coastal location x.

Tsunami activity parameters f and H * must be determined from historical tsunamis and paleotsunami data. Accordingly, substituting the values of these parameters into formulae (3) and (1), we can estimate all the necessary probabilistic values of hazard and risk associated with a tsunami.

In fact, we cannot be sure that the identified n deposits of paleotsunami are all paleotsunami N at some coastal location, belonging to the time period T and associated with the height h above sea level, because not all traces of paleotsunami are preserved.

Let us assume, that the i-th section contains n_{ik} deposits from N_k paleotsunami that actually took place during the k-th time interval. The corresponding probability P_{ik} (n_{ik}) can be estimated by the binomial distribution [17]:

$$P_{ik}(n_{ik}) = C_{N_k}^{n_{ik}} p_k^{n_{ik}} q_k^{N_k - n_{ik}}, \tag{4}$$

where $q_k = 1 - p_k$ is the probability of "erasing" the paleotsunami traces.

The likelihood function for the k-th time interval is equal to the product of the probabilities related to each i-th section [17]:

$$L_k = \prod_i P_{ik}(n_{ik}) = \prod_i \frac{N_k!}{n_{ik}! \cdot (N_k - n_{ik})!} p_k^{n_{ik}} q_k^{N_k - n_{ik}} = \prod_i \frac{N_k! q_k^{N_k}}{n_{ik}! \cdot (N_k - n_{ik})!} \left(\frac{p_k}{q_k}\right)^{n_{ik}}. \tag{5}$$

For the k-th time interval, the values of the number of paleotsunami N_k and probabilitiy p_k that maximize the value of the likelihood function L_k should be used as the maximum likelihood estimates. According to (5), these maximum likelihood values are related by a simple analytical formula:

$$p_k = \frac{\sum_{i=1}^{m} n_{ik}}{mN_k}, \quad (6)$$

where m is the number of sections. The value of the maximum likelihood estimates of paleotsunami interlayers N_k is found by a numerical method. The obtained values of the estimates of the number of tsunami deposits N_k and probabilitiy p_k of their preservation in the given time intervals, obtained by maximizing the likelihood function, are included in Table 1.

The analysis of the likelihood function (5) showed that the amount of actually identified deposits Ts1, Ts2 and Ts3 for the time intervals 2020–1907 and 1907–1855 is consistent with their distribution in 10 sections and are maximally probable. It is interesting that the estimates of the probability of preservation of tsunamigenic interlayers for these time intervals, $p_1 = 0.6$ and $p_2 = 0.2$, characterizing the conditions of their formation and preservation, differ significantly. This probability is less for the more ancient events of the period 1907–1855.

For the other two time intervals 1855–1779 and 1779–606, the numbers of identified paleotsunami interlayers Ts4 – Ts5 and Ts6 – Ts8 does not agree with their distribution in 10 sections, and the maximum of the likelihood function does not correspond to these quantities. There may be several reasons for this discrepancy. It is possible that 10 investigated sections are not enough to statistically compensate for the low probability of the preservation of traces of some ancient tsunamis. It is also possible that the conditions for the formation of tsunamigenic deposits during these longer periods of time could change significantly, which violates the requirement of uniformity of the conditions assumed in the model.

As an example, figure 3 shows the joint function of the recurrence of tsunami heights for Khalaktyrka, constructed earlier [18] by the least squares method based on the data on the historical tsunamis of 1841, 1952, and 1960, and 11 paleotsunami identified in [14] for all considered sections for the period starting from 236 BC.

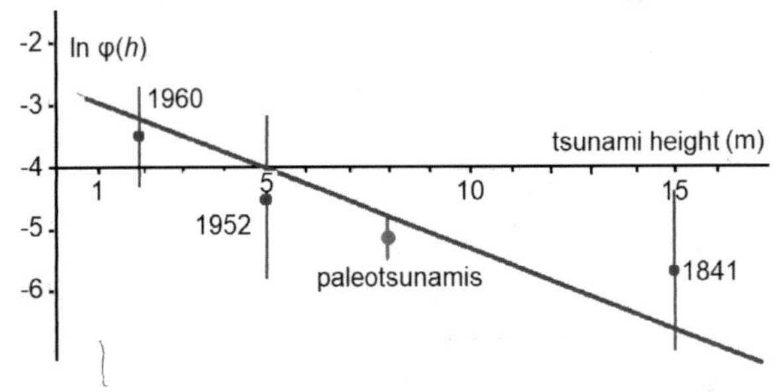

Figure 3. Empirical tsunami recurrence function for Khalaktyrka using historical and paleotsunami data [18].

Figure 3 demonstrates the significance of the paleotsunami data: exactly 11 paleotsunami correspond to one value of the recurrence function with the smallest value of the standard deviation (apriori error), compared with large standard deviations for (total) 3 large historical tsunamis.

4. Conclusion

Data on paleotsunami are very important for obtaining estimates of tsunami hazard (recurrence and possible heights) and risk with acceptable accuracy, which is necessary both for solving purely scientific problems and for planning the development of the coastal zone. However, the direct use of data on paleotsunami can lead to underestimations of the recurrency frequency, and therefore it is necessary to take into account the peculiarities of their formation and changes that occur in tsunami sediments prior to their fixation.

Using the maximum likelihood method, a model was built to estimate the real number N of paleotsunami during a certain time period T (which characterizes the frequency of tsunamis) and the probability p of the preservation of their traces based on data on the number of paleotsunami deposits in several sections associated with the same height h above ocean level.

The model was tested on data on paleotsunami on the coast near Khalaktyrka, for four time intervals set by clear deposits of key-marker volcanic tephra layers. Estimates of the number of tsunamigenic horizons for time intervals after 1855 are the most probable. For older events prior to 1855, the number of identified paleotsunami interlayers does not agree with their distribution in 10 sections, and the maximum likelihood function does not correspond to these quantities. This is explained both by the limitations of the constructed model, associated with the conditions assumed in it, and by the low probability of preserving traces of some tsunamis.

Despite some limitations of the considered method, the developed quantitative approach to assessments of the formation parameters of tsunamigenic deposits can be successfully used as a starting point for obtaining adequate quantitative assessments of tsunami hazard and risk.

Acknowledgements
The study was carried out as part of the state task for IMGG FEB RAS "Monitoring and modeling of oceanological processes, prediction of catastrophic events on the shelf and in the coastal zone", No. 121021000268-9.

References

[1] Code of rules 292.1325800.2017: *Buildings and structures in tsunami hazardous areas. Regulations of design* 2017 (Moscow: The Ministry of Construction, Housing and Communal Services) 138 p
[2] Historical Tsunami Database for the World Ocean (*Electronic Materials*) http://tsun.sscc.ru/htdbpac/
[3] Tsunami Data and Information (*Electronic Materials*) www.ngdc.noaa.gov/hazard/tsu_db.shtml
[4] Jaffe A E and Gelfenbaum G 2007 *Sedimentary Geology* **200** pp 347–361
[5] Pinegina T K, Bazanova L I, Zelenin E A et al 2018 *Pure Appl. Geophys.* **175** pp 1485–1506
[6] MacInnes B T, Bourgeois J, Pinegina T K and Kravchunovskaya E A 2009 *Geology* **37** pp 995–998
[7] Razzhigaeva N G, Ganzei L A, Grebennikova T A et al 2008 *J. of Volc. and Seismology* **2** pp 262–277
[8] Razjigaeva N G, Ganzey L A, Grebennikova T A, et al. 2014 *Pure Appl. Geophys.* **171** pp 3307–20
[9] Ganzey L A, Razjigaeva N G, Nishimura Yu et al 2015 *Russian J. of Pacific Geology* **9** pp 64–79
[10] Szczucin´ski W 2020 Post-depositional changes to tsunami deposits and their preservation potential *Geological Records of Tsunamis and Other Extreme Waves* (Elsevier) ed Max Engel et al pp 443–469
[11] Spiske M, Piepenbreier J, Benavente C and Bahlburg H 2013 *Earth Sci. Rev.* 2013 **126** pp 58–73
[12] Keating B H, Helsley C E, Wanink M and Walker D 2011 *The Tsunami Threat – Research and Technology* ner (Rijeka, Croatia: InTech) ed Nils-Axel Mör pp 389–422
[13] Pinegina T K and Bourgeois J 2020 Tephrostratigraphy and tephrochronology *Geological Records of Tsunamis and Other Extreme Waves* (Elsevier) ed Max Engel et al pp 745–759
[14] Pinegina T K, Bourgeois J, Bazanova L I, Braitseva O A and Egorov Yu O 2002 *Proc. Int. workshop. Local tsunami warning and mitigation, Petropavlovsk-Kamchatsky, September 10–15, 2002* (Moskow: Ianus-K) pp 122–131
[15] Geist E L and Parsons T *Geophys. Res. Lett.* 2008 **35** L02612
[16] Kaistrenko V 2014 *Pure Appl. Geophys.* **171** pp 3527–38
[17] Cramer H 1946 *Mathematical Methods of Statistics* (Bombay; New-York: Asia Publ. House) 575
[18] Kaistrenko V M, Pinegina T K and Klyachko M A 2003 *Proc. NATO Adv. Res. Workshop. Submarine Landslides and Tsunamis, Istanbul, Turkey, 23–26 May 2001* (Kluwer Publ., Springer) eds Yalciner A et al pp 217–228

IOP Conf. Series: Earth and Environmental Science **946** (2021) 012022 doi:10.1088/1755-1315/946/1/012022

The short-term tsunami forecast

Yu P Korolev

Institute of Marine Geology and Geophysics Far Eastern Branch of Russian Academy of Science, Yuzhno-Sakhalinsk, 693022, Russia

E-mail: yu_p_k@mail.ru

Abstract. A brief overview of the methods of a tsunami early warning in the Kuril Islands, which turned out to be ineffective during recent events, is presented. A hydrophysical method for short-term tsunami forecasting based on information about a tsunami in the ocean, used in the United States, and an express method, also using information about a tsunami in the ocean, are briefly described. The results of the retrospective forecast of the tsunami that occured on March 11, 2011, by the express method are presented.

1. Introduction

The main problem of operational tsunami forecasting in Russia is the problem of false alarms. From the moment of creation to the present, tsunami services in Russia have made a large number of false alarms (at least 75 % of the total), and their number is growing (Kuril tsunamis of 2006, 2007 and 2020, M = 8.3, 8.1 and 7.3).

False tsunami alarms, often announced too early, do not cause direct losses, but lead to a significant damage associated with stopping production in dangerous places, evacuating the population, taking ships out to the open sea, and create unjustified stressful situations for the population. Moreover, all kinds of activity in the coastal strip stops for several hours.

An example of a false alarm is the events of March 25, 2020.

An earthquake with a magnitude of 7.3 occurred on March 25, 2020 at 02:49 UTC with an epicenter 220 km east of the Onekotan Island (Kuril Islands, Russia).

8 minutes after the main shock, the Pacific Tsunami Warning Center (PTWC) issued a tsunami warning bulletin for the coast within 1000 km of the epicenter of the earthquake.

After 11 minutes a tsunami alert was announced in the Severo-Kurilsk region.

After 27 minutes Japan Meteorological Agency also warned of possible devastating tsunami waves in the Kuril Islands with an amplitude of 1 to 3 meters.

After 41 minutes after the start of the earthquake (30 minutes after the announcement of the alarm), the evacuation of the population to the safe zone was completed. About 400 people were evacuated.

The arrival of the tsunami to Severo-Kurilsk was expected at 04:04 UTC (75 minutes after the main shock).

According to visual observations, 1.2 km south of the port of Severo-Kurilsk, the first wave arrived approximately 1 hour after the earthquake. Against the background of storm waves, the tsunami height was visually estimated at 50 cm.

The tsunami alarm was canceled at 06:57 UTC (3 hours after the wave arrival).

Due to the small amplitude of the tsunami, the announced alarm turned out to be false. The prolongation of the tsunami alarm after the arrival of the first small amplitude waves was unreasonable.

All three forecasts of the tsunami on March 25, 2020 turned out to be ineffective: with the forecasted 1–3 m, the actual height was about 0.5 m.

2. Methods for early tsunami warning

The Pacific Tsunami Warning Center (PTWC) forecast is based on current regulations, magnitude criterion. The issued bulletin is informative and does not constitute a tsunami alert. The decision to declare an alarm is made by the regional centers.

The Japan Meteorological Agency (JMA), responsible for tsunami forecasting in the Pacific Northwest, is acting in accordance with the new regulations [1]. The forecast uses a database of tsunami heights at a large number of points in the ocean. The database is based on preliminary computations of tsunami waveforms from supposed seismic sources with the most probable earthquake mechanisms. The decision to declare a tsunami alert is also the responsibility of the regional centers. As practice shows, this method is not always effective.

The prediction and announcement of the tsunami alert by the Sakhalin Tsunami Warning Service is carried out on the basis of the magnitude-geographic criterion.

According to modern concepts, a tsunami alarm should be announced with a reasonable advance only at those points where the tsunami poses a real danger, and be accompanied with information about the arrival times of the first wave, the maximum wave, their amplitudes, as well as the expected time of the end of the tsunami (tsunami alarm clearing). These characteristics of a tsunami are listed in the definition of a tsunami forecast formulated by UNESCO Intergovernmental Oceanographic Commission (UNESCO IOC) in 2013 [2].

Hydrophysical methods are an effective tool for operational tsunami forecasting. Information about the formed tsunami obtained in the ocean by deep-sea bottom stations of the DART system (Deep-ocean Assessment and Reporting of Tsunamis) [3] is reliable for assessing the degree of tsunami hazard. The hydrophysical method SIFT (Short-term Inundation Forecasting for Tsunamis) [4] based on information about the tsunami in the ocean (NDBC DART Program) using a pre-computed base of synthetic mareograms provides an early numerical estimate of the amplitude, travel time, and other properties of the tsunami. The forecast is given immediately after receiving information about the passage of the first half-period of the tsunami through the registration point. The method fully meets the definition of the UNESCO IOC tsunami forecast concept.

The SIFT method has been successfully applied both for retrospective forecasting of earlier tsunamis since 1996 and for real-time forecasting of recent tsunamis [5, 6, 7, 8, 9].

The SIFT method is applicable to those points in the ocean or near the coast for which a pre-computed database is available. In particular, the method is applied to the US West Coast, the coast of Alaska and the Aleutian Islands.

For the coast of Kamchatka and the Kuril Islands, the SIFT method is not applicable due to the lack of relevant databases.

3. Express method of short-term tsunami forecast

The express method for short-term tsunami forecasting is based on the fundamental reciprocity principle. Based on this principle, the calculated relation was derived [10].

$$\zeta(s,A) = \zeta(s,M)\cdot\frac{\eta(s,A)}{\eta(s,M)},$$

where $\zeta(s,A)$ is the spectrum (Laplace transform) of the expected tsunami waveform at a given point A near the coast for which the forecast is performed, $\zeta(s,M)$ is the tsunami spectrum at a point M, where the ocean level is measured, $\eta(s,A)$ is the spectrum (Laplace transform) of the waveform at a point A from an auxiliary source, $\eta(s,M)$ is the spectrum of the waveform at a point M from the auxiliary source.

In the above relation, the ratio on the right side plays the role of a transfer function, which is used to calculate the shape of the expected tsunami at a given point near the coast (point A) from the tsunami data in the open ocean (in a point M).

The main assumption of the express method: an auxiliary source for creating the transfer function is the circular initial elevation of the free surface with the center coinciding with the epicenter of the earthquake. In this case, for the computation (forecast) of a tsunami, the only seismological information on the coordinates of the earthquake epicenter is required.

The inverse Laplace transform gives the time of arrival and the shape of the expected tsunami at a point A, which contains comprehensive information that fully complies with the definition of a tsunami forecast given by the UNESCO IOC.

4. Retrospective forecast of the Tohoku tsunami on March 11, 2011

The event on March 11, 2011, was selected as an example of using the express method.

The strongest earthquake with a magnitude of 9.0 occurred off the northeastern coast of Honshu Island on March 11, 2011.

The tsunami was recorded by almost all stations of the DART system [11], as well as by coastal tide gauges.

4.1. Retrospective forecast of the tsunami on March 11, 2011, near the Kuril Islands

A tsunami alert has been declared on all Kuril Islands. It is recommended for ships in the roads and piers to leave for a safe area. The evacuation of the population was completed 1 hour 24 minutes before the tsunami arrival in Yuzhno-Kurilsk, 2 hours 38 minutes before the wave arrival in Severo-Kurilsk, and 1 hour 18 minutes before the arrival of the wave in Kurilsk.

A retrospective forecast of the expected tsunami was made near settlements and at the location of the DART 21419 station. Their positions are shown in figure 1.

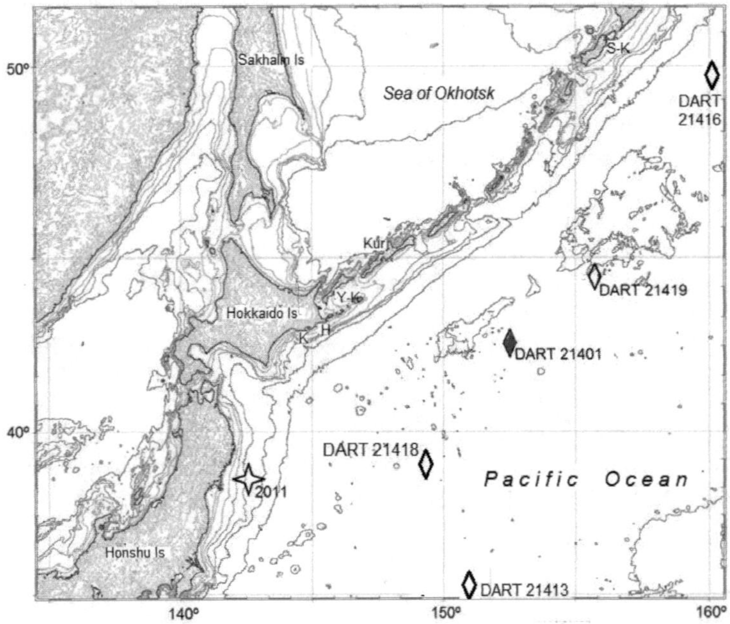

Figure 1. Scheme of the Northwest Pacific Ocean showing the earthquake epicenter and the points for which the forecast was made. The diagram indicates: S-K – Severo-Kurilsk, Kur – Kurilsk, Y-K – Yuzhno-Kurilsk, H – Hanasaki and K – Kushiro.

For the forecast, a transfer function was created using a difference grid with a step of 900 m at a latitude of 45° for 20 minutes. Taking into account the fact, that the time to obtain data on the coordinates of the earthquake epicenter of 11 min, the transfer function could be ready 31 min after the main shock of the earthquake. The forecast was made according to the data of the Russian station DART 21401. The waveform recorded by the station is shown in figure 2. The data of the station with a duration of 89 minutes from the beginning of the earthquake were used.

The calculation results are shown in figure 2.

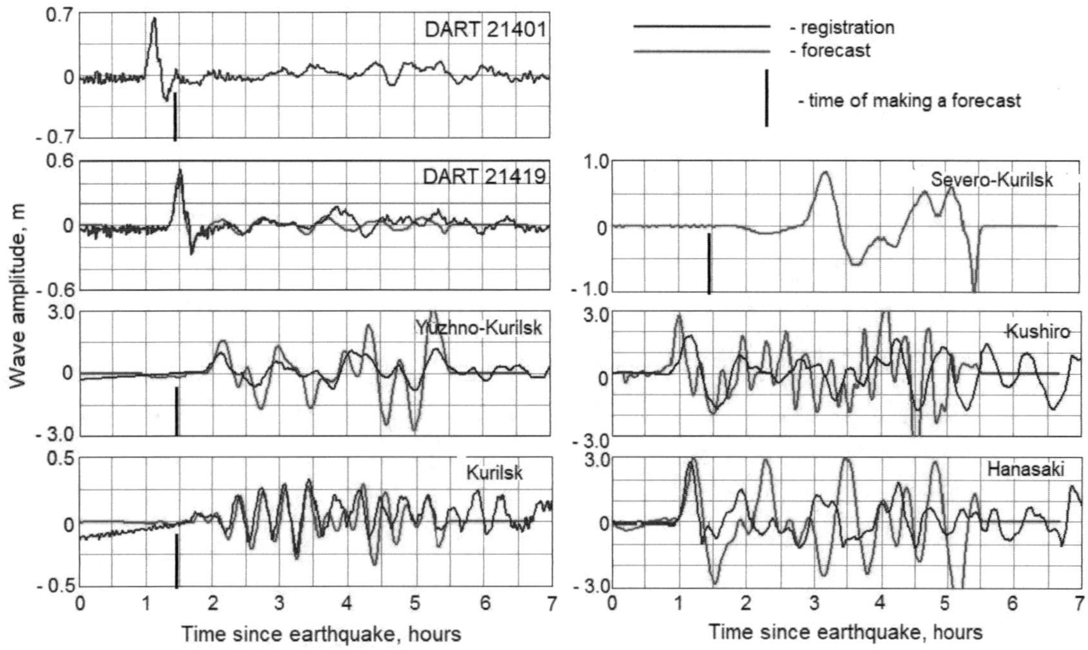

Figure 2. Results of the tsunami forecast on March 11, 2011, at settlements of the Kuril Islands and Hokkaido Island.

The predicted waveforms of the expected waves are in good agreement with the registered ones. The retrospective forecast was made after the passage of the first tsunami period through DART 21401, i.e. 89 minutes after the main shock. The forecast results are quite suitable for assessing the degree of tsunami danger and making a decision on the need to declare a tsunami alert in each settlement.

The lead time of the forecast according to DART 21401 data for Yuzhno-Kurilsk was 23 minutes, for Kurilsk – 48 minutes and Severo-Kurilsk – 75 minutes.

The lead time according to DART 21418, which is closer to the tsunami source, could be 30 min higher.

4.2. Computation of tsunami waveform in the ocean in the different directions of the source

To confirm the operability of the express method of short-term tsunami forecast, the tsunami waveforms were computed at the points of DART stations in the ocean in different directions from the tsunami source.

The locations of the stations are shown in figure 3. Diamonds with numbers indicate those stations, the computation results for which are shown below in figures 4 and 5.

The data of the Russian station DART 21401 (the first period of the wave, figure 2), located to the northeast of the tsunami source, were used. To construct the transfer function, an auxiliary circular source with a diameter of 100 km and with a center coinciding with the epicenter of the earthquake was used on a difference grid with a step of 3830 m at a latitude of 40°.

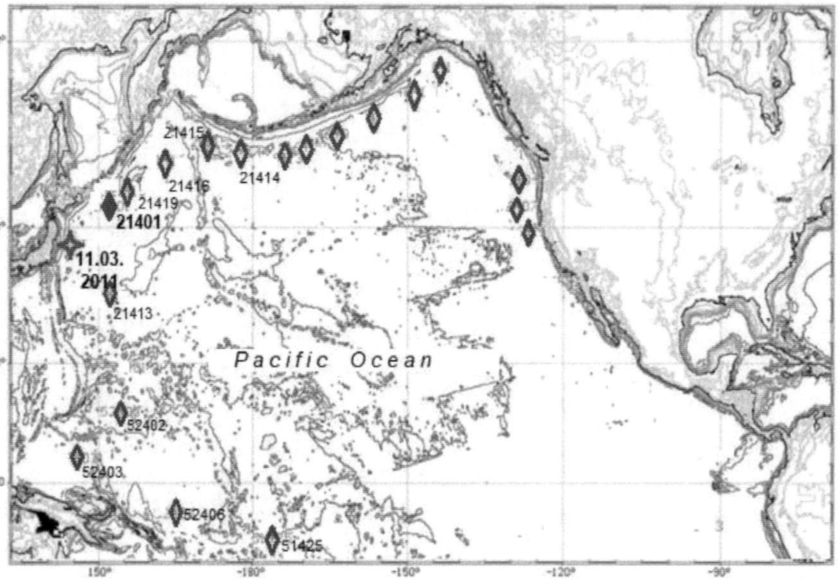

Figure 3. Scheme of the Pacific Ocean showing the earthquake epicenter and DART stations for which the computation was per formed.

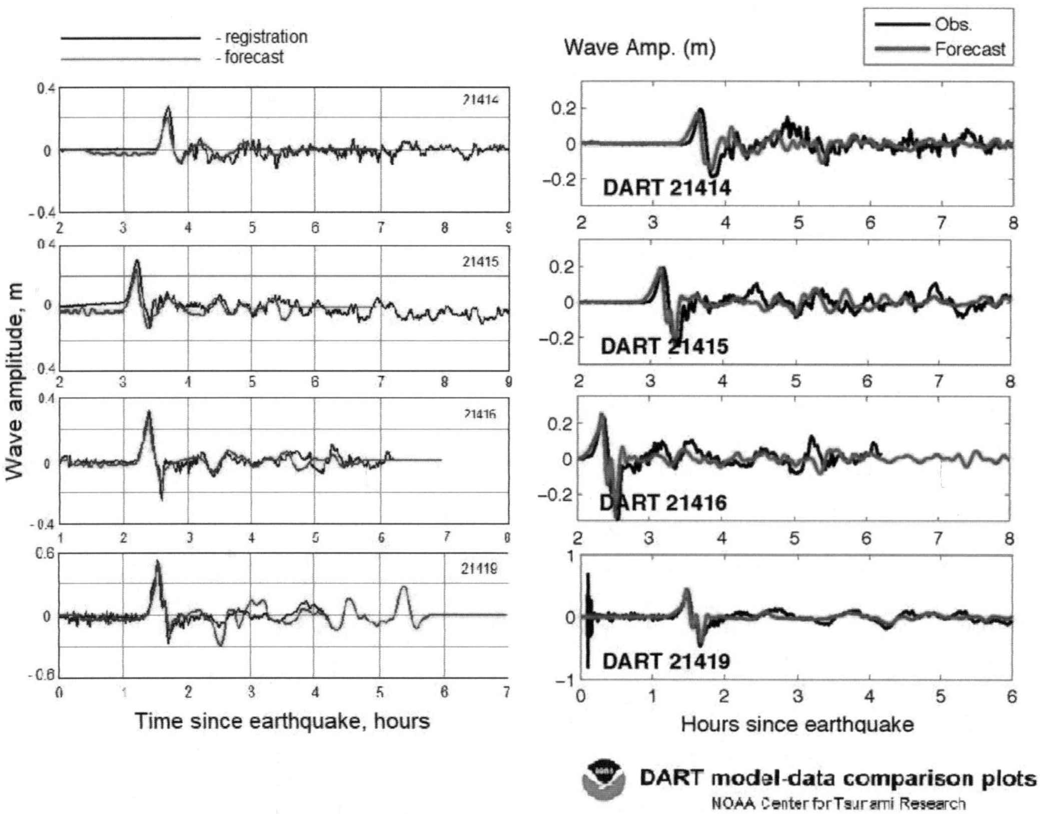

Figure 4. The results of computation the tsunami waveforms on March 11, 2011, in the direction northeast of the tsunami source (left) and similar results obtained by NOAA Center for Tsunami Research [12].

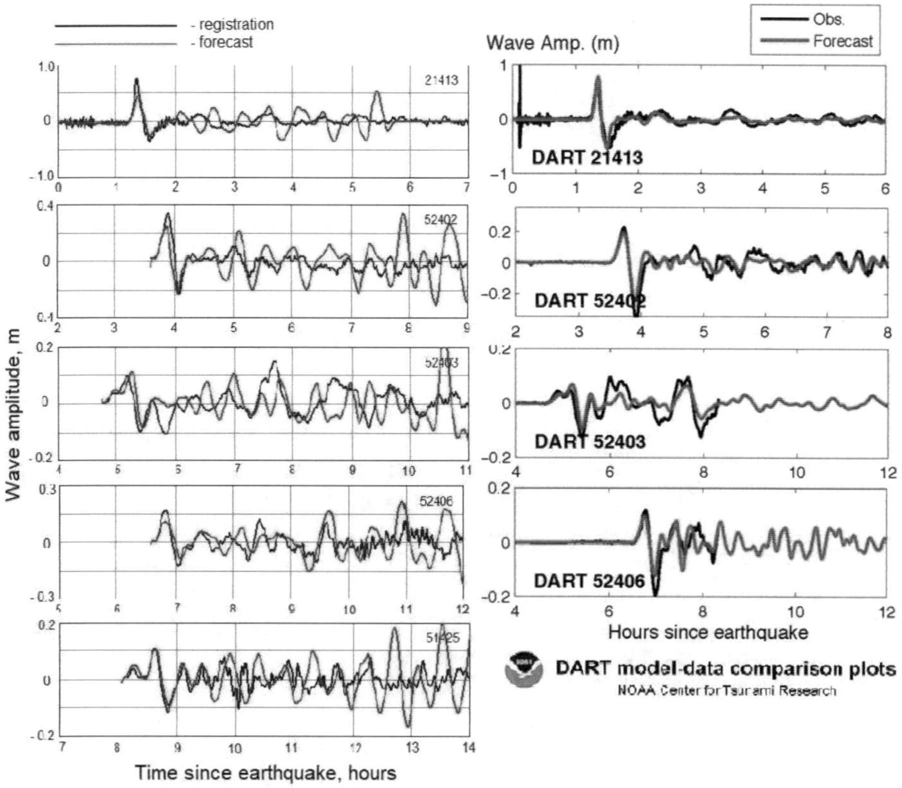

Figure 5. The results of computation the tsunami waveforms on March 11, 2011, in the direction south, southeast of the tsunami source (left) and similar results obtained by NOAA Center for Tsunami Research [12].

The presented results of the computation of the tsunami on March 11, 2011, in the ocean demonstrate good agreement between the computed and recorded waveforms, not only of the head waves, but also of subsequent waves. This indicates that the express method gives a completely adequate forecast.

The same quality of computations is for stations along the coast of the Aleutian Islands and the Alaska Peninsula, as well as along the US West Coast.

Regardless of the relative positions of the source, the level measurement point and the forecast point, the forecast quality is good, despite the fact that an axially symmetric auxiliary source was used to construct the transfer function.

Difficulties with the forecast may arise when computing near the coasts, which is associated with the inaccuracy of the bathymetric data, the coarseness of the difference grids used. The use of fast computation technology, nested grids techniques can eliminate many of the problems of online calculations.

The quality of the computation of tsunami waveforms obtained by the express method is comparable to the quality of similar computations by the NOAA Center for Tsunami Research.

5. Conclusion

The express method of short-term tsunami forecast based on tsunami data in the open ocean makes it possible to compute the waveform of the expected tsunami in real time at specified points in the ocean or near the coast.

To make a forecast, the only seismological information about the coordinates of the earthquake epicenter is need. The method does not depend on the earthquake mechanism, it can take into account additional effects in the form of subsea landslides.

The method fully complies with the definition of a tsunami forecast formulated by the UNESCO IOC.

The method does not require the creation of giant databases and can be used for tsunami forecasting in areas for which other methods (for example, SIFT) are not applicable.

The use of fast evaluation technology and method of nested grids will make it possible to forecast a tsunami immediately after receiving information about the passage of a tsunami through the DART station closest to the source with a lead time sufficient to make a decision to announce an alarm and evacuate the population.

References

[1] UNESCO/IOC 2019 *IOC Tech. Ser.* Users' Guide for the Northwest Pacific Tsunami Advisory Center (NWPTAC): Enhanced Products for the Pacific Tsunami Warning System (Paris, UNESCO) 142 (*Electronic materials*) https://unesdoc.unesco.org/ark:/48223/pf0000366546?posInSet=1&queryId=d1288da0-390e-47b1-8a51-a529b04abf93

[2] Intergovernmental Oceanographic Commission. Tsunami Glossary 2016 *IOC Technical Series* (Paris, UNESCO) *3rd edition* 85 (IOC/2008/TS/85 rev. 2)

[3] NOAA Center for Tsunami Research: DART (*Electronic materials*) http://nctr.pmel.noaa.gov/Dart

[4] NOAA Center for Tsunami Research: Tsunami Forecasting (*Electronic materials*) https://nctr.pmel.noaa.gov/tsunami-forecast.html

[5] Percival D B, Denbo D W, Eble M C, Gica E, Mofjeld H O, Spillane M C, Tang L and Titov V V 2011 *Nat. Hazards* **58** pp 567–590

[6] Tang L, Titov V V, Moore C and Wei Y 2017 Real-Time Assessment of the 16 September 2015 Chile Tsunami and Implications for Near-Field Forecast *The Chile-2015 (Illapel) Earthquake and Tsunami* (Pageoph Topical Volumes) (Birkhäuser Cham) ed C Braitenberg and A Rabinovich pp 267–285

[7] Titov V V 2009 Tsunami forecasting *The Sea* (Cambridge, MA; London, England: Harvard Univ. Press) ed E N Bernard and A R Robinson **15** pp 367–396

[8] Wei Y, Bernard E N, Tang L, Weiss R, Titov V V, Moore C, Spillane M, Hopkins M and Kanoglu U 2008 *Geophys. Res. Lett.* **35** L04609

[9] Wei Y, Cheung K F, Curtis G D and McCreery Ch S 2003 *J. Waterway, Ports, Coastal and Ocean Engineering. ASCE* **129** 2 pp 60–69

[10] Korolev Yu P 2011 *Nat. Hazards Earth Syst. Sci.* **11** pp 3081–91

[11] National Data Buoy Center (*Electronic materials*) https://ndbc.noaa.gov/dart.shtml

[12] NOAA Center for Tsunami Research: Events (*Electronic materials*) https://nctr.pmel.noaa.gov/honshu20110311/

The hindcast of recent tsunamis in Northern Pacific

P Korolev[1,*], Yu Korolev[1] and A Loskutov[1,2]

[1]Institute of Marine Geology and Geophysics Far Eastern Branch of Russian Academy of Science, Yuzhno-Sakhalinsk, 693022, Russia
[2]Laboratory of tsunamis, P P Shirshov Institute of Oceanology RAS, Moscow, 117997, Russia

E-mail: *pasha200482@mail.ru

Abstract. Three earthquakes occurred in the North Pacific in 2020, causing observable tsunamis. The tsunamis were not devastating. Numerical modelling of tsunami propagation was performed to reproduce operational forecasting (retrospective analysis) of waveforms at deep-water stations. Direct calculation of tsunami using USGS finite-fault source data on GPU was carried out. The leap-frog (Arakawa staggered grid) scheme calculation over the Pacific Ocean on a regular grid with a spatial step of 0.5 arc minutes of 1440 min (1 day) tsunami propagation was performed in approximately 90 min of computer time. With use of a hybrid cluster with several GPU accelerators and proper optimization of the simulation algorithm, this time can be reduced by tens of times. Consequently, the time for estimating the transfer function will be comparable to the travel time of a tsunami to the stations, where the forecasts data is. It will make possible to forecast the shape of a tsunami at any point with a lead time enough to decide for tsunami alert at sites where a tsunami poses a real danger. The calculation results are in good agreement with the real data of deep-ocean measurements. The quality of the forecast is comparable to the quality of calculations by other methods.

1. Introduction

The problem of the fast tsunami forecasting remains an unresolved problem for many coasts of the Pacific and Indian oceans. For several coasts (the Russian Far East, the Pacific coast of South America), the main method of operational tsunami forecasting is the magnitude-geographical criterion, which spawns unacceptable level of false alarms.

Methods based on computational fluid dynamics and using information about the generated tsunami wavefield in the open ocean (DART stations [5]) allow a better forecast [7].

Based on the forecasts, a tsunami alarm could be declared in a reasonable time only at the points where the tsunami poses a real danger and supplemented with information about the arrival times of the first wave, the maximum wave, wave amplitudes, as well as the expected time of the end of tsunami (cancelling the alarm). These characteristics of tsunamis are listed in the definition of a tsunami forecast formulated by the Intergovernmental Oceanographic Commission (IOC) of UNESCO in 2013 [1].

In 2020 several earthquakes have occurred in the North Pacific Ocean, causing the observed tsunamis. Earthquakes on March 25 with $M_w = 7.5$ near the central Kuril Islands (east of Onekotan Island), July 22 with $M_w = 7.8$ and on October 19 with $M_w = 7.6$ near the Alaska Peninsula (the area of the Shumagin Islands) [3].

Tsunamis did not cause damage, but they were the events, which might be used to test different tsunami excitation models and the effectiveness of different forecasting methods.

Purpose of the work is a development of technology for fast tsunami calculation for operational forecasting using examples of recent tsunamis.

Figure 1 shows the map of the North Pacific Ocean plotting the positions of the DART deep-water stations and earthquake epicenters.

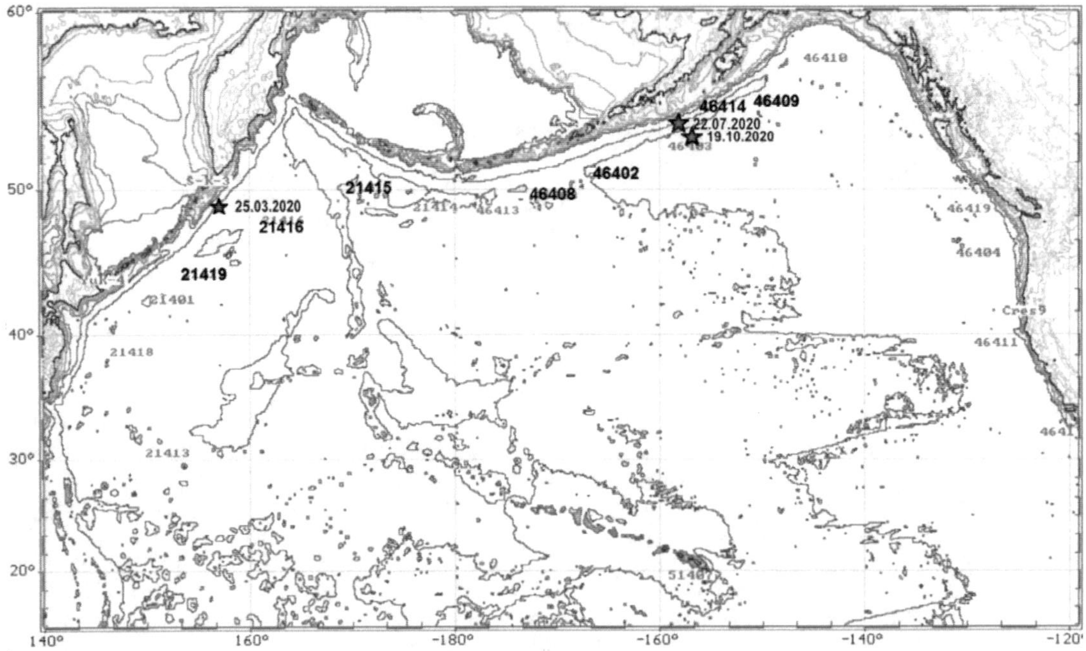

Figure 1. Positions of DART stations and earthquake epicenters in North Pacific Ocean.

2. Results

Simulation of tsunami by the fast method of operational forecasting of tsunamis on March 25, 2020, in the Northwest Pacific was performed.

For the events of July 22 and October 19, 2020, the tsunami waveforms from the sources constructed according to the USGS data [8] were calculated using GPU accelerators.

2.1. Tsunami on March 25, 2020, east of Onekotan Island

The fast method of operational forecasting of tsunamis is based on a transfer function for each point, which allows using the data from the DART station (in the present experiment, station 21416, which is closest to the epicenter of the earthquake), to calculate the shape of the expected tsunami near this point. The transfer function is built immediately after receiving information about the coordinates of the epicenter of the earthquake caused the tsunami. The transfer function is based on calculation of waveforms from a point-like source with a center coinciding with the epicenter of the earthquake at the points of registration and forecast with no earthquake magnitude data required [2]. The comparison of the results of a numerical simulation and calculations by the NOAA Center for Tsunami Research are shown in figures 2, 3.

The calculation results (red line) are in good agreement with the real tsunami measurements data (blue line). The quality of the forecast is comparable to the quality of the calculations provided by the NOAA Center for Tsunami Research [6].

For the successful implementation of the method, it is necessary to use algorithms that allow us to calculate the shape of the expected tsunami at specified points on the coast of the Kuril Islands in time not greater than 10 minutes.

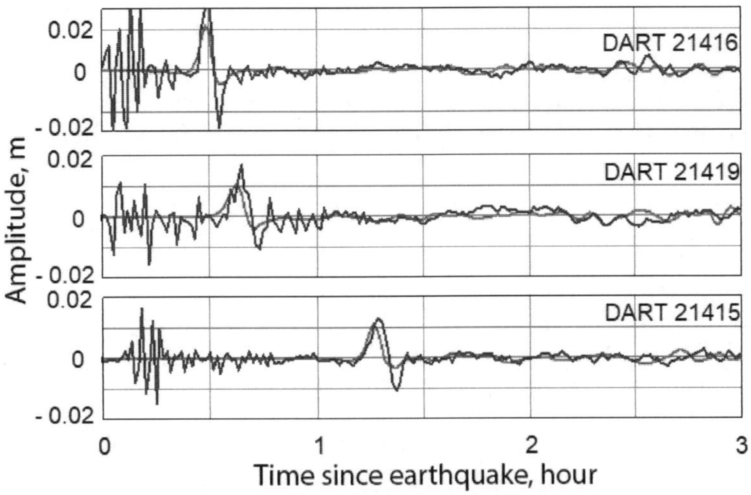

Figure 2. Results of the tsunami forecast on March 25, 2020, at DART stations near the Kuril Islands.

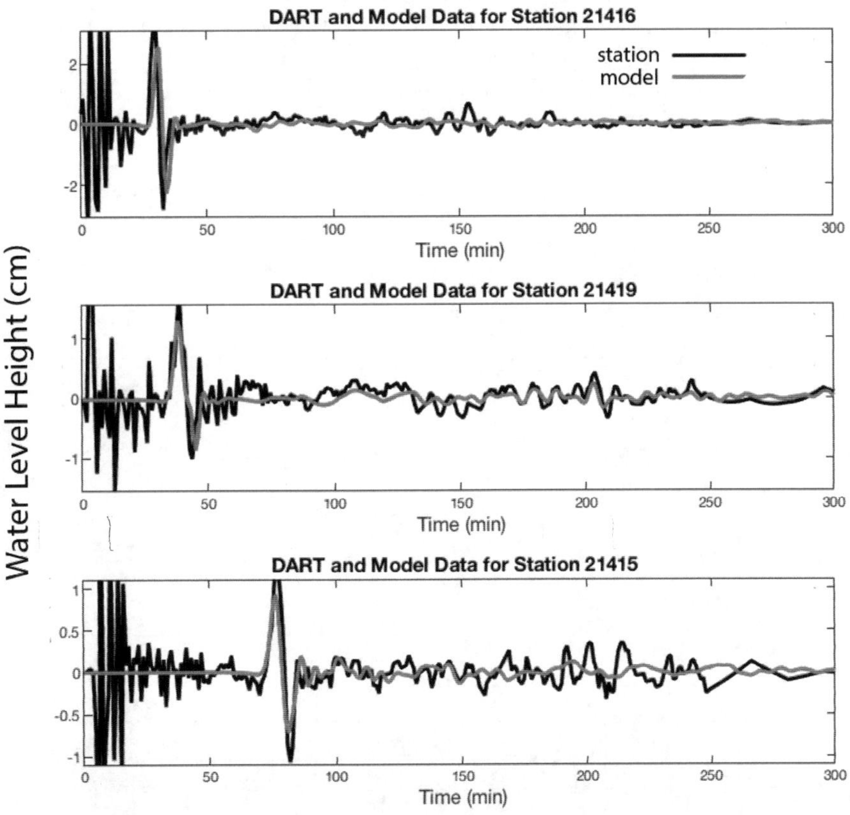

Figure 3. Results of the NOAA Center for Tsunami Research [6].

2.2. Tsunami on July 22, 2020, near Shumagin Islands

For the event on July 22, the tsunami waveforms from the USGS finite-fault sources calculated according to the [8] using GPU accelerators. The leap-frog depth-average numerical calculation of tsunami propagation over the Pacific Ocean on a regular grid with a spatial step of 0.5 arc minutes with a process duration of 1440 min (1 day) was performed in approximately 90 min of computer time on one GPU only [4].

The calculation results are represented in figure 4, 5. To verify the results, the calculations of the Center for Tsunami Research (NOAA Center for Tsunami Research) [6] are presented in figure 3 too.

Obviously, the calculation results and the results presented by the Center for Tsunami Research (NOAA Center for Tsunami Research) [6] match up pretty good. Though the discrepancy with the recorded tsunami forms indicates that calculations using seismic sources do not always give adequate results.

Evidently, the calculation results and the results presented by the Center for Tsunami Research (NOAA Center for Tsunami Research) [6] match up pretty good. Though the discrepancy with the recorded tsunami forms indicates that calculations using seismic sources do not always give adequate results.

2.3. Tsunami October 19, 2020, near Shumagin Islands

For the event of October 19, we also calculated tsunami waveforms from finite-fault seismic sources provided by USGS [8] using GPU. The same simulation parameters as for previous case were engaged. We presume that with use of a hybrid cluster with several GPU accelerators and proper optimization of the simulation algorithm, this time can be reduced by tens of times. Consequently, the time for estimating the transfer function will be comparable to the travel time of a tsunami to the stations, on which the forecasts data are based on. It will make possible to forecast the shape of a tsunami at any point with a time enough to decide for tsunami alert at sites where a tsunami poses a real danger. The calculation results are shown in figures 6, 7 where the results of the Center for Tsunami Research (NOAA Center for Tsunami Research) [6] are also shown.

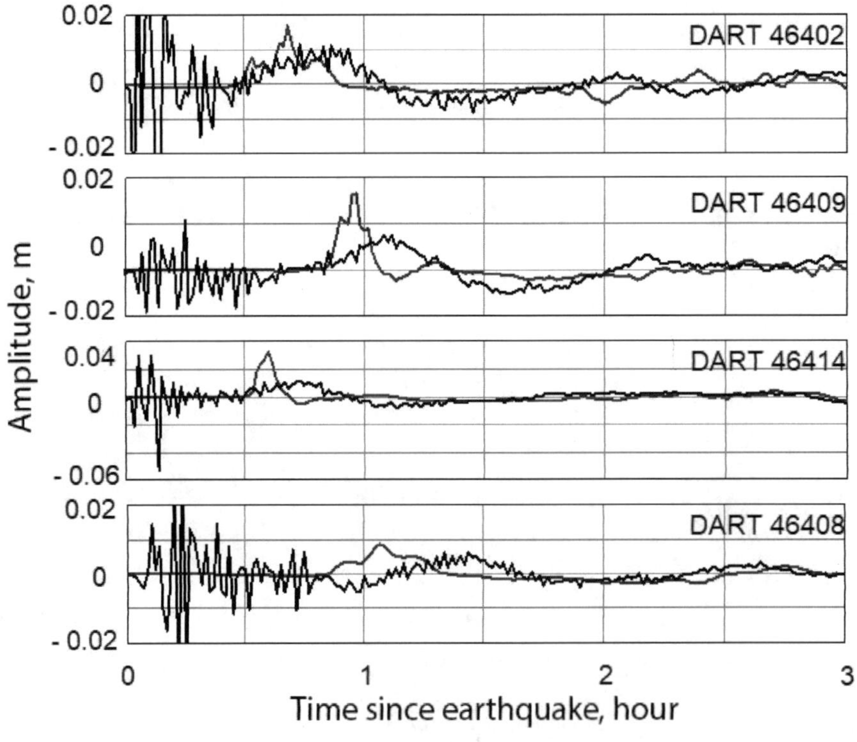

Figure 4. Results of calculations of tsunami on July 22, 2020, generated numerically using USGS finite-fault sources, compared to measurements at DART stations.

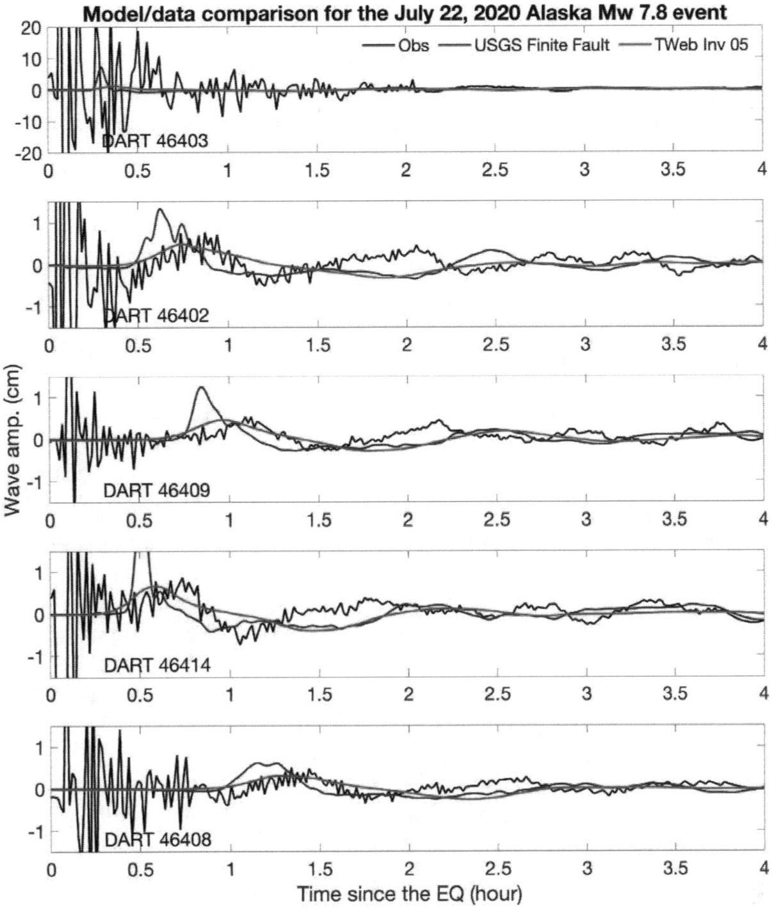

Figure 5. Results of calculations of tsunami on July 22, 2020, of the NOAA Center for Tsunami Research [6].

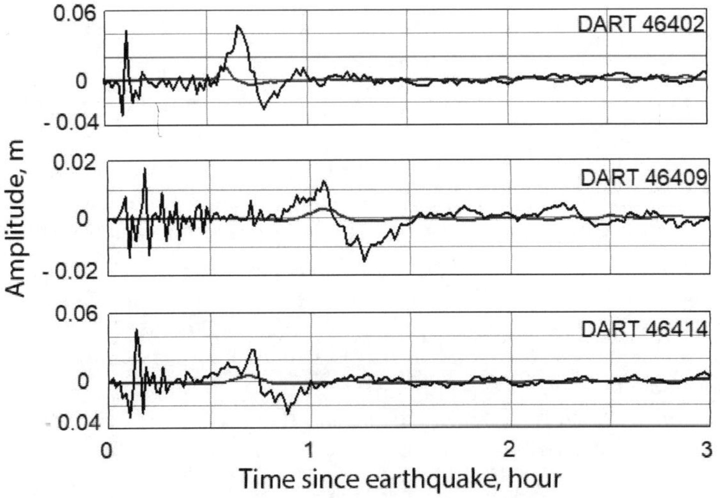

Figure 6. Results of tsunami calculations on October 19, 2020, from finite-fault seismic sources provided by USGS at DART stations.

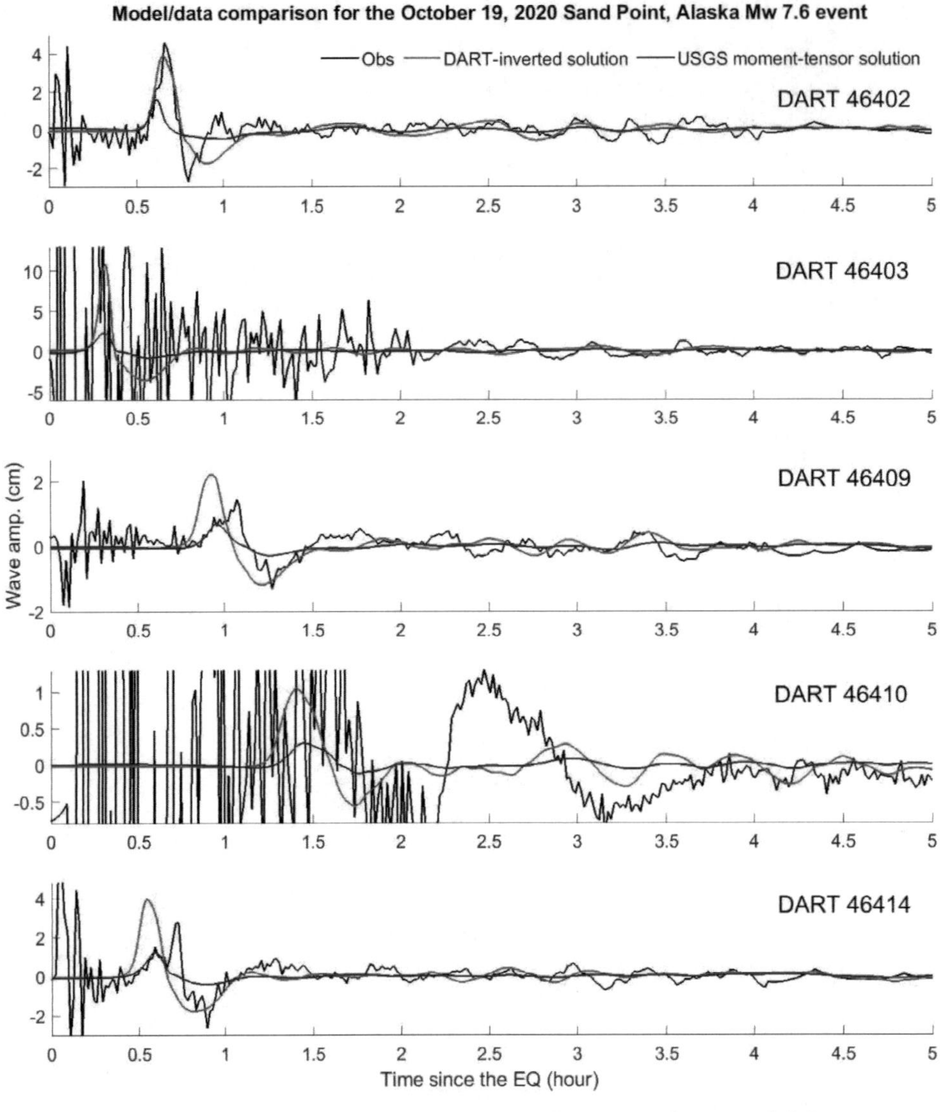

Figure 7. Results of the NOAA Center for Tsunami Research [6].

3. Conclusion

On the example of the tsunami on March 25, 2020, which occurred in the Kuril Islands region, it is shown that the fast method of operational tsunami forecast based on tsunami data in the open ocean makes it possible to calculate in real time the shape of the expected tsunami at specified points in the ocean or near the coast. To make a forecast, the seismological subsystem needs information about the coordinates of the earthquake epicenter only. The calculation results are in good agreement with the actual data on the tsunami in the ocean. The quality of the forecast is comparable to the quality of calculations by other methods.

To simulate the tsunami on July 22 and October 19, 2020, near the Alaska Peninsula, we used the calculation of waveforms using GPU accelerators. The leap-frog depth-average numerical calculation of tsunami propagation over the Pacific Ocean on a regular grid with a spatial step of 0.5 arc minutes with a process duration of 1440 min (1 day) was performed in approximately 90 min of computer time on one GPU only. The calculation results are in good agreement with the calculation results obtained by other methods.

We propose to use of a hybrid cluster with several GPU accelerators and proper optimization of the simulation algorithm, this time can be reduced by tens of times. Consequently, the time for estimating the transfer function will be comparable to the travel time of a tsunami to the stations which data the forecasts are based on. It will make possible to forecast the shape of a tsunami at any point with a lead time enough to decide for tsunami alert at sites where a tsunami poses a real danger.

References

[1] Intergovernmental Oceanographic Commission. Tsunami Glossary 2016 *IOC Technical Series* (Paris, UNESCO) *3rd edition*
[2] Korolev Yu 2011 *Nat. Hazards Earth Syst. Sci.* **11** pp 3081–91
[3] National Centers for Environmental Information: Search Tsunami Events (*Electronic materials*) https://www.ngdc.noaa.gov/hazel/view/hazards/tsunami/event-search (accessed 06/28/2021)
[4] Nickolls J, Buck I, Garland M and Skadron K 2008 *Queue* **6** pp 40–53
[5] NOAA Center for Tsunami Research: DART (*Electronic materials*) http://nctr.pmel.noaa.gov/Dart
[6] NOAA Center for Tsunami Research: Events (*Electronic materials*) https://nctr.pmel.noaa.gov/database_devel.html
[7] Titov V 2009 Tsunami forecasting *The Sea* (Cambridge, MA; London, England: Harvard Univ. Press) ed E Bernard and A Robinson **15** pp 367–396
[8] US Geological Survey 2021 *Significant Earthquakes archive* (*Electronic materials*) https://earthquake.usgs.gov/earthquakes/browse/significant.php

Transformation of baroclinic tidal waves in the conditions of the shelf of the Far Eastern seas

P D Kuznetsov, E A Rouvinskaya, O E Kurkina and A A Kurkin*

Nizhny Novgorod State Technical University n a R E Alekseev, Nizhny Novgorod, 603950, Russia

E-mail: *aakurkin@gmail.com

Abstract. This work is devoted to the study of the regimes of transformation of baroclinic tidal waves under the conditions of the Far Eastern seas within the framework of a fully nonlinear numerical model. Two sections were selected to study the features of wave dynamics – in the Sea of Okhotsk (Sakhalin Island shelf) and in the Bering Sea (near Cape Navarin). On the basis of the performed calculations, regional and seasonal features of the transformation of baroclinic waves and the structure of the flow induced by them were revealed. It is shown that the dynamics in the winter season is less intensive. The rotation effect on the formation of solibore in the studied conditions is analyzed. The obtained estimates of wave amplitudes and velocities are consistent with the published data of field observations in the Sea of Okhotsk. For the Bering Sea, the conditions are shown to be favorable for the generation of intensive internal waves, which indirectly confirms the hypothesis of influence of such waves on the formation of underwater sand dunes.

1. Introduction

One of the classical and predominant mechanisms for internal waves generation is the transformation of a tidal wave over the seafloor topography [1]. In this sense, the Far-Eastern seas, rich in various natural resources, both mineral and biological, are a favourable environment for the implementation of this scenario. They are stratified in density: most of the year have a pronounced layered structure (e.g., four layers are distinguished in the Sea of Okhotsk waters [2; 3], mainly two or three – in the Bering Sea [4]). They are exposed to a barotropic multicomponent tide, and have complex bathymetry. As a result, short-period intensive internal waves are generated, that is confirmed by the scarce data of satellite observations and field measurements (see, e.g., [5; 6] and others (a database of observations of internal waves, including those in the Far Eastern seas, is presented on the website of our laboratory: https://lmnad.nntu.ru/ru/igwatlas/)). Such waves play an important role in the ecosystem of the stratified sea shelf. And although internal waves are rarely taken into account in the design of hydraulic structures in our country, study of the typical regimes of their generation is an urgent task that has engineering applications. For quantitative assessments of dangerous dynamic effects from short-period internal waves and their further consideration in engineering surveys, it is necessary to perform regional investigation of scenarios of long baroclinic wave transformation to estimate the probability of occurrence of internal waves of extreme amplitudes. Such assessments were started for certain areas of the White Sea, the Barents Sea and the Sea of Okhotsk in [7; 8; 5]. Field and satellite observations indicate a high activity of this type of waves in some areas of the Sea of Okhotsk (e.g., see [9] for satellite sensing data) and the Bering Sea, however, the amount of such data is limited. Therefore, therefore, it is necessary to resort to numerical modeling, in order to study the scenarios of transformation of baroclinic waves in such water areas.

Within the framework of this study, we selected several vertical "sections" (corresponding to the direction of propagation of the prevailing component of the barotropic tide) in the shelf zone of the Bering Sea and the Sea of Okhotsk. The first of them was chosen on the northeastern shelf of Sakhalin Island. It crosses the Kaigansko-Vasyukanskoe-Sea oil and gas condensate field, while similar hydrological conditions are typical for the Odoptu-Sea field zone (Sakhalin-1). According to the data of field observations in this area of the northeastern shelf of Sakhalin Island, trains of short-period internal waves with amplitudes from 2 to 10 m at a depth of about 50 m were observed, while the horizontal velocities of currents induced by these waves reached 40 cm/s [10]. Manifestations of short-period internal wave trains are also observed by means of remote sensing [6].

The second cross-section is selected in the northwestern part of the Bering Sea from the Aleutian Basin to the Koryak-Kamchatka coast south of Cape Navarin. This region is poorly studied in the context of the dynamics of internal waves, because direct contact measurements of internal waves are technically difficult and economically expensive at high latitudes. However, there is a hypothesis that internal waves are responsible for the formation of large sand waves in the underwater canyons on the northern continental slope (moreover, the horizontal spatial scales of the discovered bottom dunes coincide with the lengths of the observed short-period waves (see, e.g., [11; 12])). According to satellite radar images, the zone of the Bering Sea south of Cape Navarin is characterized by high activity of short-period internal waves [13]. At the same time, this region is interesting for Russia both from an economic and an ecological point of view.

The paper presents results of numerical modeling of internal wave fields evolution in realistic conditions in the framework of program complex intended for numerical modeling of propagation and transformation of such waves in the ocean. The goal of this paper is to study regional and seasonal features of baroclinic wave transformation and the structure of the flow induced by them on the basis of the performed calculations under the conditions of the shelf of the Sea of Okhotsk and the Bering sea.

2. Initialization of mathematical model for simulation of dynamics of the internal gravity waves in the conditions of the Bering Sea and the Sea of Okhotsk

The study of internal wave dynamics was carried out by means of the program complex intended for numerical modeling of propagation and transformation of such waves in the ocean, that implements a procedure of numerical integration of fully nonlinear two-dimensional (vertical plane) system of equations of hydrodynamics inviscid incompressible stratified fluid in the Boussinesq approximation, taking into account the influence of barotropic tide [15; 14]:

$$\vec{V}_t + (\vec{V}\nabla)\vec{V} - f\vec{V} \times \vec{k} = -\nabla P - \vec{k}\rho g, \quad (1)$$

$$\rho_t + \vec{V}\nabla\rho = 0, \quad (2)$$

$$\nabla\vec{V} = 0, \quad (3)$$

$$\rho = \frac{\rho_f - \rho_0}{\rho_0}, \quad (4)$$

where $\vec{V}(u, v, w)$ is the velocity vector, ∇ is the three-dimensional vector gradient operator, subscript t denotes the time derivative, ρ_f – the density of sea water, ρ_0 – the average or characteristic density (introduced owing to the Boussinesq approximation that assumes that the density $\rho_f = \rho_0(1+\rho)$ only changes insignificantly in the basin and ρ is a nondimensional quantity that has a meaning of density anomaly), P is the pressure, g is the acceleration due to gravity, f is, as above, the Coriolis parameter and $\vec{k}$ is the unit vector along the *z*-direction. The waves propagate in the *x*-direction, *y*-axis is perpendicular to the wave motion and *z* is the vertical coordinate.

The normal to the wave propagation (cross-section) velocity is included in the model, but no variation along the *y*-coordinate is allowed. This is realised by neglecting the partial derivatives with respect to *y* in the basically three-dimensional equations (1–4). The equations are transformed to a terrain-following coordinate system (so-called sigma-coordinates) and are solved over a domain bounded below by the bathymetry *h*(*x*) (prescribed by the user) and covered by a rigid lid at the surface.

To initialize the model, it is necessary to prescribe undisturbed horizontally homogeneous density field of water masses $\rho_{mean}(z)$ as well as the velocity distribution in the barotropic tidal field in the computational domain. The steps of the numerical scheme in space and time are chosen to satisfy the Courant–Friedrich–Levy criterion for stability.

To determine water depth, we defined sections between points with coordinates 60°30'N 177°30'E and 62°30'N 177°21'E (in the Bering sea); 54°31'N 145°32' E and 53°89'N 143°16' E (in the Sea of Okhotsk) using GDEM data (figure 1). Section lengths are approximately 160 km and 220 km respectively. The direction of the section is determined in accordance with the pattern of internal gravity waves radiation from the tidal front.

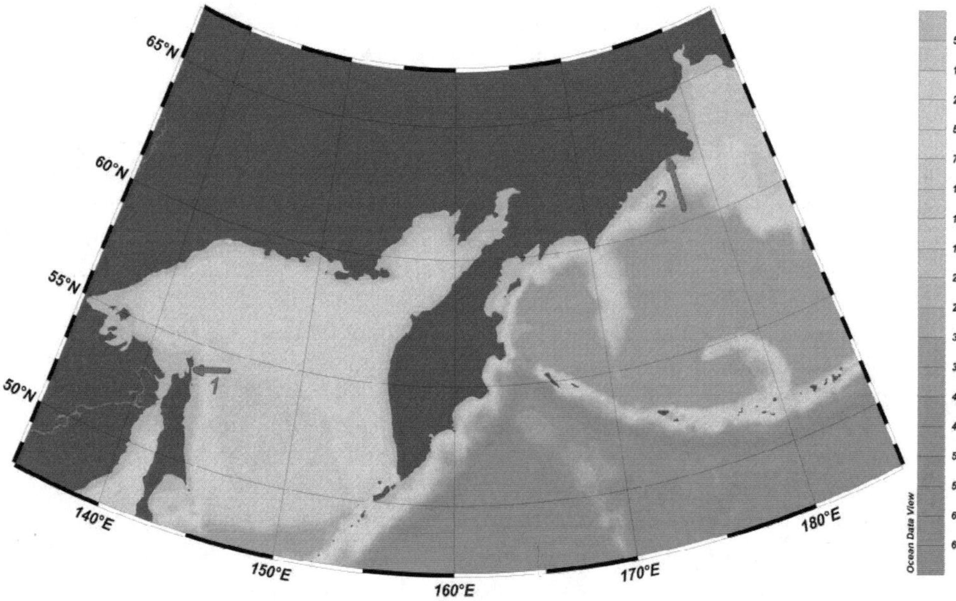

Figure 1. Geographical location of cross-sections in the Sea of Okhotsk and the Bering sea.

Since horizontally-homogeneous density field is used in our numerical model, the density profiles from the outer edges of the cross-sections have been taken, that are obtained by approximating data from GDEM climatology for winter and summer. Detailed description of algorithm of this stage is presented in [16].

An important condition for initialization of the model is barotropic tide. The horizontal velocity u_{btr} in the barotropic tide is given by:

$$u_{\text{btr}} = \frac{H}{r(x)} \sum_i U_{\max_i} \sin(\sigma_i t + \phi_i) = \sum_i \frac{Q_i}{r(x)} \sin(\sigma_i t + \phi_i), \quad (5)$$

$$Q_i = U_{\max_i} H, \quad (6)$$

where Q_i – the maximum cross flux of water for the corresponding component of the tide, $r(x) = H - h(x)$ – the local depth of the fluid, σ_i – the frequency of the tidal component, ϕ – its initial phase, and the i-index indicates the various tidal components (M2, S2, K1, O1, P1, Q1). From the continuity equation we obtain the expression for the vertical velocity of the barotropic tide

$$w_{\mathrm{btr}} = \sum_i Q_i \frac{r'(x)}{r^2(x)}(z - H)\sin(\sigma_i t + \phi_i). \tag{7}$$

The barotropic tide velocity v_{btr} in the transverse direction with respect to our section is given by:

$$v_{\mathrm{btr}} = \sum_i \frac{fQ_i}{\sigma_i r(x)}\cos(\sigma_i t + \phi_i). \tag{8}$$

In accordance with (5) – (8), the initial velocity field is initialized as follows (there are no internal waves at the initial time):

$$u = \sum_i \frac{Q_i}{r(x)}\sin(\phi_i), \tag{9}$$

$$v = \sum_i \frac{fQ_i}{\sigma_i r(x)}\cos(\phi_i), \tag{10}$$

$$w = \sum_i Q_i \frac{r'(x)}{r^2(x)}(z - H)\sin(\phi_i). \tag{11}$$

The amplitudes and phases of barotropic tide components are determined from the TPXO 8 (A TOPEX/Poseidon Global Tidal Model) model based on satellite altimetry data [17].

3. Internal waves in cross-section 1 in the Sea of Okhotsk: results and discussion

The total calculation time was 65 h, which corresponds to approximately 2.5 cycles of the prevailing diurnal component of the tide (it was noted in [1] that the full periodicity of the process is reached after approximately 2–3 tidal cycles). The fields of the vertical and horizontal components of the velocity in the computational domain, as well as the modulus of the total velocity and the density distribution at different stages of the process were constructed and analyzed. Temporal changes in the spectra of isopycnes located at different depths were analyzed. Algorithms for calculating energies (kinetic, potential and APE – Available Potential Energy – the difference between pseudo-energy and kinetic energy [18]), bottom velocities, Froude number (defined as $\mathrm{Fr} = u_{max} / c$, where c is the phase velocity of long internal waves of the lowest mode, and umax is the velocity of the barotropic tide at a given point on the path) were implemented. For the near-bottom and near-surface velocities, the exceedance probability was additionally calculated.

The initial stage of the modeled process is the formation of a long internal wave over the shelf with an amplitude of about 7 m and a length of about 15 km. The next stage of the process (after about 6 hours of the calculation time) is the steepening of the front of the long baroclinic wave. The velocities inside the baroclinic flow increase. After another 12 hours, short internal waves begin to emerge from the long wave. This can be characterized as the formation of a solibor (regimes of solibor generation for different signs of nonlinear coefficients within the framework of the Gardner equation are described in our article [19]). The horizontal velocity field above the continental slope has a complex alternating structure – a positively directed (towards the coast) "stream" stands out parallel to the slope. The maximum horizontal velocities, as well as the maximum values of the modulus of the total velocity, are achieved in the area of formation of solibor on the surface of the sea.

After about 8 hours, 4–5 short waves are discernible on the ridge of solibor. Above the deeper part of the path, the formation of baroclinic tidal waves also becomes noticeable. The evolution of a

IOP Conf. Series: Earth and Environmental Science **946** (2021) 012024 doi:10.1088/1755-1315/946/1/012024

barotropic tide into a baroclinic tide with the subsequent generation of short-period internal waves can be traced in more detail on the *x-t* diagram, for example, for an isopycn shown in figure 2, which is located at a depth of 130 m in an undisturbed state. This graph reflects the classic pattern of long-to-short wave evolution over the shelf: after about 4 semidiurnal cycles, a sequence of "solitary" waves (lower right corner of the graph) begins to emerge from the long wave, which can be identified by the "bands" (constant velocity) on the diagram.

According to the amplitude spectrum for the selected isopycn along the spatial coordinate x, one can see the "overflow" of energy into high-frequency (relative to tidal) components.

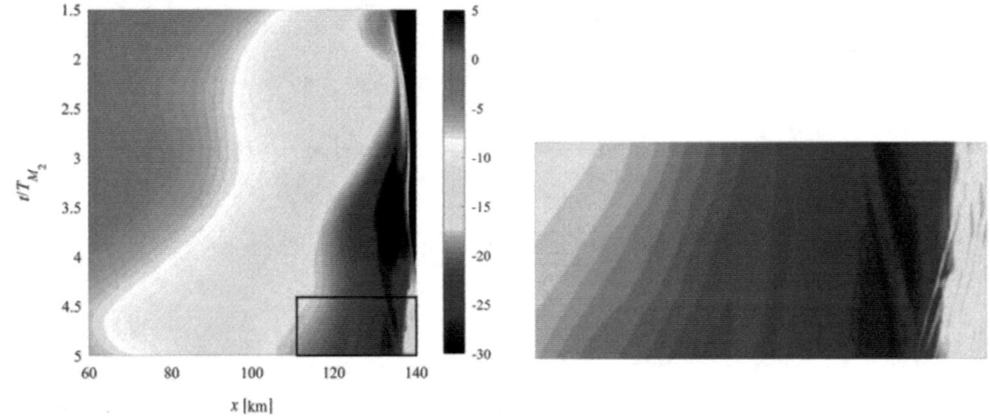

Figure 2. *x-t* diagram of the displacement of the isopycnic surface [m], which is in an undisturbed state at a depth of 130 m.

From the analysis of the induced near-bottom and near-surface velocities, it can be concluded, that the contribution of the baroclinic component is significant, especially to the near-surface velocities. In the area of the steepening of the internal wave, both the bottom and near-surface velocities exceed 0.3 m/s in absolute value (reaching even 0.4 m/s), while the barotropic component on average makes a contribution of about 0.1 m/s. Graph of the exceedance probability for near-surface velocities is shown in figure 3. It is worth noting that this result is in good agreement with the observations, published in the article [10].

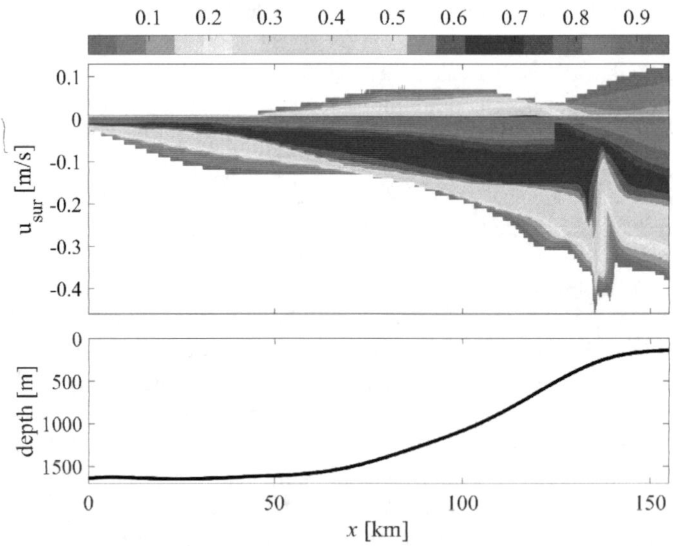

Figure 3. Distribution of the exceedance probability for the near-surface velocities.

To classify the processes of the internal wave generation it is proposed in [1] to use the Froude number magnitude. In the "supercritical" regime Fr > 1, so the generation of strong nonlinear internal lee waves by tide is possible at any latitude, whereas there is no IW generation when Fr ≪ 1 in the region north of the critical latitude. The spatial distribution of the Froude number along the simulated path indicates that no supercritical regime arises under the simulated conditions.

4. Cross-section 2 in the Bering sea: results and discussion

The total calculation time was 72 hours. The transformation regime of the baroclinic wave in section 2 is qualitatively different from that observed for section 1. The following estimates were obtained: the amplitude of long baroclinic waves can reach 20 meters (the generated short-period internal waves have amplitudes of the order of 10–15 m), and the horizontal velocity in the water column can be as high as 0.3 m/s, while the velocities of bottom flows in the upper part of the shelf reach 0.8 m/s, the Froude number exceeds 1 only in the coastal zone of the section at depths less than 150 m ("subcritical" regime is observed at a greater part of the computational domain).

It is shown that the dynamics of internal waves in the studied region also corresponds to the classical scheme of evolution of long barotropic tidal waves into baroclinic ones with the subsequent formation of wave bores. The obtained estimates illustrate dynamics of internal waves not only in the studied section, but also over a significant part of the northwestern shelf of the Bering Sea, where a similar hydrological regime takes place.

For the winter season, the generation of short-period internal waves is less intense. The generated waves have heights of up to 10 m. However, in some areas bottom velocities are still significant. In general, velocities have an amplitude of 0.2–0.4 m/s, reaching peak values of 0.6 m/s very close to the coast (see figure 4).

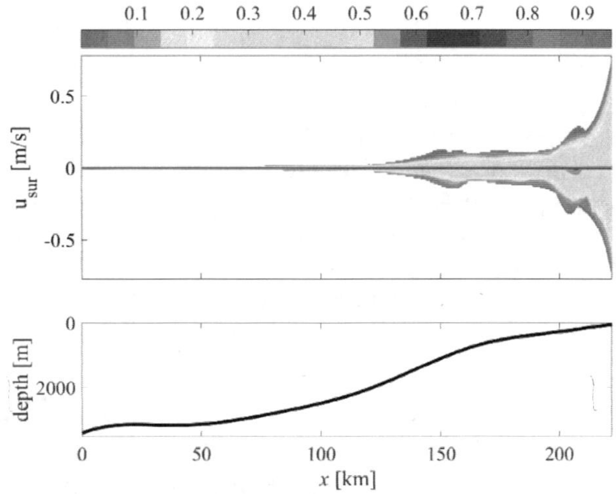

Figure 4. Distribution of the exceedance probability for the near-bottom velocities in winter.

The second cross-section is located in the region of high, though not critical latitudes (less than 74–75 °, at which, according to the predictions of the linear theory, free semidiurnal internal waves cannot exist (see, for example, [20])). We analyzed how strongly taking into account the Earth's rotation affects the dynamic picture. As expected, an increase in the Coriolis force leads to the fact that fewer short internal waves are generated from a long baroclinic wave, and their amplitudes also decrease (the regime does not change, only quantitative characteristics). It is also worth noting that in the high latitude region, intensive internal waves are often observed, and at the stage of their transformation, shoaling and breaking in the coastal zone, the Coriolis force no longer plays an important role. The simulation results for the second cross-section without rotation, with rotation corresponding to 30 °, as well as real rotation at a latitude of about 60 ° are shown in figure 5. For example, at t = 45 h, the maximum amplitude of a short period internal wave on the solibor crest is 20 m in the case of no rotation, 3 m with a rotation corresponding to 30 °, and 1 m with a real one. The number of waves in a wave train during this time period is equal to 7, 4, 2, respectively.

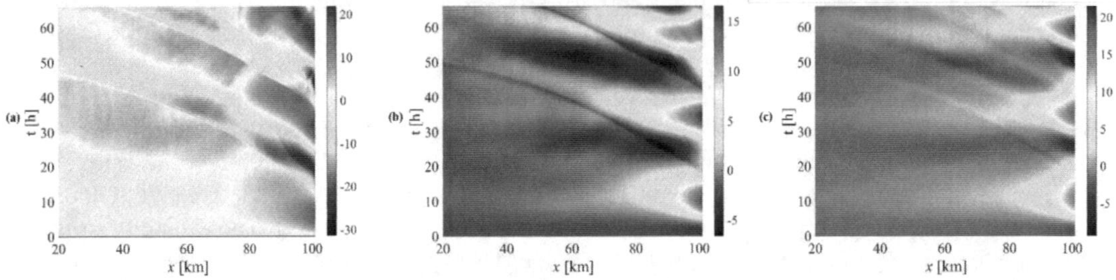

Figure 5. *x-t* diagram of the displacement of the isopycnic surface [m], which is in an undisturbed state at a depth of 200 m (a) without rotation, (b) with rotation corresponding to 30 ° and (c) 60 °.

5. Conclusion

This work is the first stage in the creation of an atlas of baroclinic wave amplitudes and associated dynamic effects for the Far Eastern seas. On the example of the selected sections in the Sea of Okhotsk and the Bering Sea, the dynamics of internal waves was simulated. It was shown that in the studied region the classical scheme of the evolution of long barotropic tidal waves into baroclinic ones with the subsequent formation of wave bores is carried out. The observed regimes differ for the selected sections. Estimates of the short period internal wave and long baroclinic wave amplitudes, as well as the velocities induced by them were obtained. For the cross-section in the Sea of Okhotsk, the internal wave amplitudes reached 5–7 m, and horizontal velocities of the currents induced by them were about 0.3–0.4 m/s both on the bottom and on the surface (that is in good agreement with measurements); for the section in the Bering Sea, the wave amplitudes exceeded 20 m, and the velocities exceeded 0.6–0.8 m/s (which theoretically confirms the hypothesis about the influence of the internal waves on the bottom topography). The influence of seasonal changes in density stratification, which leads to a decrease in the amplitudes and velocities of internal waves, was analyzed. Taking Earth rotation into account also has a qualitative effect on the dynamic picture. We calculated and analyzed such parameters as the Froude number, wave energy (APE), exceedance probabilities for near-bottom and near-surface velocities. These criteria, as well as the values of the Shields number (to determine the probability of soil erosion), will be considered in the future to analyze only extreme wave phenomena.

Acknowledgments

This study was initiated in the framework of the state assignment in the field of scientific activity (topic No. 0728-2020-0007 "Wave climate of the stratified sea shelf: nonlinear dynamic processes and their impact on the coastal zone and hydraulic structures") and the grant of the President of the Russian Federation (SP-1225.2019.5).

References

[1] Vlasenko V, Stashchuk N and Hutter K 2005 *Baroclinic Tides: Theoretical Modeling and Observational Evidence* (Cambridge: Cambridge University Press) 372 p
[2] Moroshkin K V 1966 *Water masses of the Sea of Okhotsk* (Nauka) 66 p
[3] Kruts A A and Luchin VA 2013 Vertical water structure in the Okhotsk Sea *Izv. TINRO* **175** pp 234–253
[4] Dudkov S P 2010 Mezhgodovye izmeneniya prostranstvennogo sootnosheniya tipov vertikal'noj stratifikacij vod na severo-zapadnom shel'fe Beringova morya letom 2005–2008 gg. *Izv. TINRO* **162** pp 306–323

[5] Zimin A V and Svergun E I 2018 Short-period internal waves in the shelf areas of the white, Barents and Okhotsk seas: Estimation of the extreme heights occurrence and dynamic effects in the bottom layer *Fund. i Prikladn. Gidrofiz.* **11** 4 pp 66–72

[6] Mitnik L M and Dubina V A 2007 *Proc. Envisat Symp.* Spatial-temporal distribution and characteristics of internal waves in the Okhotsk and Japan Seas studied by ERS-1/2 SAR and Envisat ASAR pp 23–27

[7] Svergun E I and Zimin A V 2017 Forecast of the occurrence of intense internal waves in the White and Barents Seas according to expeditionary research *Fund. i Prikladn. Gidrofiz.* **10** 2 pp 13–19

[8] Svergun E I and Zimin A V 2020 Characteristics of Short-Period Internal Waves in the Avacha Bay Based on the In Situ and Satellite Observations in August-September, 2018 *Morskoy Gidrofizicheskiy Zhurnal* **36** 3 pp 300–312

[9] Mitnik L and Dubina V 2019 The Sea of Okhotsk: Scientific Applications of Remote Sensing *In book: Remote Sensing of the Asian Seas* pp 159–175

[10] Rutenko A N and Sosnin V A 2014 Hydrodynamic processes on the Sakhalin shelf in the coastal Piltun area of the grey whale feeding and their correlation with atmospheric circulation *Russ. Meteorol. Hydrol.* **39** pp 335–349

[11] Karl H A Cacchione D A and Carlson P R 1986 Internal-wave currents as a mechanism to account for large sand waves in Navarinsky Canyon head, Bering Sea *Journal of Sedimentary Research* **56** 5 pp 706–714

[12] Lagerloef G S E and Muench R D 1987 Near-inertial current oscillations in the vicinity of the Bering Sea marginal ice zone *Journal of Geophysical Research* **92** C11 pp 11789

[13] Svergun E I and Kozlov I E 2021 Characteristics of short-period internal waves in the Bering sea in summer 2019 from Sentinel-1 data *Sovremennye problemy distantsionnogo zondirovaniya Zemli iz kosmosa* **18** 3 pp 269–276

[14] Tyugin D Yu, Kurkin A A, Pelinovsky E N and Kurkina O E 2012 Increase of productivity of software package for modeling of internal gravity waves IGW Research with the help of Intel® Parallel Studio XE 2013 *Fund. i Prikladn. Gidrofiz.* **5** 3 pp 89–95

[15] Lamb K 1994 Numerical experiments of internal wave generation by strong tidal flow across a finite amplitude bank edge *J. Geoph. Res.* **99** C1 pp 843–864

[16] Rouvinskaya E A, Kurkina O E and Kurkin A A 2011 Modeling of internal weather in the ecosystem of the stratified sea shelf *Ecological Systems and Devices* **6** pp 8–16

[17] Egbert G D and Erofeeva S Y 2002 Efficient inverse modeling of barotropic ocean *tides J. Atmos. Oceanic Technol.* **19** 2 pp 183–204

[18] Lamb K 2008 On the calculation of the available potential energy of an isolated perturbation in a density-stratified fluid *Journal of Fluid Mechanics* **597** pp 415–427

[19] Kurkina O, Rouvinskaya E, Talipova T, Kurkin A and Pelinovsky E 2016 Nonlinear disintegration of sine wave in the framework of the Gardner equation *Physica D.* **333** pp 222–234

[20] Morozov E G and Paka V T Internal waves in a high-latitude region *Oceanology* 2010 **50** 5 pp 668–674.

Comparative analyses of changes in the ice cover of the Tatar Strait in relation to 1961-1990 and 1991-2020 climate normals

D V Zarubina[1*], **V M Pishchalnik**[1] **and V A Romanyuk**[2]

[1] Sakhalin State University, Yuzhno-Sakhalinsk, Russian Federation
[2] RN-SakhalinNIPImorneft LLC, Yuzhno-Sakhalinsk, Russian Federation

E-mail: [*]dariadorofeyeva26@gmail.com

Abstract. Current warming in the Northern Hemisphere has encompassed not only the surface atmosphere, but also manifested itself in a change in the state of the sea ice cover, as an integral part of the climate system. One of the indicators of the warming was a reduction in ice cover observed in the water area of the Far Eastern seas since the late 1980s. The prospective areas of hydrocarbon deposits discovered in the Tatar Strait, as well as the existing needs for the transportation of hydrocarbons along the shipping routes in the strait, led to the emerging interest in researching the ice regime of the Sea of Japan in the modern climatic era. The proposed paper analyses the changes in ice cover of the Tatar Strait in relation to 1961–1990 and 1991–2020 climate normals. The regularities of the long-term variability of ice cover within non-overlapping 30-year periods were revealed. The analysis of the dynamics of ice cover anomalies, calculated for each normal separately, was carried out. An assessment of the distribution of winter types in the periods under study was carried out. The characteristic features of variations in the intra-seasonal course of ice cover and the dates of the onset of ice phases in different types of winters were identified.

1. Material and Methods

According to the provisions of the World Meteorological Organization guidelines for the calculation of climate normals [1], non-overlapping 30-year time periods starting from 1901 were taken as climatological standard normals. The period from 1961 to 1990 was defined as the standard reference period for long-term assessment of climate change. The purpose of this work was to make a comparative analysis of changes in the elements of the ice regime of the Tatar Strait for the periods corresponding to the standard reference period 1961–1990 and the standard climate normal 1991–2020 (figure 1).

From 1960 to 1991, calculation of the area of ice extent was carried out according to the data obtained from the maps of regular aircraft ice reconnaissance [2, 3]. From 1971 to the present, the main source of information was satellite imagery of the ice extent from spacecrafts carried out in the microwave range. Accuracy assessment of calculated data for aeronautical observations did not exceed 5% [2], for satellites – 2% [4; 5].

The calculation of the ice cover of the Tatar Strait was carried out using ICE software package [6, 7] according to the data of colour-coded schematic maps of ice concentration [8]. The calculations were carried out within the boundaries of the strait, determined by the Main Directorate of Navigation and Oceanography of the Ministry of Defence of the Russian

Federation [9]. The value of ice cover for the season was calculated by averaging the values of the ice extent area on the 15th day of each month for the period from December to April.

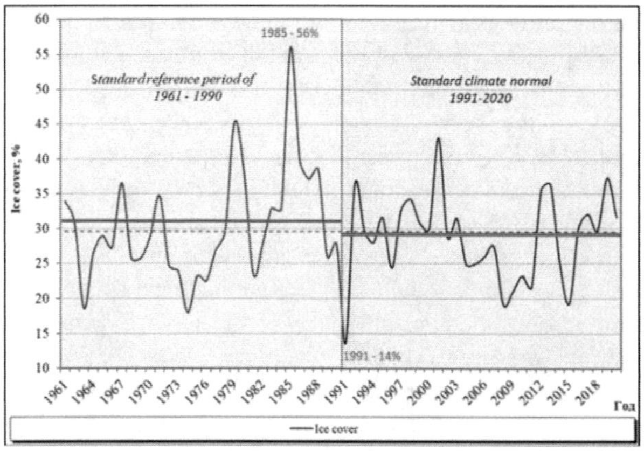

Figure 1. Variations in ice cover in the Tatar Strait in the period 1961–2020.

Red solid line denotes the level of ice cover corresponding to the standard reference period of 1961–1990, blue solid line – to the standard climate normal 1991–2020. Linear trend over 60 years is highlighted with black dashed line.

The analysis of ice cover anomalies was carried out by the method of integral curves, which allowed one to identify the characteristic patterns of its long-term variability [10]. The integral curves were calculated by sequential algebraic addition of ice cover anomaly values.

The typification of winters was determined by the anomaly of ice conditions using 0.8 σ criterion (σ is the standard deviation) [11], according to which extremely severe winters (ESv ≥ 1.2 σ) were identified; as well as severe (+0.4 σ ≤ Sv <+1.2 σ); moderate (–0.4 σ <M<+0.4 σ); soft (–1.2 σ <S≤ –0.4 σ); extremely soft (ES ≤ –1.2 σ) winters.

2. Results

The average value of ice cover of the Tatar Strait for the standard reference period was 31 % or 32.7 thousand km². The maximum value was observed in 1985, and the minimum – in 1974, 56 and 18 %, respectively. The range of fluctuations in its values was 38 % (figure 2). The analysis of the long-term variability of ice cover in the standard reference period revealed a general tendency of its increase by 3 % for every 10 years (a significant linear trend at the 95 % level), which corresponds to an increase in the ice area by ~10 thousand km² over a 30-year period. At the same time, for the entire reconstructed series of observations in the Tatar Strait from 1882 to 2018 [12], a general tendency for a decrease in the area of ice extent at a rate of 0.7% over 10 years was established.

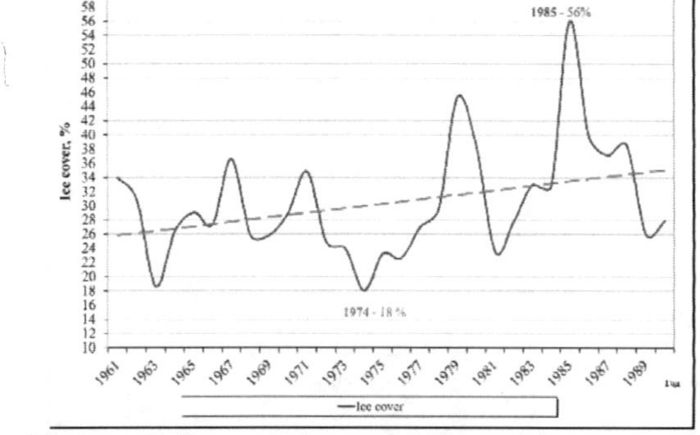

Figure 2. Long-term variability of ice cover of the Tatar Strait for 1961–1990.

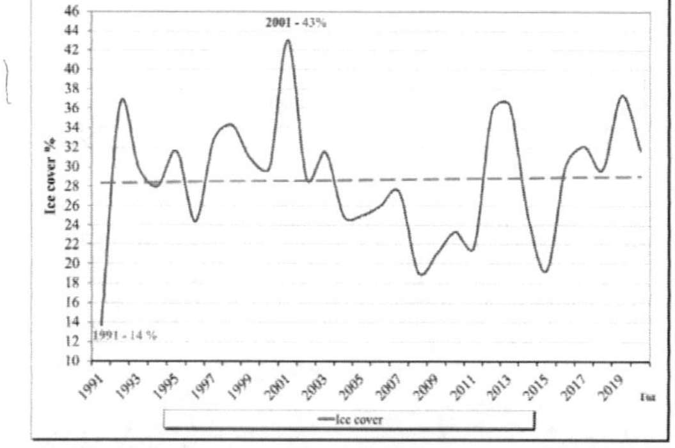

Figure 3. Long-term variability of ice cover of the Tatar Strait for 1991–2020.

The average value of ice cover in the standard climate normal 1991–2020 was 29 % or 30.9 thousand km² (figure 3). The difference between the mean values of ice cover for both normals was less than the accuracy of calculations. The range of fluctuations in ice cover was 24 %, which was 1.6 times less than the same value in the standard reference period. There was no linear trend slope. The maximum ice cover was recorded in 2001, when its value was 43 %. The minimum ice cover of 14 % was recorded in 1991, which corresponded to the absolute minimum in the entire history of observations [12].

It was established that long-term fluctuations in the ice cover's seasonal variability in the Tatar Strait and the Sea of Okhotsk were in phase only in two isolated periods: 1894–1920 and 1985–2009. In the current warming period, the general downward trend of the ice cover in the Tatar Strait commenced in 1985, five years later than in the Sea of Okhotsk [12]. Those discrepancies happened due to the morphometry of the studied water areas (the wind tunnel effect in the strait) and the particular conditions of destruction of the ice cover in southern latitudessouth of 48°N. In the Tatar Strait, the ice cover was destroyed due to the active thermal effect of the western flow of the warm Tsushima Current, while in the southern part of the Sea of Okhotsk, the processes of ice accumulation in the "ultimate ice trap of the Sea of Okhotsk" prevailed over the processes of ice destruction under the influence of the eastern branch of Tsushima Current flowing through the La Perouse Strait.

Figure 4 and figure 5 show the integral curves of ice cover anomalies calculated for the Tatar Strait relative to the standard reference period 1961–1990 and the standard climate normal 1991–2020. The configurations of the integral curves of ice cover anomalies in the periods corresponding to the considered climate normals characterize the tendencies of changes in the direction of ice processes. There was no dependency of the course of the integral curve on the normal selected for the calculation, which indicates the relative stability of ice formation processes in the strait over the past sixty years.

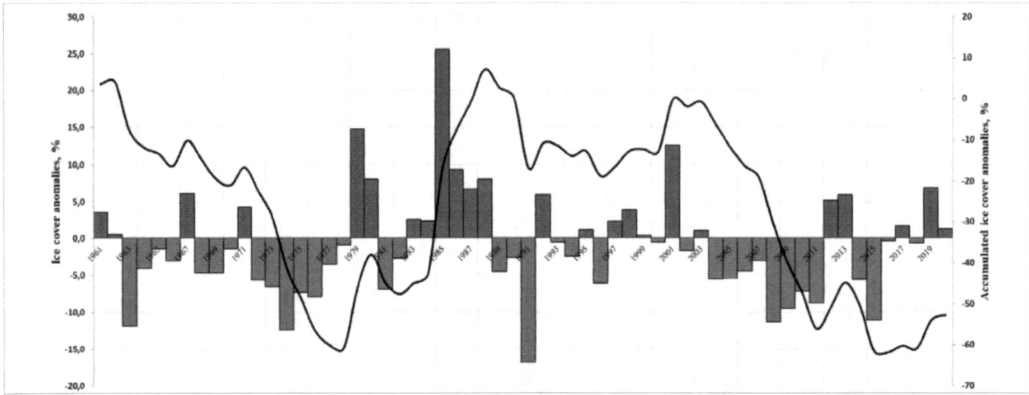

Figure 4. Integral curve of ice cover anomalies in the Tatar Strait relative to the standard reference period 1961–1990.

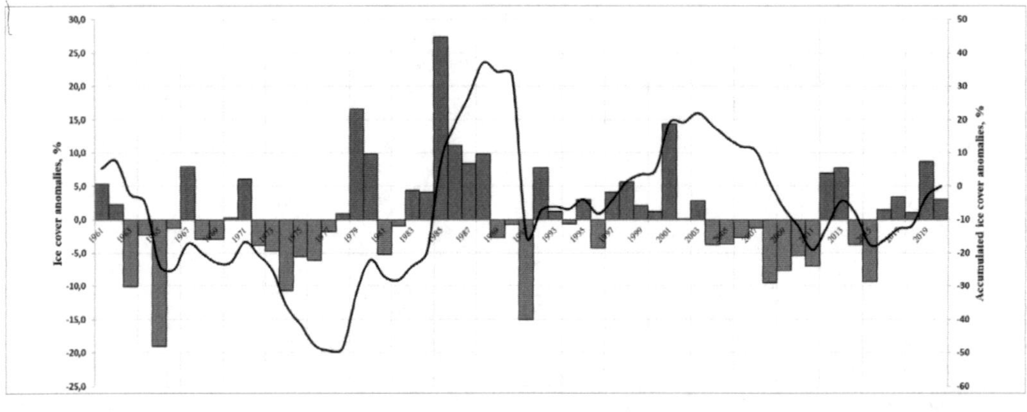

Figure 5. Integral curve of ice cover anomalies in the Tatar Strait relative to the standard climate normal 1991–2020.

The typification of winters according to the anomalies of the average ice cover values for a season in the Tatar Strait was carried out relative to the average value of ice cover for a season in the period 1961–2020 (figure 6, 7 and 8). The results of the analysis for the selected gradations of winter types showed that in the 60-year period (figure 6), the moderate (M) type of winters prevails – 33 %. The next most frequent type was the soft (S) type (28 %). The ratio of severe (Sv) winters accounted for 20 %. Extreme types accounted for 8 % in the category of extremely soft (ES) and 10 % in the category of extremely severe (ESv) winters. Comparative analysis of 30-year consecutive periods (figure 7 and figure 8) allowed us to ascertain a decrease in the frequency of ESv winters by 6 % in 1991–2020. The ratio of ES and S ice cover anomalies taken together in the standard reference period 1961–1990 and in the period of the standard climate normal 1991–2020 was the same – 18 %. The frequency of Sv winters remained at the same level. The proportion of M winters in 1991–2020 increased by 7 %.

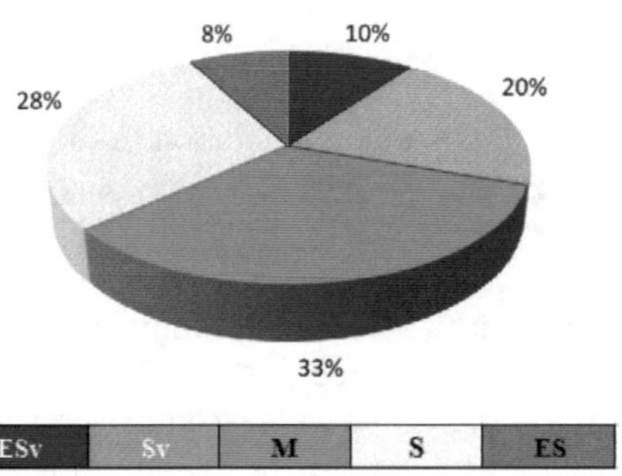

Figure 6. Frequency of occurrence of winter types in the Tatar Strait for the period from 1961 to 2020.

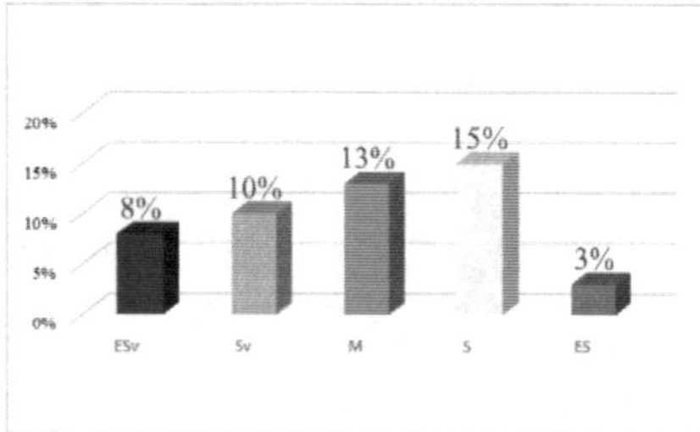

Figure 7. The frequency of occurrence of winter types in the Tatar Strait for the period from 1961 to 1990.

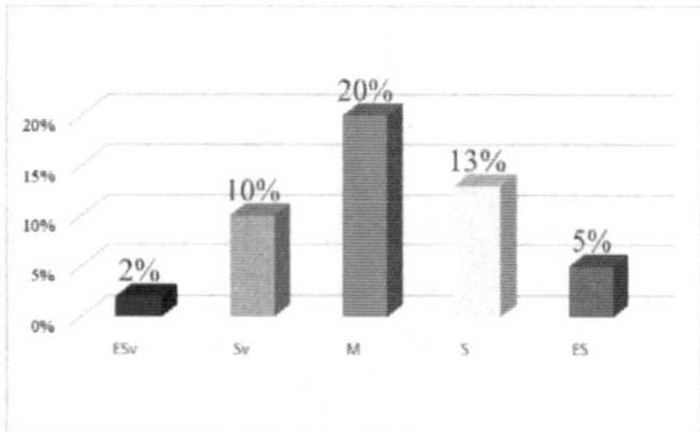

Figure 8. The frequency of occurrence of winter types in the Tatar Strait for the period from 1991 to 2020.

The analysis of temporal distribution and frequency of ice cover anomalies performed using the created calendar of ice cover anomalies (Table 1) allowed us to detect the fact that the vast majority of ESv and Sv anomalies in ice cover were observed at the end of the standard reference period 1961–1990. The period corresponding to the standard climate normal 1991–2020 was characterized by a predominance of moderate winters, evenly represented throughout the entire 30-year series.

Table 1. Typification of winters based on anomaly of ice conditions in the Tatar Strait from 1961 to 2020.

Value of anomaly	15,9	-6,8	-0,5	1,6	-2,0	5,2	-3,2	-4,7	-1,2	-0,3	-13,5	0,8	-2,0	4,6	4,6	3,6	2,1	10,4	8,6	6,5	7,9	-6,1	-6,8	4,7	10,2	-0,6	-2,5	0,2	-7,8	-2,1
Year	1961	1962	1963	1964	1965	1966	1967	1968	1969	1970	1971	1972	1973	1974	1975	1976	1977	1978	1979	1980	1981	1982	1983	1984	1985	1986	1987	1988	1989	1990
Value of anomaly	-15,9	6,8	0,3	-1,8	2,6	-5,2	5,2	4,7	1,2	0,3	13,5	-0,8	2,0	-4,6	-4,6	-3,6	-2,1	-10,4	-8,6	-6,3	-7,9	6,1	6,8	-4,7	-10,2	0,6	2,5	0,2	7,8	2,1
Year	1991	1992	1993	1994	1995	1996	1997	1998	1999	2000	2001	2002	2003	2004	2005	2006	2007	2008	2009	2010	2011	2012	2013	2014	2015	2016	2017	2018	2019	2020

ESv	Sv	M	S	ES

Figure 9 shows the intra-seasonal variation of the average ice cover values corresponding to different climate normals. At the stage of ice formation, no significant differences in the course of the curves were observed. Relative to the standard reference period 1961–1990, during the period of the

IOP Conf. Series: Earth and Environmental Science **946** (2021) 012025 doi:10.1088/1755-1315/946/1/012025

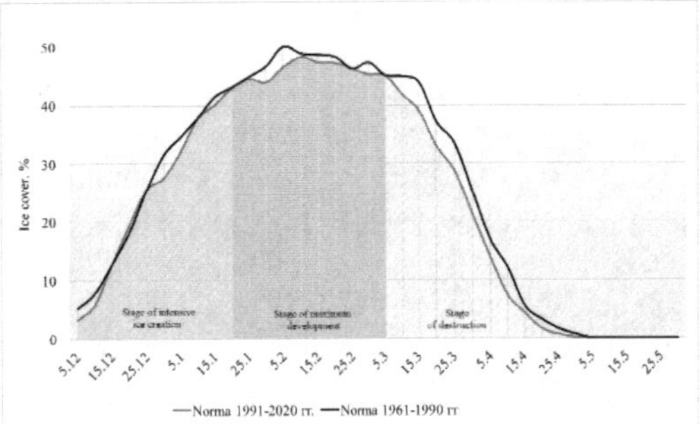

Figure 9. The intra-seasonal course of ice cover during the period of the current and the reference climate normal.

standard climate normal 1991–2020 there was a 5-day delay of the onset of the maximum ice cover, and the stage of destruction of ice cover occurred 10 days earlier.

The first appearance of ice extent in the strait was observed on average on November 13, the average date for stable ice formation was November 15. The early dates and late dates of the first and the stable appearance of ice were scattered from 15 to 20 days. The first ice clearing of the strait on average was recorded on April 25. The difference between the dates of the earliest and the latest initial clearing of the water area from sea ice was 45 days (April 5 and May 20, respectively). The average date for final clearing of ice was April 27. In moderate types of winters, the earliest final clearing of the studied water area from ice occurred on April 5, and the latest – on May 20. The accuracy of calculating the onset of ice phases using pentad maps was +/-2.5 days.

The average duration of the ice season was 159 days for soft winters, 164 days for severe winters and 165 days for moderate winters (average duration – 163 days). The shortest ice season in the last 30 years was recorded in the winter of 1992/93, when its duration was 141 days. The longest one was 1998/1999 ice season, when ice remained in the studied water area for 181 days.

The dates of the beginning of ice formation and the final clearing of the strait in the reference climate normal were established according to the data of marine hydro-meteorological stations. According to long-term average data, in 1961–1990 the duration of the ice period was 190 days. Analysis of the available data allowed us to conclude that the duration of ice season in the period corresponding to the standard climate normal 1991–2020 decreased by almost a month, namely by 27 days, from 190 to 163 days. The difference in the duration of ice season, as function of the types of winters, became less substantial and was only 5 days. At the same time, in the period corresponding to the reference normal, it could increase or decrease by a month compared to the average long-term dates, depending on the severity of winter [2].

3. Conclusion

Based on the analysis of variations in long-term ice cover fluctuations in the Tatar Strait in the periods corresponding to the standard reference period 1961–1990 and the standard climate normal 1991–2020 the following was established:

1. A reduction in the average ice cover over the past 30 years by 2 % (from 31 % to 29 %), which was within the accuracy of the calculations. Linear trend of sea ice cover in the Tatar Strait during the standard reference period 1961–1990 had a positive trend. The slope of the linear trend during the period of the standard climate normal 1991–2020 was close to zero.
2. The range of fluctuations in the value of ice cover in the standard climate normal 1991–2020 amounted to 24 %, which was 1.6 times less than the same value in the standard reference period 1961–1990.
3. Relative to the standard reference period 1961–1990, during the period of the standard climate normal 1991–2020, there was a 5-day delay in the stage of the onset of the maximum ice cover, and the stage of destruction of ice cover occurred 10 days earlier. The duration of ice season during the period of the standard climate normal 1991–2020 decreased by almost a month from 190 to 163 days.
4. The ratio of soft and extremely soft ice cover anomalies in aggregate in the standard reference period 1961–1990 and during the period of the standard climate normal 1991–2020 remained at

the level of 18 %. In 1991–2020, the proportion of moderate winters increased by 7 %, and the frequency of extremely severe winters decreased by 6 %. The period corresponding to the standard climate normal 1991–2020 was characterized with predominance of moderate winters, evenly represented throughout the entire 30-year series.

5. The nature of the course of the integral curves testified to the relative stability of ice formation processes in the strait over the past sixty years.

References

[1] *Rukovodyashchie ukazaniya VMO po raschetu klimaticheskih norm* 1970 (VMO: Zheneva) 1203 32 p

[2] Yakunin L P, Plotnikov V V, and Petrov A G 2003 Ledovye usloviya (Ice conditions) *Gidrometeorologiya I gidrokhimiya morei* (SPb: Gidrometeoizdat) **8** 1 pp 347–394

[3] Plotnikov V V, Chetyrbotskiy A N and Gordeychuk T V 2010 Otsenka sostoyaniya ledyanogo pokrova Yaponskogo morya *Meteorol. i gidrol.* 3 pp 46–55

[4] Mitnik L M, Trusenkova O O and Lobanov V B 2015 Distancionnoe radiofizicheskoe zondirovanie okeana i atmosfery iz kosmosa: dostizheniya i perspektivy (obzor) *Vestnik DVO RAN* 6 pp 21–22

[5] Pishchalnik V M, Bobkov A O, Minervin I G and Romanyuk V A 2017 Analysis of Variations in the Ice Regime in Individual Regions of the Rea of Okhotsk during the Warming Period *Herald of the Russian Academy of Sciences* (Pleiades Publishing Ltd.) **87** 5 pp 429–440

[6] Pishchalnik V M, Bobkov A O, Minervin I G and Romanyuk V A 2015 Svidetel'stvo o gosudarstvennoy registratsii programmy dlya EVM 2015660472 (Certificate of state registration of computer programs 2015660472) Software complex for calculating the ice cover in the Sea of Okhotsk and the Sea of Japan according to remote sensing data (LED PC)

[7] Pishchalnik V M, Bobkov A O, Minervin I G and Romanyuk V A 2015 *IX-th Vserossiyskiy Symp. Fizika geosfer* Osnovnye printsipy raboty programnogo kompleksa "LED" (Basic Principles of Operation of the Program «ICE») (Vladivostok: Dalnauka) pp 556–561

[8] Yaponskoe meteorologicheskoe agentstvo (*Electronic Materials*) http://www.data.jma.go.jp

[9] *Granitsy oceanov i morey (Boundaries of seas and oceans)* 2000 (Sankt-Petersburg: GUNiO MO) pp 127–137

[10] Girs A A and Kondratovich K V 1978 *Metody dolgosrochnyh prognozov pogody* (L.: GMI) 344 p

[11] Spichkin V A 1987 Opredelenie criteriya krupnoi anomaliy *Trudy AARI* (Sankt-Petersburg) **402** pp 15–20

[12] Pishchalnik V M, Dorofeeva D V, Minervin I G, Shumilov I V and Nikulina I V 2019 Year-to-year dynamics of the ice cover anomalies in the Tatar Strait for the period from 1882 to 2018 *Izv. TINRO* **196** pp 114–122

IOP Conf. Series: Earth and Environmental Science **946** (2021) 012026 doi:10.1088/1755-1315/946/1/012026

Hydroacoustic study of the basin of Biryuzovoe Karyernoe lake (Korsakovsky district, Sakhalin Island)

D N Kozlov*, **F A Romanyuk and R V Zharkov**

Institute of Marine Geology and Geophysics Far Eastern Branch of Russian Academy of Science, Yuzhno-Sakhalinsk, 693022, Russia

E-mail: *kozlovdn@bk.ru

Abstract. For the first time, the paper presents the results of a detailed hydroacoustic study of the basin of the Biryuzovoe Karyernoe Lake (Novikovo village, Korsakovsky District, Sakhalin Region), formed after the conservation of the germaniferous coal deposit of the same name. On the basis of 12 hydroacoustic profiles, a detailed bathymetric scheme of the lake was compiled, its main morphometric parameters were calculated, and the morphological appearance was described. Preliminary conclusions have been made about the rates of sedimentation within the basin of the lake, and perspective points for monitoring the dynamics of bottom and slope sedimentation have been established. Based on the results of the chemical analysis of the lake waters, a conclusion was made on the prospects of its use for tourist and recreational purposes.

Keywords: Biryuzovoe Karyernoe lake, Sakhalin, hydroacoustic research, morphometry, bathymetric map, geochemistry of waters, sedimentation

1. Introduction

Biryuzovye Karyernye lakes are a group of unique anthropogenic reservoirs located in the vicinity of the Novikovo village (Korsakovsky district, Sakhalin region) (figure 1).

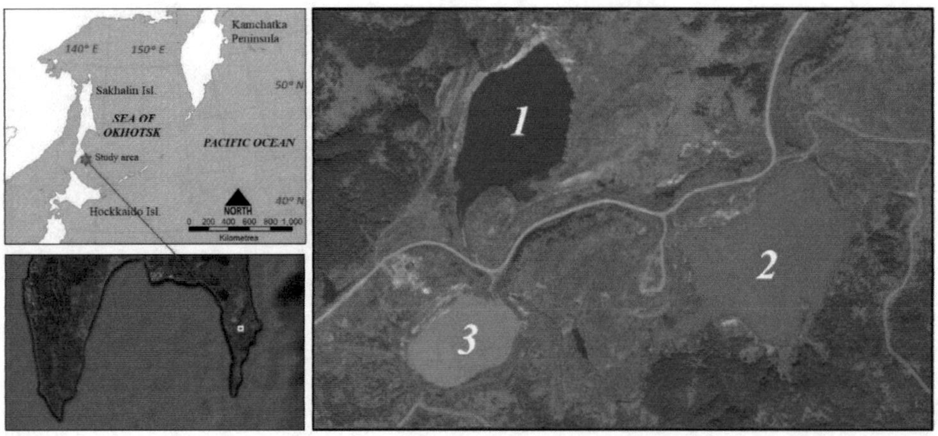

Figure 1. Geographic location of the research object. Some of the largest Karyernye lakes are marked with numbers: 1 – Biryuzovoe, 2 – Utinoe, 3 – Zelenoe.

These lakes are 6 large and 14 small reservoirs, which were formed after the flooding of open pits, where industrial production of germaniferous coals was carried out from the 1960s to the early 2000s. It should be noted that, from a theoretical point of view, a more accurate name for these water bodies is not lakes, but watered quarries, however, on various websites, in documents and in media publications [1, 2] they are called lakes, so we consider it quite acceptable to name and designate them using both versions. Today, Biryuzovye Karyernye lakes are one of the most visited water bodies in the south of Sakhalin [3] due to the uniqueness of their morphology and specific water color (figure 2), associated with the content of dissolved sulfates. Despite the popularity of these water bodies, data on the morphology of the basins and modern processes occurring in their water area are not available in the open press. Such gaps in the cadastre of water bodies cause the existing and very urgent problem of assessing water resources in the Far East as one of the least developed, and at the same time, the most promising regions for the development of the country. In addition, the recording and updating of information on the state of lake basins and the rate of sedimentation allow solving fundamental issues of lithogenesis and quantifying vertical movements in sedimentation basins. The lack of data in the scientific literature on the chemistry of waters of certain objects also complicates their development. In this regard, a group of specialists from the IMGG FEB RAS conducted a hydroacoustic study of the basin of the most visited and deepest lake, called the Biryuzovoe Karyernoe lake, from September 16 to 17, 2021, and took samples of its waters. The objectives of the study aims to obtain a high-precision digital bathymetric scheme of the lake and its subsequent analysis to identify promising areas for monitoring sedimentation processes, assess the chemistry of waters.

Figure 2. Biryuzovoe Karyernoe lake. Echo-sounding survey. Photo by R.V. Zharkov, 2021.

2. Materials and methods

A hydroacoustic survey of the basin of Lake Biryuzovoe Karyernoe was carried out in mid-July 2021. The work was carried out using a proven methodology [4–6] using a Lowrance LMS-527c DF iGPS echo-sounder with synchronous satellite referencing of profiles; the echo-sounder emitter frequency was 200 kHz, the shooting step was 1 m. The equipment was installed on a light motor boat CatFish 240. Data processing was performed using software packages Lowrance Sonar Viewer 2.1.2 (echo-sounder data export with georeference SR-ORG Projection 8230 – Lowrance Mercator), QGIS v. 3.20.3 (data rebinding in WGS 84, measurement of surface morphometric characteristics of the lake) and Surfer v. 13.0 (building of a bathymetric scheme, calculation of the morphometric characteristics of the lake basin). The calculation of the morphometric characteristics of the lake was carried out according to generally accepted methods [7–10]. Chemical analysis of water was carried out according to standard methods in a testing laboratory center Federal Budgetary Institution of Health Care «Center of Hygiene and Epidemiology in the Sakhalin Region» (Yuzhno-Sakhalinsk, Russia). In a water sample, odor (at a temperature of 20 °C and 60 °C), pH, hydrogen sulfide, silicon, sulfates, chlorides, hydrocarbonates, sodium, potassium, calcium, magnesium, and total mineralization were determined.

3. Results and discussion

Based on the results of the echo-sounding survey, 12 profiles of various lengths were made. The total number of measurement points that were obtained during the survey was about 7 thousand. Based on these data, an accurate bathymetric scheme with an interactive component was compiled; it is possible to change the detail over a wide range by setting the step of isobaths from detailed (from 1 to 5 m) to generalized (from 10 to 20 and more m). In addition to the bathymetric scheme, a 3D modeling of the lake basin was carried out on the basis of these data, also with an interactive component for visualizing the survey results (figure 3).

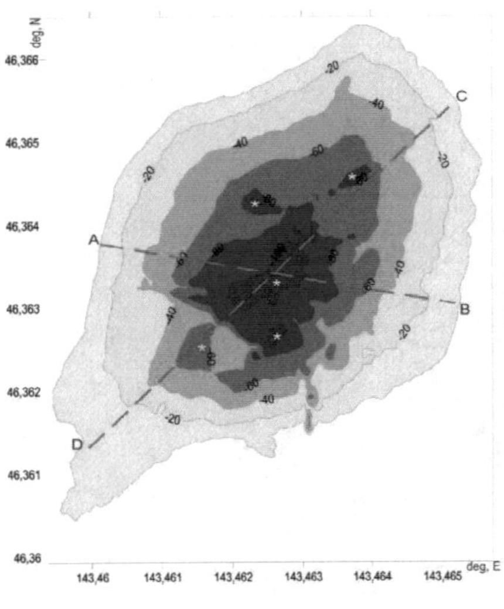

Figure 3. Bathymetric map of Biryuzovoe Karyernoe lake, isobaths are given at 20 m intervals. The dotted line in the diagram marks the location of the AB and CD profiles, white asterisks mark promising sites for sampling bottom sediments.

Calculations based on data of GIS and the echo-sounder navigator have shown that today the Biryuzovoe Karyernoe lake has the following morphometric characteristics: mirror area is 0.21 km^2, abs. mirror height above sea level is 45 m, the coastline tortuosity coefficient is 1.34, the maximum length is 0.78 km, the maximum width is 0.4 km, the volume is 8.6 10^{-3} km^3, the maximum depth is 106 m, the length of the coastline is 2.17 km. Some additional parameters were also calculated: catchment area – 0.9 km^2, area index – 0.23, specific catchment area – 4.29, average width – 0.269 km, average depth – 41 m, capacity coefficient – 0.39, openness index – 0.005.

The data obtained in the course of interpretation, calculations and modeling allow us to draw a number of conclusions:

- today, the Biryuzovoe Karyernoe lake with a maximum elevation of 106 m is the deepest reservoir on Sakhalin (formally, while being a watered quarry), traditionally, lake Tunaicha was the leader here with a maximum depth of about 42 m;
- analysis of bathymetric profiles (figure 4) has shown that the serpentine observed along the perimeter of the above-water part of the basin is clearly traced with depth, down to the bottom

of the basin. In many places in the western part of the lake, the serpentine ledge is practically absent (at a depth of about 60 m). It is possible that the walls of the basin are more susceptible to erosion here;

- on the received echograms there are no significant, clearly distinguishable thicknesses of bottom sediments, which corresponds to the age of the basin. For a relatively short period of time (15 years) with low rates of erosion, sedimentation processes proceed rather slowly;
- analysis of profiles and bathymetric maps made it possible to identify a number of areas of prospective monitoring of the thickness and composition of bottom sediments. In the future, it is possible to use these areas for monitoring the rates and specifics of sedimentation;
- the calculated morphometric parameters (area index, specific catchment area, capacity coefficient and openness index) and the morphological appearance of the basin of Biryuzovoe Karyernoe lake suggest a significant similarity with volcanogenic crater lakes in the Kuril-Kamchatka region, such as Lake Kipyashcheye (Kunashir Island), Glazok (Ketoy Island), Verkhnee (Kamchatka Peninsula), Stubel (Kamchatka Peninsula).

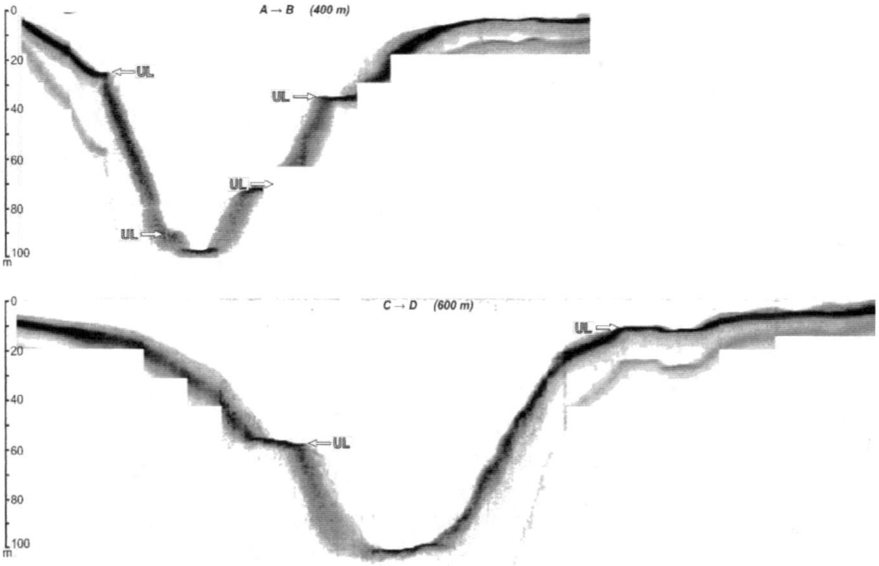

Figure 4. Bathymetric profiles of the Biryuzovoe Karyernoe lake AB and CD, the position of the profiles is given in figure 3. UL – underwater ledge.

In addition to hydroacoustic surveys, a physicochemical analysis of the lake water for the content of the main macronutrients was also carried out. Laboratory studies have shown that water is odorless at temperatures of 20 °C and 60 °C, refers to slightly alkaline (pH 7.8), fresh (mineralization 0.22 g/l), hydrocarbonate-sulphate sodium-calcium. The content of iron, hydrogen sulfide and silicon in the water is insignificant. A similar, but more intense color is found in Lake Biryuzovoe in the caldera of Zavaritsky volcano (Shiashkotan Island), the waters of which are mineralized (up to 4 g/l) and are classified as chloride-sulfate sodium in composition [11].

4. Conclusion

Biryuzovoe Karyernoe lake today can be considered the deepest of the surveyed water bodies of Sakhalin. A high-precision bathymetric scheme with georeferencing has been compiled for it, a number of morphometric calculations have been performed, they are necessary to describe the morphology of the bottom of the basin of the reservoir. To assess the rates of sedimentation processes within the lake bottom, several promising areas for long-term monitoring were noted.

At present, it can be argued that no significant bottom sediments have been found according to hydroacoustic studies. Chemical analysis has shown that the lake waters are weakly alkaline, fresh, hydrocarbonate-sulphate sodium-calcium. The water is suitable for swimming, and its turquoise color, especially bright in sunny weather, is due to the sulphate content [11]. Considering the great interest of residents and guests of Sakhalin Island to the Biryuzovoe Karyernoe lake, recreational and ecological tourism can be developed here.

References

[1] Turister 2020 *Blue (Biryuzovye) Lakes on Sakhalin* (*Electronic materials*) https://www.tourister.ru/world/europe/russia/city/novikovo/lakes/37699

[2] Ministry of Tourism of the Sakhalin Region 2019 *Description of the route "Cape Eustathia and Biryuzovye Lakes on Sakhalin"* (*Electronic materials*) https://tourism.sakhalin.gov.ru/mys-evstafiya-i-biryuzovye-ozera/

[3] *Places and routes for tourism and hiking in Sakhalin, Korsakovsky district* (*Electronic materials*) https://idilesom.com/sakh/places?district=7&sortby=views

[4] Kozlov D N 2015 *Crater lakes of the Kuril Islands* (Yuzhno-Sakhalinsk: Sakhalin Regional Museum, IMGG FEB RAS) p 12–15

[5] Kozlov D N, Degterev A V and Zarochintsev V S 2018 Koltsevoe caldera lake: current state and structure of the basin (Onekotan Island, Kuril Islands) *Geosystems of Transition Zones* **4** pp 359–364

[6] Kozlov D N, Lebedeva E V and Zharkov R V 2019 Structure of volcanic lake Klyuchevoe and lake Stubel basins (Ksudach caldera, Kamchatka) *Geography and Natural Resources* **4** pp 153–164

[7] Vereshhagin G Yu 1930 Methods for the morphometric characterization of lakes *Proceedings of the Olonets Scientific Expedition* **1** pp 3–114

[8] Chebotarev A I 1953 *Land hydrology and river flow calculations* (Leningrad: Hydrometeoizdat) 562 p

[9] Muraveyskiy S D 1960 Essays on the theory and methods of lake morphometry *Rivers and Lakes* (Moscow) pp 91–125

[10] Ufimtsev G F, Timofeev D A, Simonov Yu G, Siridonov A I, Siliverstov Yu, Borsuk O A, Lastochkin A N, Rozhdestvensky A, Loskutov Yu I, Koshkarev A V, Antonov S I, Chichagov V, Zhindarev L A and Agafonov B P. 2004 *Relief morphology* (Moscow: Sci. world) p 184

[11] Markhinin E K and Stratula D S 1977 *Hydrothermal Systems of Kuril Islands* (Moscow: Nauka) p 227

Materials for the database on the largest lakes of the Kuril Islands (morphometry, genesis and geographic distribution)

D N Kozlov

Institute of Marine Geology and Geophysics Far Eastern Branch of Russian Academy of Science, Yuzhno-Sakhalinsk, 693022, Russia

E-mail: kozlovdn@bk.ru

Abstract. This paper presents generalized modern data on the location, morphometry, and genesis of the largest lake basins of the Kuril Islands, obtained in a series of expeditions of the laboratory of volcanology and volcanic hazard of the Institute of Marine Geology and Geophysics FEB RAS in 2005–2020. The data is supplemented with modern geoinformation open access resources. A sample of the largest of 1099 Kuril lakes has been made according to the areal criterion, thus, 20 objects have been included in the list of studied water bodies, which are represented by 7 volcanic and 13 lagoon lakes. The considered lakes are clearly divided according to genesis, area and height of the mirror, maximum depth, as well as the ratio of the main morphometric characteristics. This information is part of a database currently being created on the largest lakes in the region.

Keywords: Kuril Islands, volcano, lake, lagoon, database, morphometry

1. Introduction

Research and assessment of water resources, their rational use are important global problems, which every year become more and more urgent for Russia. One of the main water facilities are lakes, therefore, it is necessary to constantly update data on their morphometry, spatial distribution, classification, genesis, etc., this will allow us to use our water resources most efficiently.

Considering this issue on the example of the Sakhalin Region, it can be noted that many Sakhalin lakes have been studied relatively well, while the Kuril ones have been studied either very superficially or in the context of specific issues (fishery resources, volcanological researches). In addition, most of the available information has lost its relevance and needs to be updated; the existing materials have been published in a series of articles in specialized and scientific journals, and several scientific works. In this regard, the author has set the task of generalizing the available information, creating and publishing a database on the lakes of the Kuril Islands.

2. Materials and methods

The first step in this direction was the study of the largest lakes, since they have the greatest water storage and potential for use. A review of morphometric parameters, genesis and geographical location of the largest lakes was carried out, observation results were collected and updated from expeditionary studies of the laboratory of volcanology and volcanic hazard of the IMGG FEB RAS for the period from 2005 to 2020 [1–3], as well as comparison with the materials from open access geoinformation resources (Google Earth, SAS.Planet) and literature data [4–7]. Field studies of volcanic reservoirs were carried out using the successfully tested method of digital bathymetric surveying with satellite referencing of profiles [1]; earlier, based on the results of the interpretation of these data, bathymetric schemes and tables of morphometric parameters of lakes were compiled, which served as the basis for creating a database.

3. Results and discussion

There are 17219 lakes in the Sakhalin region, of which 1099 are located on the Kuril Islands, the largest are volcanic lakes, as well as some lagoon ones. In order to generalize the sample of all Kuril lakes, water bodies were selected that satisfy the criteria S≥1 km² and correspond to small reservoirs according to the morphometric classification by P.V. Ivanov [9]. Using such a sample, 20 small and medium-sized water bodies of the region were selected. They are represented by 7 volcanic and 13 lagoon lakes (figure 1), all of them are clearly divided by origin, area and height of the mirror, as well as by maximum depth (table 1).

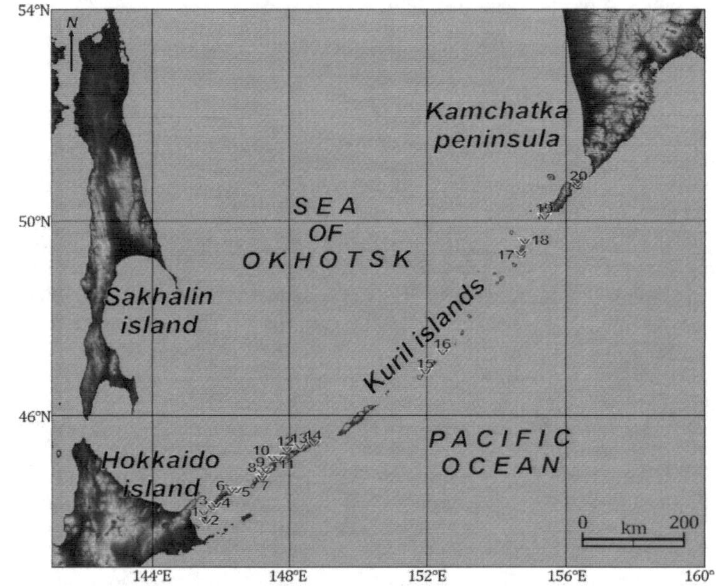

Figure 1. Location map of the largest lakes of the Kuril Islands: 1 – Goryacheye, 2 – Veslovskoye, 3 – Peschanoye, 4 – Lagunnoye, 5 – Krugloye, 6 – Dlinnoye, 7 – Krasivoye, 8 – Dobroye, 9 – Lesozavodskoye, 10 – Kuibyshevskoye, 11 – Blagodatnoye, 12 – Lebedinoye, 13 – Sopochnoye, 14 – Slavnoye, 15 – Biryuzovoye, 16 – Malahitovoye, 17 – Koltsevoye, 18 – Chernoye, 19 – Zerkal'noye, 20 – Bol'shoye.

If we consider the selected reservoirs in more detail, then we can highlight a number of important parameters, significant similarities and differences. The considered 20 lakes make up only 1.8 % of the total number of Kuril lakes, but their total area is 80.4 km², or almost 70 % of the total area of all lakes. This once again focuses the attention of the scientific community on the role of these reservoirs in the structure of water resources of the Kuril Islands. The average maximum depth of the considered lakes is 56.1 m, for lagoon lakes this indicator is 8.9 m, and for volcanic ones – 113 m. The geographical distribution of the lakes is very uneven, 14 of them are located in the Southern Kuriles, 4 in the Northern Kuriles and 2 in the Middle Kuriles, while the largest of the considered reservoirs is the Koltsevoye caldera lake located on the Northern Kuriles (Onekotan Island). According to the morphometric areal classification, the Koltsevoye Lake belongs to the middle lakes, and the other to the small ones.

Table 1. The largest lakes of the Kuril Islands.

№	Island	Name	Coordinates	Alt. of water surf., m a.s.l.	Area of water surf., km²	Max. depth, m	Genesis	Hydrol. regime
1		Goryacheye	43°52′ N 145°30′ E	128	3.1	62.3	V	DR
2		Veslovskoye	43°43′ N 145°33′ E	1	1.2	1	L	DR
3		Peschanoye	43°55′ N 145°36′ E	5	7.4	21.5	L	DR
4		Lagunnoye	44°03′ N 145°45′ E	1	3.5	23.4	L	DR
5		Krugloye	44°22′ N 146°24′ E	8	3.26	5.1	L	EN
6	Kunashir	Dlinnoye	44°24′ N 146°27′ E	1	2.58	2.8	L	DR
7		Krasivoye	44°37′ N 147°12′ E	82	5.8	50	V	DR
8		Dobroye	44°44′ N 147°14′ E	6	2.6	1.2	L	DR
9		Lesozavodskoye	44°46′ N 147°13′ E	9	1.45	2	L	EN
10		Kuibyshevskoye	45°03′ N 147°39′ E	5	1.4	11	L	DR
11		Blagodatnoye	45°01′ N 147°42′ E	4	4.06	15.7	L	DR
12		Lebedinoye	45°13′ N 147°54′ E	1	1.04	3	L	DR
13		Sopochnoye	45°18′ N 148°24′ E	3	1.33	21.5	L	DR
14	Iturup	Slavnoye	45°21′ N 148°44′ E	164	2.86	4	V	DR
15	Simushir	Biryuzovoye	46°54′ N 151°57′ E	50	3.2	87	V	EN
16	Ketoy	Malahitovoye	47°19′ N 152°27′ E	648	1.5	110	V	DR
17		Koltsevoye	49°20′ N 154°43′ E	385	26	369	V	EN
18	Onekotan	Chernoye	49°34′ N 154°50′ E	72	5.8	110	V	EN
19	Paramushir	Zerkal'noye	50°03′ N 155°25′ E	1	1.25	6	L	DR
20	Shumshu	Bol'shoye	50°45′ N 156°15′ E	2	1.08	1.8	L	DR

Note. V – volcanic, L – lagoon, EN– endorheic, DR – drainage.

Volcanic lakes occupy an area of 48.26 km² (60 % of the total area of 20 lakes), their depth varies from several tens to several hundred meters, the average is 113 m. They are characterized by relatively high levels of the mirror – from 50 to 648 m above sea level. Lagoon lakes occupy an area of 32.15 km² (40 % of the total area of 20 lakes), the depth of reservoirs is small – from 1 to 23 m, on the average 8.9 m. The absolute height of the lake surface above sea level. m. is from 1–5 to 8–9 m.

4. Conclusion

The information obtained using the morphometric parameters, genesis and geographical distribution of the largest lakes of the Kuril Islands were analyzed and summarized in order to fill the database. They can be summarized in the following form:

1) There are 20 lakes on the Kuril Islands with a surface area of more than 1 km² each (1.8 % of the total number of lakes in the region). The area of these 20 lakes is 80.4 km² (69.9 % of the area of all Kuril lakes). The total area of volcanic lakes is 48.26 km², the maximum depths reach from several tens to several hundred meters, they are characterized by high elevations of the mirror above sea level (50–648 m). Lagoon lakes cover an area of 32.15 km² and have relatively shallow depths and minimum level markings;
2) According to the morphometric areal classification, all lakes with the exception of Koltsevoye Lake belong to small lakes, Koltsevoye lake is medium;

3) Of the 20 considered lakes, 14 are located in the Southern Kuril Islands, 4 – in the Northern Kuril Islands (the largest reservoir of the region is located here) and 2 in the Middle Kuril Islands; 7 lakes are of volcanic genesis, and 13 are of lagoon genesis;
4) The type of water exchange in the largest lakes does not depend on their geographic location and genesis. In volcanic lakes, the role of hydrothermal waters in feeding is often significant;

The tasks of further research include filling, updating and publishing the database on the Internet, as well as clarifying the list of objects, expanding the set of morphometric and limnological criteria, including data on the volumes of water bodies, their average depths, coastline ruggedness coefficients, etc.

Acknowledgements
This work was supported by the Russian Science Foundation (project No. 21-17-00049).

References
[1] Kozlov D N 2015 *Crater lakes of the Kuril Islands* (Yuzhno-Sakhalinsk: Sakhalin Regional Museum, Institute of Marine Geology and Geophysics FEB RAS) 112 p
[2] Kozlov D N, Degterev A V and Zarochintsev V S 2018 Koltsevoe caldera lake: current state and structure of the basin (Onekotan Island, Kuril Islands) *Geosystems of Transition Zones* **4** pp 359–364
[3] Kozlov D N The largest lakes of the Kuril Islands: morphometry and geographical distribution (materials for the database) *Geosystems of Transition Zones* **4** pp 506–513
[4] Korsunskaya G V 1958 *The Kuril island arc* (M.: Glavnoye izd-vo geogr. Literatury) 222 p
[5] Gorshkov G S 1967 *Volcanoes of the Kuril island arc* (Moscow: Nauka) 287 p
[6] Zelenov K K, Kanakina M A Biryuzovoe lake (Zavaritsky caldera) and changes in the chemistry of its waters as a result of the 1957 eruption *Bul. volc. Station* **32** p 33–44
[7] Zlobin T K et al. 2004 *Kuril Islands* (Yuzhno-Sakhalinsk: Sakhalin. Book Publ.) 227 p
[8] Ivanov P V 1948. Classification of the lakes around the world by size and their average depth *Bull. LGU* **20** p 29–36

Accumulative landforms in valleys with gas-hydrothermal manifestations of the Kuril-Kamchatka region

E V Lebedeva[1] and R V Zharkov[2]

[1] Institute of Geography RAS, Moscow, Russia
[2] Institute of Marine Geology and Geophysics Far Eastern Branch of Russian Academy of Science, Yuzhno-Sakhalinsk, 693022, Russia

E-mail: [1]ekaterina.lebedeva@gmail.com, [2]rafael_zharkov@mail.ru

Abstract. Within the watercourse valleys of geothermal zones, where manifestations of gas-hydrothermal activity are noted, the processes of relief formation are accompanied by chemical and thermal effects. Thermal waters contribute to the formation of specific accumulative landforms both directly in the channel and in the bottom and on the sides of river valleys. Similar phenomena have been studied in river valleys draining the slopes of the Mendeleev volcano (Kunashir Island), Baransky volcano (Iturup Island), Mutnovsky volcano and Uzon-Geyzernaya caldera (Kamchatka). The typification of accumulative relief forms is carried out, the influence of gas-hydrothermal manifestations on fluvial and slope processes is determined.

1. Introduction

In the river valleys of the geothermal zones of the territories of modern volcanism, there are often areas of active gas-hydrothermal manifestations that accompany volcanic activity, but can also persist after its attenuation. The locations of the outlets of thermal waters and solfataras within the volcanic edifices are usually determined by systems of radial and ring faults. River valleys on the slopes of volcanoes and at their feet are also usually confined to fault zones; therefore, gas-hydrothermal manifestations are often observed in the bottoms of erosional gullies and valleys. To date, few works have been published on the structure and development of such valleys; the influence of gas-hydrothermal waters on the formation of fluvial relief has been poorly studied. The purpose of these studies is to use the example of valleys on the slopes of several volcanoes in the Kuril-Kamchatka region to classify the forms of accumulative relief in the conditions of manifestations of gas-hydrothermal activity, as well as to determine the influence of the latter on the features of geomorphological processes in the valleys.

2. Methodology

The study is based on the results of field observations, sampling of alluvial and sinter sediments, waters of thermal springs and mud pots in the watercourse valleys on the slopes of the Baransky volcano (Iturup Island), Mendeleev volcano (Kunashir Island), as well as in the basins of rivers Falshivaya (slopes of the Mutnovsky volcano), Shumnaya and Geyzernaya (Uzon-Geyzernaya caldera) in Kamchatka. Analyzes were carried out in the laboratories of the IEC SB RAS (Irkutsk), FEGI FEB RAS (Vladivostok), POI FEB RAS (Vladivostok), «Primorgeologiya» (Vladivostok), IVS FEB RAS (Petropavlovsk-Kamchatsky) and IMGG FEB RAS (Yuzhno-Sakhalinsk).

3. Results

The formation and compositional characteristics of gas hydrotherms in volcanic areas, including the Kuril-Kamchatka region, have long attracted the attention of researchers and are widely discussed [1–10]. Gas-hydrothermal manifestations of the studied region include steam-gas outlets (mainly solfataras), mud pots and thermal springs (figure 1). In areas of high level of gas-hydrothermal activity, in the so-called solfatara fields, as a rule, carbonic, sulfate hydrothermal fluids with a complex cationic composition, having an average temperature of 80–100 °C, are observed. With distance from the center of volcanic edifices and the extinction of volcanic activity, thermal springs of various physical and chemical properties are encountered.

Figure 1. The geographical position of the Kuril-Kamchatka region (a); solfatara of the Mendeleev volcano, Kunashir Island (b); mud pot in the valley of the Geyzernaya river, Uzon-Geyzernaya caldera, Kamchatka (c); thermal springs in the Kipyashchaya river, Baransky volcano, Iturup Island (d).

3.1. Typification of accumulative landforms in the watercourse valleys with gas-hydrothermal manifestations

Often, watercourses draining the slopes of volcanoes originate from the solfatara fields or cross them. Thus, within the limits of valleys, not only outlets of thermal waters, but also solfataras and mud pots can be observed. Therefore, we can distinguish several groups of landforms, which owe their appearance mainly to steam-gas outlets, or to the springs of thermal waters of various composition, as well as their specific interaction with the bedrocks.

3.1.1. Landforms associated with steam-gas outlets. The formation of sublimates in the form of deposits of native sulfur occurs around the solfatara. The composition of solfatara gases is predominantly carbon dioxide (more than 90 %), with some content of SO_2, H_2S, HCl, etc. Precipitation occurs as a result of a sharp decrease in temperature and pressure, specific cone-shaped structures are formed around long-lived solfataras (figure. 1, b), the height of which sometimes reaches 1–2 meters, and the average diameter is up to 1–2 m and more.

3.1.2. Landforms associated with the activity of mud pots. Mud pots are funnels and small basins (usually the first meters in diameter) filled with a liquid clay mass (figure. 1, c). The chemical composition of the water from squeezed mud mass is, as a rule, similar to the composition of the waters of the surrounding thermal springs. Acidic or slightly acidic sulphate mud pots with a mud mass temperature of up to 100 °C are most commonly found in solfatara fields and thermal sites [7, 9, 11, 12]. They are often located on the floodplain, and therefore, the removal of clay material by water flows can be observed there during floods or the rainy seasons.

In some cases, they are confined to the low terraces of streams, but, for example, in the valley of the river Geyzernaya, they are located on a terrace-like surface with elevations of about 30 m above the river. Usually mud pots, the diameter of which ranges from 0.5 to 3–5 m, are framed by the ring *of mud material* a few tens of cm high and up to 0.5–1.5 m wide; sometimes along their periphery, tongues of clay masses with a length of a few meters, less often tens of them, can be traced. In some areas near the pots, there are small groups of small *«mud volcanoes»* with a cone height of 20–30 cm.

3.1.3. Landforms created with the participation of thermal springs. Three types of accumulative formations associated with the inflow of mineralized waters into the streams are confined directly to the river channels: *channel troughs*, *festoons* and *armored steps of waterfalls*. All of them are formed in the areas of rapids, where water is saturated with oxygen, and the transition of soluble compounds into an insoluble form occurs and they precipitate. The same can be observed at other geochemical barriers: when streams merge or flow into water bodies with a different chemical composition of water. All this also leads to the formation of dense crusts in the channel and on the surfaces of the floodplain and terraces, to the cementation of alluvial deposits.

The most widespread microrelief forms are *channel troughs* – gutter-like sections of the channel formed by layered formations on rapids and in areas of narrowing. Their length is a few meters, and their width is from 0.2 to 1.0 m. They were found in almost all valleys of the studied watercourses with hydrothermal manifestations. The results of the analyzes have shown that they can be composed mainly of silica (SiO_2 36–74 %), and oxides and hydroxides of Fe and Mn (up to 30 % or more), sometimes carbonates.

Festoons are platforms of 0.4–0.7 m, often with a rounded edge, sometimes bordered by a small rim, stepwise located on rapids-waterfall sections of the channel. In the study area, we have found them only in the middle course valley of the Kipyashchaya river (Baransky volcano, Iturup Island). The Kipyashchaya river has a salinity of 4 g/l, sulphate-chloride calcium-sodium composition with content SiO_2 of more than 300 mg/l, pH 1.4 [11]. Judging by the results of silicate analysis, the festoons in this section of the channel were formed by a process of the deposition of siliceous sinter (up to 93 %).

The most pronounced example of *armored steps* of a waterfall was observed in the lower part of the valley of the Kislyi brook (Mendeleev volcano, Kunashir Island). Lighter evorsion hollows with a diameter of up to 0.5–0.7 m and a depth of up to half a meter with a rounded pebble and small-boulder material on the bottom and small channel troughs are stand out well against the background of 5 steps of the waterfall covered with a dense brown crust. The results of geochemical analysis showed that the armoring of the steps was due to the deposition of siliceous sinter (SiO_2 – 79–86 %) with the participation of iron oxides and hydroxides (3–6 %). The thickness of the dense crust varies from 5 to 10 cm. Armoring of the waterfall steps is also observed in the Geyzernaya river basin (waterfalls Igrushka, Troinoy, etc.), and Kipyashchaya river (Iturup Island), but such well-defined steps are rare.

Laminated sinter formations are formed in the valleys near thermal springs and geysers: *shields (or aprons)*, *walls* and *sinter terraces*. *Shields* and *walls* made of geyserite are most widely represented in the valley of the river Geyzernaya in the area of the Geysernoye thermal field, where chloride sodium springs with a mineralization of 1.8–2.2 g/l dominate. Here, *cone-shaped or isometric structures* from geyserite, a dense layered siliceous agglomerate, are also formed around long-lived thermal springs and geysers. Their height can reach 1–2 m and more. Large *cones* are often framed by sinter sub-horizontal or inclined formations – the so-called *shields (or aprons*), which overlap the adjacent areas in the form of a blanket. They are composed mainly of geyserite. Geyserite *walls* are ledges or sub-vertical sections of the valley sides below the outlet of thermal springs, covered with sinter crusts of varying thickness; their area sometimes reaches several tens of square meters. One of the most large-scale formations in river valleys with hydrothermal manifestations can be *sinter terraces* – terrace-like surfaces formed by deposits of thermal waters at the outlets in the sides of the valley. Good examples are observed on the slopes of the Mendeleev volcano: in the valley of the Kislyi brook at the outlet of the springs of the Lower group and in the valley of the Zmeinyi brook at the Stolbovskie springs. Thermal waters in the middle reaches of the Kislyi brook valley (at an absolute height of about 120–130 m) flow out in the floodplain and in the ledge of a 5-meter terrace, mainly at the left side. Their waters are ultra-acid, nitrogen, chloride-sulfate sodium with pH 1.6–2.1, containing SiO_2 up to 140–250 mg/l with a total salinity of 2.5–4.0 g/l [2, 11, 13]. A sinter terrace with a height of about 1.5 m and a length of up to 10–15 m, composed of silica (33 %), oxides and hydroxides of Fe (23 %) and SO_3 (14 %), has been near the ledge of a 5-meter terrace. The temperature of the Stolbovskie thermal springs, which flow out along the left side of the Zmeinyi brook in its middle reaches, is about 80 °C, the soil above them is warmed up to 40–60 °C. Neutral (pH 6.7–7.0), nitrogen, chloride-sulfate sodium hydrothermal fluids flow out here through small griffins, the walls of which are covered with a snow-white coating of sulfur formed due to of the activity of sulfur bacteria [2, 11], which oxidize gaseous sulfur compounds (H_2S and SO_2). At the site of the main water outlet, a small sinter terrace with a height in the central part of about 1.5 m above the brook was formed, upstream and downstream its surface decreases to 0.5 m, its total length is about 7–10 m. This sinter formations consist of silica – up to 42 %, CaO – from 21 % to 52 %, oxides and hydroxides of Fe – up to 7 %, and Al – up to 12 %.

3.2. Specificity of the accumulation of alluvial deposits in the valleys

It should be noted that alluvial deposits in the study area are formed as a result of erosion and processing of hydrothermally altered volcanic and volcanic-sedimentary rocks, both as a bedrock, and displaced and processed by glacial, slope, mudflow processes, as well as chemogenic sediments. Both in the process of moving in the channel of the watercourse, and during deposition and burial, the pebble material is exposed to the aggressive effects of thermal waters and continues to erode, which is clearly seen in the analysis of the boulder component of the alluvium of the Kipyashchaya river draining the slopes of the Baransky volcano (Iturup Island), carried out by [14].

Our analysis of pebble material from watercourses on the Kunashir Island (Lesnaya river, Kislyi brook, Valentiny brook) has also shown that *weathered by hydrothermal solutions* sometimes altered to saprolite material dominates there: the content of altered pebbles ranges from 43 % to 96 %. The depth of weathering crust of pebbles is on average 2 cm. The fragments of floodplain and terraces covered with layered *dense crusts* and/or composed of alluvial deposits weathered by hydrothermal processes, and their erosional fragments of various sizes in channels occur in the considered valleys. Moreover, there are two types of alluvium processing by mineralized waters. Strong *cementation* of alluvial deposits is more often observed, as a rule, with ferruginous and ferromanganese solutions.

3.3. Slope processes and mudflow accumulation in valleys

Hydrothermally altered rocks, weathered to clays, are easily eroded by surface waters, sometimes badlands are formed at these areas, but more often the formation of slips and landslides occurs. Plastic moist clays are an ideal substrate for the formation of numerous displacements on the slopes, as a result of which local multilevel terraces are formed on the sides of the valleys. Due to this fact, the valleys themselves in such areas expand significantly, so they acquire beads-like shape in plan [15–17]. On the slopes of the valley of the Kislyi brook in its upper reaches, for example, we have found out up to 4 levels of such terraces, which are large blocks of displaced hydrothermally altered rocks, which indicates a high activity of landslide processes.

The observations carried out in the valley of the Geyzernaya river also clearly show that its sides are characterized by active development of various gravitational slope processes [17–19], which often form *mudflows*. In the bottom of the valley, there is an accumulation of slope material displaced as a result of rockfalls and landslides, with periodic blocking of the valley and the formation of temporary *dams* and dammed lakes, in which the accumulation of river sediments occurs. Further redeposition of the slope and alluvial material occurs during the destruction of dams with the formation of mudflows. We have already considered this process in more detail earlier [17].

4. Conclusions

Hydrothermal zones of areas of active volcanism are geosystems, in the formation of which endogenous energy is actively involved. Within such areas, the physical and chemical properties of rocks change greatly, specific forms of meso- and microrelief are created, and vegetation changes. In geomorphological terms, the watercourse valleys with gas-hydrothermal manifestations differ significantly from valleys without them. First of all, due to the fact that they are exposed to chemical and thermal effects, their waters become warm, highly mineralized, and specific accumulative processes take place in the channels and valleys. The mineralization of waters and the presence of geochemical barriers, where the precipitation of compounds from aqueous solutions takes place, is of great importance for the development of forms of accumulative relief.

As a result of complex interactions with thermal waters in river valleys, the following occurs:

1) *the formation of specific accumulative forms* of micro- and mesorelief both in the channels and on the sides of the valleys due to the deposition of new mineral formations;
2) *processing of alluvial deposits and bedrocks with hydrothermal solutions*, which leads to a radical change in their properties and, accordingly, in the characteristics of fluvial processes: there is an intensification of erosion of rocks weathered to clay in some areas, and in others, on the contrary, erosion slows down, rapids are often formed in the case of cementation of sediments;
3) *activation of slope processes* in areas of gas-hydrothermal manifestations with the formation of rockslide bodies and landslide masses, due to which the accumulation of displaced slope material occurs in the bottoms of the valleys, followed by its redeposition by mudflow processes.

Acknowledgements

Field research was carried out within the framework of the state assignment of the Institute of Geography of the Russian Academy of Sciences No. 0148-2019-0005, the generalization of materials was supported by the grant of the Russian Science Foundation No. 21-17-00216.

References

[1] Ellis A J and Mahon W A.J 1964 Natural hydrothermal systems and experimental hot water/rock interaction *Geochemica et Cosmochimica Acta* **28** pp 1323–57

[2] Markhinin E K and Stratula D S 1977 *Hydrothermal Systems of Kuril Islands* (Moscow: Nauka) pp 212

[3] Giggenbach W F 1984 Mass transfer in hydrothermal alteration systems – a conceptual approach *Geochemical et Cosmochimica Acta* **48** pp 2693–2711

[4] Taran Yu A, Pilipenko V P and Rozhkov A M 1986 Geochemistry of hydrothermal solutions and gases of the Mutnovskaya hydrothermal system *Geothermal and geochemical studies of high-temperature fluids* (Moscow: Nauka) p 207

[5] Rychagov S N 1993 Hydrothermal system of Baransky volcano (Iturup Island): model of geological structure *Volcanology and Seismology* 2 pp 59–75

[6] Chudaev O V 2003 *Composition and conditions of formation of modern hydrothermal systems in the Russian Far East* (Vladivostok: Dalnauka) p 203

[7] Bortnikova S B, Bessonova E P, Gora M P et al. 2013 *Gas hydrotherms of active volcanoes in Kamchatka and the Kuril Islands: composition, structure, genesis* (Novosibirsk: INGiG SO RAS) p 282

[8] Degterev A V, Kozlov D N, Romanyuk F A, Zharkov R V and Rybin A V 2018 The state of Berutarube volcano in 2017 (Iturup Island, Kuril Islands) *Geosystems of Transition Zones* **2** 4 pp 386–391

[9] Zharkov R V 2020 Active Volcanoes and Thermal Springs of Kunashir Island (Russia) *IOP Conference Series: Earth and Environmental Science* **459** 2 022039

[10] Zharkov R V 2020 Thermal manifestations of the Ebeko volcano (Paramushir Island, Kuriles) and their recreation and tourism potential *Geosystems of Transition Zones* **4** 4 pp 514–525

[11] Zharkov R V 2014 *Termal springs of the South Kuril Islands* (Vladivostok: Dalnauka) p 378

[12] Bragin I V, Chelnokov G A and Kharitonova N A 2019 Geochemistry of thermal springs at Baransky volcano, Southern Kuriles (Russia) *Environ. Earth Sci.* **78** 3 pp 79–89

[13] Kalacheva E G, Taran Yu A, Kotenko T A et al. 2017 The hydrothermal system of Mendeleev volcano, Kunashir Island, Kuril Islands: the geochemistry and the transport of magmatic components *Volcanology and Seismology* **11** 5 pp 335–352

[14] Ladygin V M, Frolova Yu V and Rychagov S N 2014 Transformation of effusive rocks under the influence of acid leaching by surface thermal waters (geothermal system of Baransky volcano, Iturup Is.) *Volcanology and Seismology* 1 pp 20–37

[15] Lebedeva E V 2018 Sequences of catastrophic geomorphic processes in the river valleys of volcanic regions *Geomorphology* 4 pp 38–55

[16] Lebedeva E V 2019 Kinds of impacts of volcanic and post volcanic activity on fluvial relief *Geomorphology* 4 pp 49–66

[17] Lebedeva E V, Sugrobov V M, Chizhova V P and Zavadskaya A V 2020 The valley of the river Geyzernaya (Kamchatka): hydrothermal activity and features of relief forming *Geomorphology* 2 pp 60–73

[18] Dvigalo V N, Melekestsev I V 2009 Geological and geomorphological consequences of catastrophic rockfall and rockfall – landslides processes in the Kamchatka Valley of Geysers (according to aerial photogrammetry) *Volcanology and Seismology* 5 pp 24–37

[19] Pinegina T K, Delemen' I F, Droznin V A et al. 2007 Kamchatka Valley of Geysers after the catastrophe on 3 June *Vestnik DVO RAN* 1 pp 33–44

IOP Conf. Series: Earth and Environmental Science **946** (2021) 012029 doi:10.1088/1755-1315/946/1/012029

Search for hydrogeochemical indicators of the genetic relation between mud volcanism and oil and gas fields

O A Nikitenko* and V V Ershov

Institute of Marine Geology and Geophysics Far East Branch of Russian Academy of Sciences, Yuzhno-Sakhalinsk, 693022, Russia

E-mail: *nikitenko.olga@list.ru

Abstract. The paper reports the results of a comparative analysis of the chemical and isotope composition ($\delta^{18}O$ and δD) of mud volcanic waters and formation waters from oil and gas fields. Studies show that the waters discharged by mud volcanoes in most cases are very similar to formation waters. The most characteristic geochemical traits of both waters are elevated concentrations of hydrocarbonate ions, iodine, boron, bromine, and a low content of sulfate ions.

1. Introduction

Mud volcanism is a geological phenomenon associated with the processes of focused discharge of subsurface fluids in fault zones of sedimentary basins. Mud volcanoes are an important element in the transformation of sedimentary deposits – high temperature and pressure induce sediment diagenesis and catagenesis in mud volcanic systems, and drive the maturation of the organic matter contained therein [1]. Transformation of organic matter at high depths can be a major cause of mud volcanoes formation. It is believed that as organic matter matures under clay cap rocks, hydrocarbon gas accumulates there and anomalously high formation pressure forms in mud volcanic chambers. This supposition is largely corroborated by a consistent distribution of mud volcanoes within large oil and gas basins [2]. The possibility of finding industrially exploitable hydrocarbon reserves at mud volcanism occurrences is evidenced by the case of Lokbatan volcano in Azerbaijan. In 1933, a wildcat was drilled at the foot of Lokbatan volcano, discovering an oil and gas field, which is in operation still now [2]. That said, some researchers remark that by far not all oil and gas regions have mud volcanoes. For example, no mud volcanoes have been found in large oil and gas provinces in the Near East, Siberian Platform, Volga-Ural and Timan-Pechora basins [3]. A presumable reason for that is the age and lithological characteristics of the reservoir rock. The oil reservoir rocks for the oil and gas basins with mud volcanoes are Cenozoic (rarely Cretaceous) terrigenous-clay strata. The reservoir rocks in the oil and gas basins lacking mud volcanoes are ancient (Paleozoic and Precambrian) carbonate, terrigenous-carbonate, and evaporate sediments.

Mud volcanoes are natural wells that inform about the processes taking place in the Earth's interior. Research into the genetic relation between mud volcanoes and oil and gas fields are of high practical importance, as it provides answers for the feasibility of estimating the deep hydrocarbon potential and for the development of new geological prospecting methods. Immediate relation to hydrocarbon fields can be identified through geochemical studies of subsurface fluids discharged by mud volcanoes. Evidence directly pointing to mud volcanoes associated with hydrocarbon fields is oil impurities in emissions of mud volcanoes. Not all mud volcanoes in oil and gas regions, however, produce oil seepage [2]. For example, in the South Caspian oil and gas basin, which harbors the largest mud volcanic province worldwide, oil seepage happens only in 40 out of several hundred mud volcanoes. Hence, the absence of oil impurities in the material discharged by a mud volcano is not per se a negative indicator of the region's hydrocarbon potential.

Another fact in favor of the relation between mud volcanism and the hydrocarbon potential is the geochemical characteristics of gaseous emissions from mud volcanoes. Analysis of worldwide data on the chemical and isotope composition of mud volcanic gases shows mud volcanoes to emit large amounts of hydrocarbon gases, the prevalent one is thermogenic methane [4]. Researchers also point out that many gaseous emissions from mud volcanoes contain carbon dioxide produced through anaerobic biodegradation of petroleum.

The hydrogeological settings are known to significantly affect oil and gas fields, since groundwater always accompanies oil and gas reservoirs [5]. Active participation of groundwater in hydrocarbon synthesis, migration, and destruction processes is the basis for the application of hydrogeochemical indicators for estimate of the hydrocarbon potential. Mud volcanic waters is often similar to formation waters, which is another indicator of a genetic relation between mud volcanoes and oil and gas fields. It is therefore reasonable to inquire whether the geochemical characteristics of the groundwater discharged by mud volcanoes can be used to estimate the hydrocarbon potential of the region?

2. Materials and methods

This paper gives a comparative analysis of the chemical and isotope composition of waters on terrestrial mud volcanoes and formation waters from oil and gas fields made on the basis of published hydrogeochemical data. The study made use of consolidated reports on water composition from over 170 mud volcanoes situated in different parts of the world – Azerbaijan, Georgia, India (Andaman Islands), Italy, China, Russia (Kerch Peninsula, Taman Peninsula, Sakhalin Island), USA (Alaska), Taiwan, Trinidad and Tobago, Turkmenistan, Japan [6–8]. Specific geochemical characteristics of formation waters are highlighted through the cases of several large oil and gas regions of Russia and other countries [9–16].

3. Results and discussion

The most characteristic geochemical features of the groundwater accompanying hydrocarbon fields are elevated total dissolved solids, and Cl–Ca or HCO_3–Na type water [5]. Formation waters types here are given in accordance with the classification suggested by V.A. Sulin and used in petroleum hydrogeology. The origin of Cl–Ca-type brines is believed to be associated with the geochemical evolution of sedimentation seawaters. The origin of HCO_3–Na-type formation waters, on the other hand, is associated with hydrochemical inversions in the sedimentary column. The causes of such inversions are still being debated. Mud volcanic waters, in turn, chiefly belong to the HCO_3–Na type according to Sulin's classification or, less often, to the Cl–Mg type. The synthesis and analysis of worldwide data on the chemical and isotope composition of the waters on terrestrial mud volcanoes show that the primary original source of mud volcanic waters is sediment-buried seawaters [6, 8].

The genesis of natural waters is often identified relying on their isotope composition. Comparisons and analysis of the isotope parameters mud volcanic waters and formation waters demonstrate that both of them feature a wide variation of $\delta^{18}O$ and δD values (fig. 1). In most cases, $\delta^{18}O$ and δD values are almost identical for mud volcanic waters and formation waters. Supposedly, the diversity of the isotope composition of mud volcanic waters is largely shaped by a combination of the following processes: mixing of the original sedimentation seawaters with the waters formed through dehydration of clay minerals; dilution by meteoric waters; oxygen-18 isotope exchange with the water-hosting rocks [6]. As a result, most mud volcanic waters are significantly enriched in oxygen-18 and have a lower deuterium content than seawaters. The tendency for formation waters from oil and gas fields is generally similar. Like mud volcanic waters, they exhibit a marked positive oxygen shift (fig. 1). Formation waters enrichment in oxygen-18 is largely due to isotope exchange with the reservoir rock

[10–12, 14, 16]. Ongoing processes of isotope exchange in the "water–rock" system in reservoir conditions indicate that the subsurface is hydrologically rather closed, which is a good sign for the integrity of oil and gas fields. Formation waters enrichment in oxygen-18 can also be due to the input of dehydration waters. Also, the isotope composition of formation waters is significantly influenced by the input of meteoric waters, which is the reason for a reduction in deuterium concentration in the formation waters. This process usually takes place in early stages of sedimentary basin evolution.

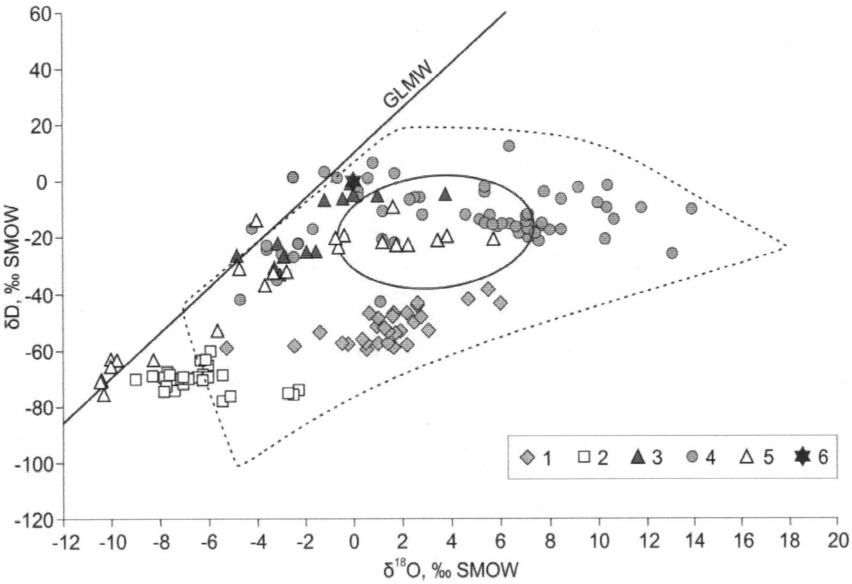

Figure 1. The ratio of $\delta^{18}O$ and δD of formation waters in different oil and gas areas: 1 – the Lunnan oilfield, China [11]; 2 – the oilfields in Novosibirsk region, Russia [14]; 3 – the Carabobo oilfield, Venezuela [10]; 4 – oil and gas fields of Eagle Ford Group, Texas, USA [12]; 5 – oilfield of San Joaquin Valley, California, USA [16] and Seawater – 6. GLMW – Global Meteoric Water Line. The dotted line marks the area that corresponds to the ratio of $\delta^{18}O$ and δD in mud volcanic waters in different regions [6]. The ellipse marks the area of the most typical $\delta^{18}O$ and δD values for mud volcanic waters.

An important geochemical feature of the groundwater associated with oil and gas fields is that it is low content of sulfate ions [5]. For example, SO_4^{2-} concentrations in formation waters of oil and gas fields in the Yenisei-Khatanga trough range from 2 to 98 mg/l, while their total dissolved solids content is 2.2 to 21.1 g/l [13]. The low SO_4^{2-} content in formation waters is a result of sulfate reduction processes involving anaerobic bacteria in the presence of organic matter. Mud volcanic waters also feature low SO_4^{2-} concentrations. They are significantly depleted in SO_4^{2-} compared to seawaters. We estimate the average SO_4^{2-} content in mud volcanic waters to be 20.5 mg/l [8]. Such a low SO_4^{2-} content in mud volcanic waters can be a consequence of anaerobic methane oxidation accompanied by sulfate reduction.

Bacterial sulfate reduction in groundwater rich in organic matter is a reaction that produces CO_2. Dissolution of CO_2 in groundwater, in turn, causes a rise in HCO_3^- concentrations. This is believed to be one of the reasons for elevated HCO_3^- content in formation waters [17]. Rather high HCO_3^- concentrations are found also in mud volcanic waters. According to our estimates, average HCO_3^- content in mud volcanic waters is 2 g/l [8]. Analysis of worldwide data, on the other hand, shows that 20% of samples have HCO_3^- concentrations over 5 g/l, and the maximum HCO_3^- concentration record is 15 g/l. CO_2 input to mud volcanic chambers can come from oil biodegradation, as corroborated by high $\delta^{13}C$–CO_2 values in gaseous emissions from mud volcanoes [4]. It is possible also that mud volcanic waters get enriched in HCO_3^- as a result of anaerobic methane oxidation, since methane is a dominant component of mud volcanic gases. Besides, mud volcanic chambers can get CO_2 from mantle degassing and metamorphic CO_2 generated during thermal degradation of carbonates.

Mud volcanic waters are characterized by an elevated content of trace elements such as iodine, boron, and bromine. These trace elements in groundwater are direct indicators of the hydrocarbon potential. Groundwater enrichment with iodine is attributed to organic matter transformation in the process of sedimentary rock formation, considering that iodine, being originally a biophilic element, is a constituent of the organic matter in water-hosting rocks. The most significant rise in iodine content in groundwater is believed to happen during the organic matter thermal maturation stage [18]. In content of chlorine and iodine concentrations, waters of mud volcanoes are very similar to formation waters (fig. 2). Iodine concentrations in mud volcanic waters range from 0.02 to 92.5 mg/l. The range of iodine concentrations in formation waters from the oil and gas fields examined in this paper is 0.75 to 81 mg/l.

Analysis of worldwide data shows boron and bromine concentrations in mud volcanic waters to vary widely: from 0.3 to 915 mg/l for boron (75.6 mg/l on average), from 0.1 to 376 mg/l for bromine (44 mg/l on average). Enrichment of mud volcanic waters in these trace elements is often associated with the interaction processes in the "water-rock" system. Namely, high boron content in mud volcanic waters can be explained by smectite transformation into illite, during which large amounts of boron-rich water is discharged into the pore space. The said trace elements in mud volcanic waters can be of biogenic origin, considering that the sedimentation process involves boron and bromine accumulation in the organic matter contained in the sedimentary deposits. The destruction of the organic matter in sedimentary deposits under high temperatures and pressures leads to elevation of bromine and boron concentrations in the groundwater of sedimentary basins [19, 20]. The thermal settings for smectite illitization largely coincide with the temperatures at which hydrocarbons are generated as organic matter is thermally matured. Note that boron and bromine concentrations in formation waters are comparable to those in mud volcanic waters.

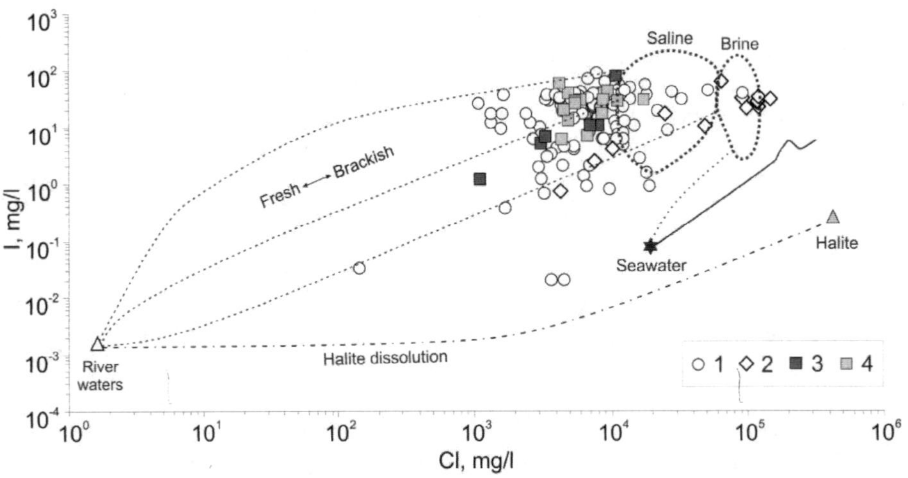

Figure 2. The ratio of Cl and I concentrations of mud volcanic waters (1) in different regions [7, 8] and formation waters in different oil and gas areas (2 – the Timan-Pechora oil and gas province, Russia [9, 15]; 3 – the Yenisei-Khatanga regional trough, Russia [13]; 4 – Sakhalin Island, Russia) plotted on a diagram from [18].

The rates of diagenetic and catagenetic processes in sedimentary deposits and the organic matter contained therein are primarily controlled by the thermal conditions, which are directly related to the depth at which these processes are happening. The generation temperatures of mud volcanic waters calculated based on hydrochemical geothermometry can be used as proxies of the bedding depth of the reservoirs that feed mud volcanoes with water. The generation temperatures of mud volcanic waters in different regions of the world, calculated from a geothermometry dataset (Mg–Li, Na–Li, K–Mg, Na–K), mainly vary from 50 to 110 °C. Judging by the calculated temperatures, mud volcanic chambers are bedded at 1.0 to 3.5 km depths. These are also the depths of the oil and gas generation zone. Let us remark that peak intensities of liquid and gaseous hydrocarbon generation occur at different temperatures (and, hence, at different depths). The principal oil generation zone corresponds to a temperature range of 95 to 175 °C, whereas the main gas generation zone occurs at higher temperatures – 105 to 220 °C [21]. In that case, where the thermal flux is, for example, 60 mW/m^2, the oil generation zone lies at a depth of 1 to 4.5 km, and the gas generation zone is at 1.7 to 7.5 km depth, i.e. is deeper bedded. Gases emitted from some mud volcanoes were found to contain methane rich in $\delta^{13}C–CH_4$, which is formed in late stages of catagenesis. Hence, mud volcanoes can serve as a source of information on the gas content of deeply buried deposits.

4. Conclusion

Comparative analysis of the chemical and isotope composition of mud volcanic waters and formation waters from oil and gas fields shows them to have similar hydrogeochemical characteristics. In our opinion, the similarity observed between mud volcanic and formation waters is largely due to their shared geological environments – within large sedimentary basins, where groundwater composition is significantly influenced by the processes of hydrostatic compaction of sedimentary deposits, thermal dehydration of clay minerals, organic matter maturation, and interactions in the "water–rock–gas" system. Let us specially emphasize that generation temperatures of mud volcanic waters and depths calculated through hydrochemical geothermometry largely coincide with those of oil and gas generation zones. Thus, the characteristics and patterns of mud volcanic waters and formation waters generation described in this paper suggest that there exists at least a paragenetic relation between mud volcanoes and oil and gas fields.

References

[1] Kholodov V N 2019 *Lithol. Min. Resour.* **54** 1 pp 38–52
[2] Aliev Ad A, Guliev I S, Dadashev F G and Rakhmanov R R 2015 *Atlas of the world mud volcanoes* (Baku: Nafta-Press) p 321
[3] Kholodov V N 2002 *Lithol. Min. Resour.* **37** pp 197–209
[4] Ershov V V and Bondarenko D D 2020 *Geoecology. Engineering geology. Hydrogeology. Geocryology* **3** pp 23–35
[5] Kartsev A A *Waters of oil and gas fields of the USSR* 1989 (Moscow: Nedra) p 382
[6] Nikitenko O A and Ershov V V 2017 *Bulletin of KRAESC. Earth Sciences* **2** 34 pp 49–60
[7] Nikitenko O A and Ershov V V 2020 *Astrakhan Bulletin for Environmental Education* **1** 55 pp 26–30
[8] Nikitenko O A and Ershov V V 2021 *Geochemistry* **66** 10 pp 887–903
[9] Balashova E Y and Farnosova E N 2017 *Journal Advances in Chemistry and Chemical Technology* **5** pp 76–78
[10] Boschetti T, Angulo B, Quintero F, Volcan L and Casalins A *Geol. Acta* **16** 3 pp 257–64
[11] Chen J, Liu P, Yu C, Zhang B and Xiao Z 2013 *Geofluids* **13** pp 381–94
[12] Engle M A, Doolan C A, Pitman J A, Varonka M S, Chenault J, Orem W H, McMahon P B and Jubb A M 2020 *Chem. Geol.* **550** 119754
[13] Novikov D A 2017 *Geodynamics & Tectonophysics* **8** 4 pp 881–901
[14] Novikov D A, Pyrayev A N, Chernykh A V, Dultsev F F and Ryzhkova S V 2021 *Bulletin of the Tomsk Polytechnic University. Geo Assets Engineering* **332** 2 pp 59–72
[15] Zytner Yu I and Chibisova V S 2013 *Petroleum Geology – Theoretical and Applied Studies* **8** 3 pp 1–19
[16] McMahon P B, Kulongoski J T, Vengosh A, Cozzarelli I M, Landon M K, Kharaka Y K, Gillespie J M and Davis T A 2018 *Appl. Geochem.* **98** pp 127–140

[17] Poroshin V D and Mulyak V V 2004 *Methods of hydrochemical data processing and interpretation to control development process of belorussian oil fields* (Moscow: Nedra) p 220
[18] Özdemir A 2018. *Journal of Sustainable Engineering Applications and Technological Developments* **1** 2 pp 103–150
[19] Krasintseva V V 1968 *Hydrogeochemistry of chlorine and bromine* (Moscow: Nauka) p 196
[20] Williams L B, Hervig R L, Wieser M E and Hutcheon I 2001 *Chem. Geol.* **174** pp 445–461
[21] Pepper A S and Corvi P J 1995 *Mar. Pet. Geol.* **12** 3 pp 291–319

IOP Conf. Series: Earth and Environmental Science **946** (2021) 012030 doi:10.1088/1755-1315/946/1/012030

Mud volcanism as a dangerous phenomenon for oil and gas facilities

G V Ustyugov* **and V V Ershov**

Institute of Marine Geology and Geophysics Far Eastern Branch of Russian Academy of Science, Yuzhno-Sakhalinsk, 693022, Russia

E-mail: *gen.ustyugov@mail.ru

Abstract. The research dwells on the danger of mud volcanism for human economic activity, namely, oil and gas production. We performed quantitative assessment of mud volcanoes activities, using Azerbaijan and Kerch-Taman region as examples. Average annual number of mud volcanoes eruptions is 3–4 for Azerbaijan and 1–2 for Kerch-Taman region. We estimate the catalogues of mud volcanic eruptions for those areas to be 52 % and 39 % complete, respectively. Mud volcanoes eruptions are quite frequent. In both regions, over 50 % of all recorded eruptions occur within ten years of the latest eruption. Analysis of mud volcanic eruptions catalogues shows that the volume of breccia ejected during an eruption is practically not related to how long the mud volcano was quiescent. Analysis of potential impact of seismicity on mud volcanic activity shows that the probability of mud volcanoes responding to an earthquake is 6 % and 10 % for Azerbaijan and Kerch-Taman region, respectively.

1. Introduction

A mud volcano is a natural fluid dynamic system with intensive transport of substances (such as gas, water, and sedimentary rocks) and energy from the bowels of the Earth to the surface. Mud volcanoes are often found in areas of large oil and gas deposits. Mud volcanism is sometimes viewed as one of the signs indicating presence of hydrocarbons in the area [1]. A lot of engineering structures related to oil and gas production are designed and operated close to onshore and offshore mud volcanoes [2]. Mud volcanism is a dangerous geological phenomenon posing a serious threat to engineering infrastructure and natural ecosystems, and it may have serious socioeconomic and environmental consequences [3–6]. A relatively recent and vivid example is eruption of LUSI mud volcano back in 2006 near well BJP-1. The well was drilled with some violations of the process, which is considered to be one of the possible reasons for the eruption, which caused huge financial damage to Indonesian economy and dealt irreparable damage to the ecosystem of Java [4]. In addition to damage to expensive oil and gas facilities, there's a risk of a massive oil spill onshore and offshore. Response to such spills can be more expensive than all operating costs [2].

Oil companies put a lot of emphasis on security in areas of potential operations. Namely, they need to determine the most dangerous zones in areas of mineral resources production, as well as the scope of potential emergencies [2]. Therefore, at every stage of oil and gas fields development comprehensive measures must be taken to identify, monitor, and study mud volcanoes in the development area. This is necessary to predict potential emergencies and prevent consequences. These tasks get more difficult when developing offshore oil and gas fields, because underwater mud volcanoes are harder to find than onshore ones. In these cases, we can use various methods to register and study fluid dynamic systems [5, 7]. Evidently, analysis of geological risks from mud volcanic activity requires knowledge of quantitative patterns in mud volcanoes eruptions.

2. Materials and methods

Statistical analysis of mud volcanoes eruptive activity was performed using Azerbaijan and Kerch-Taman region as examples. Azerbaijan is the largest mud volcanic province on Earth (over 350 volcanoes). Kerch-Taman region is among the top ten regions on Earth in terms of the number of mud volcanoes (over 80) [1]. There is a rather detailed catalogue of mud volcanoes eruptions for Azerbaijan, with data from 1810 to 2014 [1], updated with data up to 2018 [8]. The resulting database contains information on over 400 eruptions of 93 volcanoes. There is a similar catalogue for Kerch-Taman region with data from 1794 to 2016 [9], also updated with data up to 2018 [10–14]. The resulting catalogue contains data on over 100 eruptions of 30 volcanoes. We made a joint analysis of eruptions and earthquakes catalogues to study how seismicity influences mud volcanic activity [15].

Accurate and reliable quantitative assessment is only possible with a representative dataset. It is evident that the used mud volcanoes eruptions catalogues are incomplete due to lack of regular observations (especially before the middle of the 20th century). The chances of missing an eruption depend on its strength. The probability of a strong eruption not being recorded is lower [2]. It is also not very likely that an eruption will go unnoticed if the mud volcano is close to a settlement [16]. Therefore, completeness of mud volcanoes catalogues has its own importance and will be discussed separately further on.

3. Results and discussion

Currently over 2500 mud volcanoes were found on Earth, over 1400 of them offshore. The largest volcanoes and the most powerful eruptions are typical for mud volcanoes located in oil and gas areas [1]. In recent years people actively research and develop the Arctic region which is very promising in terms of oil and gas production. Currently there is one known underwater mud volcano in the Arctic, called Haakon Mosby, between the shores of Norway and Spitzbergen [7]. Certain geologic conditions lead us to think that, in fact, there are way more mud volcanoes in the Arctic waters than is currently known. Offshore oil and gas fields in the north-east of Sakhalin Island are being actively developed. Piltun-Astokhskoye, Dagi and Chayvo fields located in close vicinity of Dagi mud volcano are of special interest here [17]. Many mud volcanoes were found in the Black Sea and in the Sea of Azov. An increased number of mud volcanoes can be found in the Kerch Strait. This needs to be considered when ensuring navigation safety, because up to 10000 marine vessels go through the Kerch Strait every year [6]. Azerbaijan also actively develops its oil and gas fields close to numerous onshore and offshore mud volcanoes [1].

Major hazards are mud slides, gas emissions, spontaneous combustion of methane, deformation of the land surface and sea bottom, rockfalls and landslides, cracks in the eruptive center and around the volcano, appearance of zones with anomalous formation pressures, and volcanic earthquakes [5, 18]. Quite often wells drilling is accompanied by accidental gas discharge with burning methane. Islands and sand banks generated by underwater mud volcanoes eruptions are the main threats to marine traffic. Strong single gas outbursts from underwater mud volcanoes are dangerous too because they can cause the vessel to careen sharply and capsize due to changed water density. In 1953, a gas outburst from an underwater mud volcano caused the Kayo Maru 5, Japanese science vessel, to go down, killing 31 persons [3]. One of the types of mud volcanoes are mud diapirs, domed piercing structures with a core made up of pliant clays in the central part. Those clays are pushed upwards, raising, breaking, and bursting through the overlying rock. The core of the diapir does not always reach the land surface. Physical modelling indicates that the thicker the overlying rock (the deeper the heart of the mud volcano), the more pressure is required for the subsoil fluids to migrate to the surface. If the pressure was not sufficient, instead of a mud volcano we may end up with a mud diapir which did not reach the surface [19]. Mud diapirs which did not reach the sea bottom are also called blind mud volcanoes, which goes to show how hard to find such natural objects are [20]. However, the mud diapirs are a potential threat for human economic activity. Drilling operations in area of mud diapir may cause development of a new mud volcano, in the form of a strong eruption. Mud volcanoes may also activate as a result of strong earthquakes. Earthquakes are considered to be a potential trigger for mud volcanic eruptions [16, 21].

Analysis of mud volcanic eruptions catalogues for Azerbaijan and Kerch-Taman region revealed some patterns. Since 1946 until now, Azerbaijan has 3–4 mud volcanoes eruptions a year (the average value is 3.62). Before 1946 the annual average number of mud volcanoes eruptions was significantly lower, from 0.35 (1810–1846) to 1.89 (1900–1946). We suppose the main reason was that some eruptions were not recorded, because there was no regular monitoring of mud volcanoes at that time. Otherwise, we'd have to believe that intensity of mud volcanic eruptions increased by several times over two centuries. This period is too short through the geological time for a significant increase of long geological processes to occur. Most likely before 1946 Azerbaijan had the same 3–4 eruptions a year. We estimate completeness of Azerbaijan mud volcanic eruptions catalogue as 52 %. Kerch-Taman region has 1–2 mud volcanoes eruptions a year since 1977 until now (the average value is 1.45). Before 1977, the average number of mud volcanic eruptions was significantly lower (0.32 times a year), which is probably not accurate because not all of them were recorded. Following the same logic, we estimate the completeness of mud volcanoes eruptions catalogue here to be about 39 %.

We analyzed the relation between the volume of sedimentary rock thrown out during mud volcanoes eruptions in Azerbaijan and Kerch-Taman region and the number of recorded eruptions with a certain volume of discharge (figure 1). In areas of strong and medium eruptions, we see a clear linear trend towards increased number of eruptions while the eruptions volume log decreases. However, when we get to weak eruptions, we see a drastic decrease in the number of eruptions. Yet most likely weak eruptions are more numerous than strong ones. The probable reason for this is that many eruptions were not recorded. If we extrapolate the resulting linear trend onto the weak eruptions, we can estimate the number of eruptions that were not recorded, as well as the total volume of mud volcanoes emissions over a selected period.

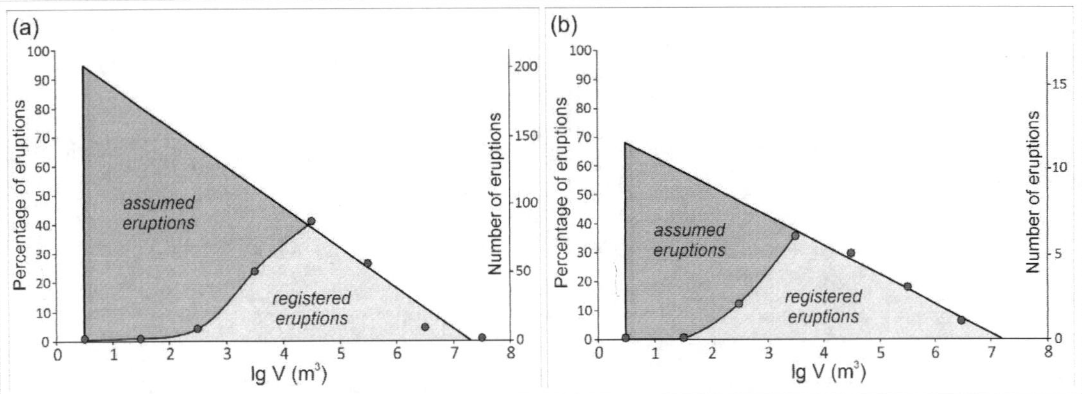

Figure 1. Relation between the number of recorded eruptions of mud volcanoes and the volume of ejected sedimentary rocks: a – Azerbaijan, b – Kerch-Taman region. The right scale shows the number of eruptions of a given volume of ejected sedimentary rocks. The left scale shows the percentage of eruptions of a given volume from the total number of recorded eruptions.

Active mud volcanoes are not as destructive as magma volcanoes, but they erupt far more frequently [18]. The time intervals between eruptions of each mud volcano vary greatly. However, both in Azerbaijan and in Kerch-Taman region we saw a common trend: when the quiescent time of mud volcanoes increases, there's a rapid decline in the number of eruptions. In both regions, 51-54 % of all recorded eruptions occur within the first decade after the mud volcanoes becomes quiescent. We should note that unrecorded eruptions we spoke of extend the time periods between eruptions, thereby increasing the share of eruptions with a long quiescent period.

Area of impact from mud volcanoes may go as far as 4–5 kilometers from the eruption center [18]. For instance, a lot of sedimentary rock is thrown out during an eruption. We can expect that the longer an active volcano remains quiescent, the more emissions we can expect on its next eruption. However, the available data does not support this theory. In over two centuries of observations, all scopes and sizes of emissions, from thousands to millions cubic meters, were recorded in Azerbaijan and Kerch-Taman region, after all kinds of quiescent periods (from one year to decades) (figure 2). There are many instances where a mud volcano just a year after its previous eruption threw out way more breccia than after many years of being quiescent, and many cases where things were just the other way around.

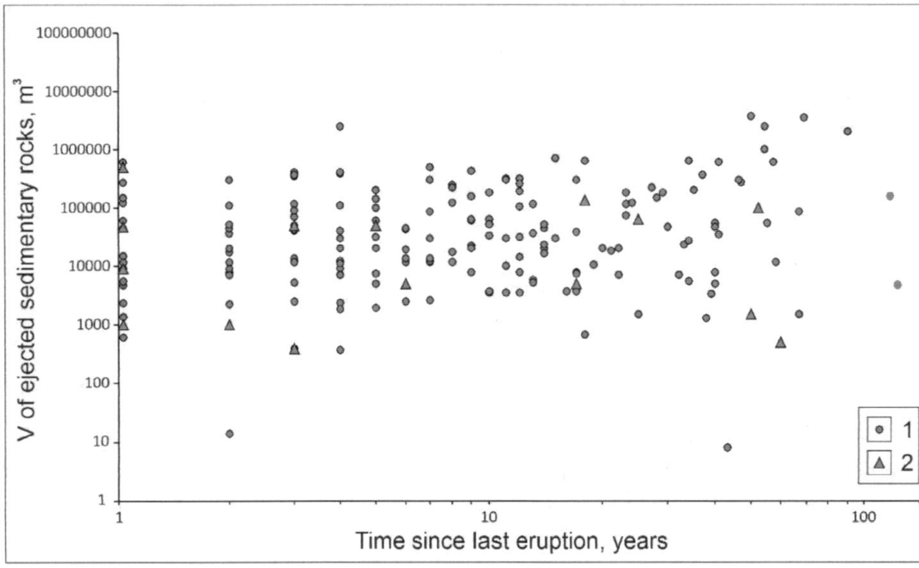

Figure 2. Volumes of breccia ejected during eruptions by mud volcanoes in Azerbaijan (1) and the Kerch-Taman region (2), in relation to the time interval since the last eruption.

Mud volcanoes activity can change due to earthquakes. The fastest responses of mud volcanoes were on the same day as the earthquake that triggered it. At the same time, there are cases where eruptions occur only two or three years after a seismic event [22]. There are quantitative assessments of threshold values of seismic impact on various fluid dynamic systems, including mud volcanoes. The following formula is quite common:

$$\log R_{max} = 2.05(\pm 0.10) + 0.45M$$

where M is the earthquake magnitude, R_{max} is the maximum distance of epicenter from the fluid dynamic system where trigger effect can be observed [21, 23]. The R value in the formula is presented in meters. With this formula we determined cases of Azerbaijan and Kerch-Taman region mud volcanoes potentially responding to earthquakes. Our calculations indicate that in Azerbaijan in 1810–2018 earthquakes should have provoked 572 mud volcanic eruptions. In fact, 35 eruptions were recorded, so only 6 % of cases were a response. The mud volcano was considered to have responded to an earthquake if it erupted within two years of the earthquake. Using longer time intervals is not reasonable, because mud volcanoes typically erupt frequently [15]. Therefore, in a longer time interval, mud volcanoes will erupt for natural reasons, irrespectively of earthquakes. We should point out there were instances of anomalous reaction of mud volcanoes to earthquakes. For instance, after an earthquake of November 25, 2000 (M = 6.4) 19 mud volcanic eruptions was recorded in Azerbaijan, out of the potential 84, therefore, probability of mud volcanoes responding to an earthquake was about 23 %. If we remove this anomaly from our dataset, the probability of Azerbaijan mud volcanoes responding to an earthquake will be only about 3 %. In Kerch-Taman region in 1794–2018, earthquakes should have provoked 31 mud volcanoes eruptions. In fact, three eruptions were recorded, so the response ratio was 10 %.

4. Conclusion

Mud volcanoes are often found in areas of large oil and gas deposits, which poses a threat to oil and gas production facilities. Mud volcanic eruptions are frequent, and the impact area can be as large as several kilometers.

Quantitative analysis of mud volcanoes activity using Azerbaijan and Kerch-Taman region as examples shows that annual average number of eruptions in these regions is 3.62 and 1.45, respectively. The time intervals between eruptions of each volcano vary greatly. However, more than 50 % of all mud volcanoes eruptions in both regions occur within the first decade after the previous eruption.

Analysis of eruptions catalogues for Azerbaijan and Kerch-Taman region shows that the volume of sedimentary rock thrown out by eruptions is not really related to the time the mud volcanoes were quiescent. In over two centuries of observations, all scopes and sizes of emissions, from thousands to millions cubic meters, were recorded in Azerbaijan and Kerch-Taman region, after all kinds of quiescent periods (from a year to decades).

Seismic events may impact mud volcanic activity serving as a trigger. At the same time, this impact is not the main factor of mud volcanoes eruptions; eruptions without any preceding earthquake are far more prevalent. Our calculations indicate that in Azerbaijan and Kerch-Taman region mud volcanoes respond to earthquakes in 6 % and 10 % of all potential cases, respectively. And since the mud volcanic eruptions catalogues are far from complete, we can suppose that the actual probability of response can be twice that much.

References

[1] Aliev A A, Guliev I S, Dadashev F G and Rakhmanov R R 2015 *Atlas of the World's Mud Volcanoes* ed Ak A Alizadeh and A D Ismailzade (Baku: Nafta-Press) p 322
[2] Bagirov E and Lerche I 1999 *Impact of Natural Hazards on Oil and Gas Extraction: the South Caspian Basin* (New York: Springer Science+Business Media) p 353
[3] Bogoyavlensky V I and Bogoyavlensky I V 2017 *Proc. of the State Oceanographic Institute* **218** pp 191–210
[4] Bogoyavlensky V I, Bogoyavlensky I V and Kargina T N 2017 *J. Drilling and Oil* **11** pp 18–27
[5] Glazyrin E A 2017 *Modern Problems of Geology, Geophysics and Geoecology of the North Caucasus* ed I A Kerimov (Moscow: IHNST RAS) **7** chapter 2 pp 39–48
[6] Shnyukov E F and Maslakov N A 2009 *Geol. and Min. Res. of the World Ocean* **2** pp 81–91
[7] Bondur V G and Kuznetsova T V 2015 *Earth and Space Exploration* **4** pp 30–43
[8] Baloglanov E E, Abbasov O R and Akhundov R V 2018 *European J. of Nat. History* **5** pp 12–26
[9] Shnyukov E and Yanko-Hombach V 2020 *Mud Volcanoes of the Black Sea Region and their Environmental Significance* (Cham: Springer Nature Switzerland AG) p 494
[10] Bogatikov O A, Voitov G I, Sobisevich L E, Sobisevich A L, Naumenko-Bondarenko I I, Puzich I N and Korobeinik G S 2003 *Reports of the Academy of Sciences* **390** pp 805–808
[11] Ivanova E A, Masberg I V, Chekalov V P and Gulin M B 2017 *Ecology: Rational Nature Management and Life Safety* ed Shakhanova A V (Maykop: ASU Publishing House) pp 143–146
[12] Izmailov Ya A and Gusakov I N 2017 *Geol. and Min. Res. of the World Ocean* **2** pp 77–80
[13] Olenchenko V V, Shnyukov E F, Gaskova O L, Koh S N, Sokol E V, Bortnikova S B and Eltsov I N 2015 *Reports of the Academy of Sciences* **464** pp 214–219
[14] Shnyukov E F, Sobolevsky Yu V, Gnatenko G I, Naumenko P I and Kutniy V A 1986 *Mud*

Volcanoes of the Kerch-Taman Region: Atlas ed E F Shnyukov (Kiev: Naukova Dumka) p 152
[15] Ustyugov G V and Ershov V V 2019 *VSU Bull. Geol. Series* **4** pp 6–13
[16] Mellors R, Kilb D, Aliyev A, Gasanov A and Yetirmishli G 2007 *J. of Geophys. Res.* **112** B04304
[17] Lazebnaya M A and Lobzhanidze N E 2018 *Issues of Geology and Integrated Development of Natural Resources in East Asia* ed Y V Polyakova (Blagoveshchensk: The Royal Word) **2** pp 9–13
[18] Mironyuk S G 2010 *Georisk* **3** pp 20–28
[19] Wan Z, Zhang J, Lin G, Zhong S, Li Q, Wei J and Sun Y 2021 *Geofluids* **2021** pp 1–16
[20] Shnyukov E F and Kobolev V P 2020 *Geol. and Min. Res. of the World Ocean* **16** pp 49–65
[21] Manga M, Brumm M and Rudolph M L 2009 *Marine and Petroleum Geol.* **26** pp 1785–98
[22] Babayev G, Tibaldi A, Bonali F L and Kadirov F 2014 *Nat. Hazards* **72** pp 789–808
[23] Wang C, Wong A, Dreger D S and Manga M 2006 *Bull. of the Seismological Society of America* **96** pp 355–363

Specificity of the products of Tyatya volcano eruption in 1973

O V Veselov*, A I Kazakov and D N Kozlov

Institute of Marine Geology and Geophysics Far Eastern Branch of Russian Academy of Science, Yuzhno-Sakhalinsk, 693022, Russia

E-mail: *o.veselov@imgg.ru

Abstract. Based on 198 determinations of the complete silicate analysis of Tyatya volcano (Kunashir Island), we compared the data of the latter (1973) and all previous eruptions to reveal their differences and determine the evolution of Tyatya volcanism in the Holocene. The application of the estimation of the thermobaric parameters of the upper mantle magma chamber of the volcano made it possible to determine the temperature decrease in the magma chamber in 1973. An assumption was made about the polycyclic magmatic activity of the Tyatya volcano. The results of statistical processing of data on the thickness of tephra and the size of its fragments from the 1973 eruption are presented. Polynomial regressions of various degrees are applied. Models of tephra distribution based on three-dimensional trend analysis have been constructed. The results of the study are recommended for the creation of information databases on the petrochemistry of eruption products and the distribution of pyroclastics from volcanoes in the Kuril Island arc.

1. Introduction

The objectives of this paper are to compare the results of the analysis of the main oxides of the products of the last and all previous eruptions of Tyatya volcano to determine their evolution, assess the thermobaric characteristics of the products of the volcano's magmatic activity and the main features of the areal distribution of ash-stone deposits of the volcanic eruption in 1973.

Tyatya stratovolcano, one of the most impressive volcanoes of the Great Kuril Ridge, is in the north of Kunashir Island (figure 1). The height of its central cone is 1819 m, the height of the somma is 1485 m, and the base diameter is up to 18 km. At the ring ridge of the caldera, the diameter is up to 2.5 km. The exact age of the volcano's appearance has not yet been determined. However, based on the results of the description of the sections of the volcano's paleosols, in which samples were taken for radiocarbon and geochemical analyzes of pyroclastics, it can be concluded that the volcano began to form at the end of the Late Pleistocene [1]. During its formation, its height probably exceeded 2 km. As a result of one of the powerful explosions, which occurred, probably at the beginning of the fourth century AD, the upper part of the volcano collapsed with the formation of a caldera, in which a cone later grew, reaching the height of more than 1800 m. The last manifestations of volcanic activity of the volcano were observed in 1812 and 1973.

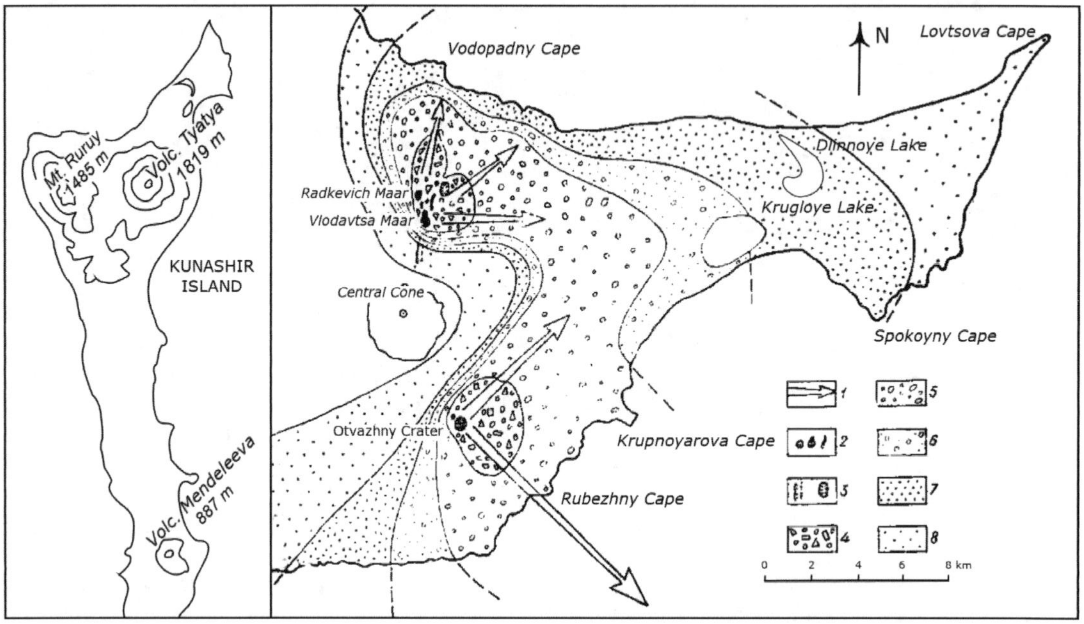

Figure 1. The scheme of the Tyatya volcano, craters, maars, zones of distribution of erupted material [2]. 1 – direction of propagation of erupted material and its predominant distribution; 2 – craters, maars, cracks with explosion funnels; 3 – funnels of the explosion of previous eruptions; 4 – zones of forest damage by stone, slag and hot ash; 5 – zone of ash, lapilli, slag bombs with a layer thickness of more than 30 cm; 6 – zone of ash and coarse-grained material with a layer thickness of 15–30 cm; 7 – zone of ash and fine-grained material with a layer thickness of 3–15 cm; 8 – ash zone with a layer thickness of less than 3 cm.

The 1973 eruption began on July 14 and lasted two weeks. On the first day, explosions occurred on the northern slope of the volcano with the formation of maars and rupture cracks at an altitude below 500 m. Further, powerful explosions formed two craters at an altitude of about 400 m on the southeastern slope of the volcano, from which the main amount of the explosive material was ejected. Two cones of deposits have grown over the funnels – Pogranichnik and Otvazhny. The most powerful ejections of pyroclastics occurred from the Otvazhny. Ash clouds, carried over with north-western winds, reached the Kuril deep-water trench. The mechanism of lateral breakthroughs of magmatic products of the volcano during the 1973 eruption is described in detail in [3]. It occurred at the intersection of ring and radial faults in the body of the volcano. According to publications [2, 4], the total volume of ejected material was at least 200 million m^3. By estimate of later studies [5], the volume of stony-ash deposits on the slopes of Tyatya volcano reached 110 million m^3.

2. Geological and petrological characteristics of the eruption products

According to the literature, at least 198 determinations of the complete silicate analysis of the Tyatya volcano igneous rocks are known, including the explosive products of 1973 account for > 42 % [4, 3, 6, 7, 2, 8, 9, 10, 11, 12].

Analytical data for all groups of determinations of basic oxides conform to the requirements for their geochemistry. This made it possible to compare the petrochemical characteristics of the rocks of the 1973 explosions with all the products of previous eruptions to reveal their differences. Variational diagrams of the dependence of basic oxides on SiO_2 were constructed (figure 2).

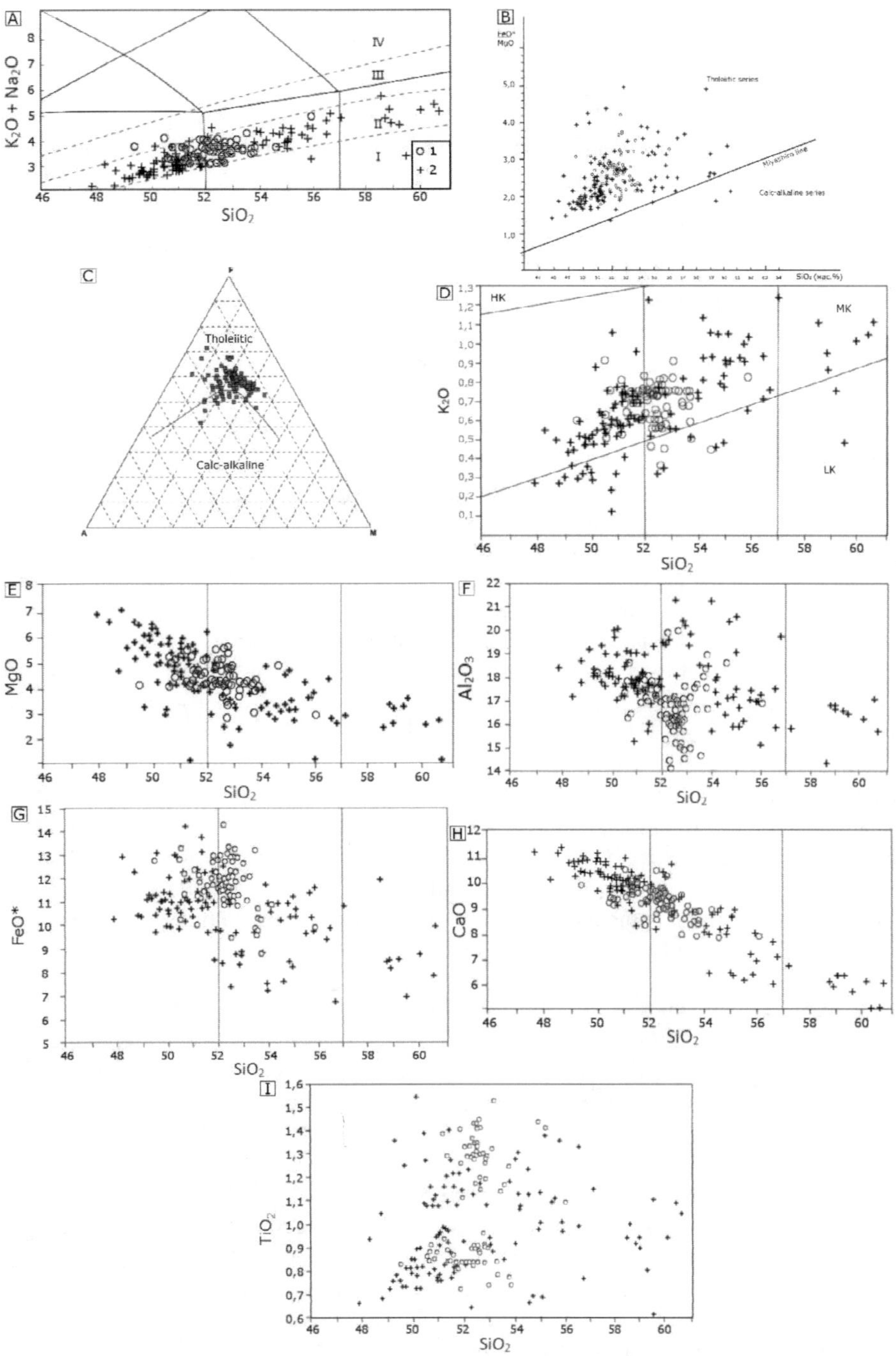

Figure 2. Classification diagrams of the Tyatya volcano igneous rocks (Kunashir Island): 1 – the eruption of 1973, 2 – eruptions until 1973 (content of oxides indicated in wt. %): A – TAS diagram of (SiO_2 – (K_2O + Na_2O)) [13] and the scheme of alkali seriality [10], alkalinity series: I – low, II – medium, III – increased, IV – high; B – diagram SiO_2 – FeO* / MgO according to A Miyashiro's criterion [14]; C – AFM diagram (K_2O + Na_2O – FeO* – MgO) [15] with trends in the differentiation of the tholeiitic series of the volcano; D – diagram SiO_2 – K_2O; E – diagram SiO_2 – MgO; F – diagram SiO_2 – Al_2O_3 ; G – diagram SiO_2 – FeO*; H – diagram SiO_2 - CaO; I – diagram SiO_2 – TiO_2.

In general, the rocks of all Tyatya eruptions are enriched in alumina, iron oxide, and depleted in magnesium oxide in comparison with typical continental basaltoids. The products of the 1973 eruption are mainly represented by basaltic andesites: 65 of 86 samples are basaltic andesites (75 %), while in previously igneous rocks, 67 of 112 samples are basalts, 35 are basaltic andesites, and 10 are andesites. The wide range of composition of rocks erupted before 1973 indicates a prolonged evolution of the volcano's magma chambers. In classification diagram SiO_2 – (Na_2O + K_2O) (wt. %) supplemented series of alkalinity (figure 2A), volcanic rocks are in the field of medium alkalinity under the rules of terminology Commission of Petrographic Committee RAS. As can be seen from the analysis diagram SiO_2 – FeO* / MgO, volcanic rocks of Tyatya volcano eruptions are in tholeiites field according to criterion A Miyashiro [14] (figure 2B). This conclusion is supported by the AFM diagram (figure 2C). The finding of the Tyatya igneous rocks in the tholeiite field indicates that its basaltic magmatism belongs to the Fennerow type of magma differentiation. Basaltoids of this type are characterized by intense accumulation of iron in the residual melt, which, in the opinion of petrologists [16], is associated with a large proportion of plagioclase in the separating phenocrysts. The increased content of ferrous monoxide in the melt (FeO / Fe_2O_3 > 1), as a rule, indicates the reducing conditions for the formation of magmas and the low fugacity of oxygen. Therefore, the gas components H_2 and CO are oxidized to H_2O and CO_2, forming independent systems that leave the melt and do not significantly affect the evolution of basaltic magma. The low water content in the igneous rocks of Tyatya volcano (the average value in 82 % of the samples is 0.08 wt. %) confirms this conclusion. From an analysis of the diagrams SiO_2 – (Na_2O + K_2O), SiO_2 – FeO* / MgO, SiO_2 – MgO, SiO_2 – CaO (figure 2A, B, E, H), which clearly demonstrates the solid concentration of figurative points, it follows that the relatively constant thermobaric conditions of the formation of magmas of the Tyatya volcano and the unity of the genesis of the rocks of the volcano during the activity of its upper mantle chamber in the Holocene. However, the geochemistry of the products of the previous and the last eruptions differs markedly, which can be seen in several Harker diagrams (figure 2B, F, G, I). It follows from the analysis of these diagrams that the rocks of the 1973 eruption are enriched in iron, titanium, reduced content of aluminum and to a small extent in calcium in comparison with the basaltoids of previous eruptions, which suggests a decrease in the temperature parameters of the main magma chamber of Tyatya volcano. The basalts and basaltic andesites of the early stages of volcanic activity have a higher alumina composition, which indicates more high temperatures of magmas of the early stages of magmatic activity of the volcano (figure 2F).

3. Thermobaric characteristics of the Tyatya volcano magma chambers

The almost dry composition of the Tyatya volcano igneous rocks makes it possible to use the formulas obtained in experiments with dry samples of ultrabasic and basic rocks to determine the thermobaric parameters of its magmatic melts, these results confirm the conclusion that the MgO content in the magmatic melt, relative to most other rock-forming oxides, most reliably correlates with temperature and pressure (depth) of the magma chamber.

In [17], the relationship between the conditions of existence of magmas before their crystallization and the petrochemical characteristics of igneous rocks was studied through the ratio of various petrogenic elements, in particular, through the coefficient ω = MgO / (CaO + Na_2O + K_2O) (mol. %). Based on a statistically reliable data set of basic oxides of Cenozoic volcanic rocks, a close relationship between the ω ratio, magma composition and magma changes has been confirmed. For this ratio, a regression equation was derived, which determines the depth of the magma chamber (km):

$$H = 24.17 + 22.28\,\omega. \quad (1)$$

Based on the improvement of laboratory methods for studying the separation of the melt from the substrate, which served as natural peridotites, equations were obtained that relate the melting temperature of the upper mantle substance and the concentration of SiO_2 and MgO in them. How it was indicated below, these regression equations derived from experimental data indicate that the MgO content in the magma melt relative to other rock-forming oxides most reliably correlates with the temperature and pressure (depth) of the magma chamber, if the melt is in equilibrium with olivine. Representation in several publications [18, 19, 20, 21, 22] of relatively simple empirical relations between SiO_2, MgO, alkalis and thermobaric parameters magma chambers in subcrustal of the upper mantle volcanic zones allows to solve the problem of calculating the temperature and pressure characteristics of the magma chamber before the eruption of Tyatya volcano to assess the evolution of the temperature regime of its magma chambers. These regression equations are:

$$T_1(°C) = 1091.7 + 21.5\ MgO\ (wt.\ \%); \tag{2}$$

$$T_2(°C) = 1019.8 + 1384.2\ MgO/(MgO + SiO_2)\ (wt.\ \%). \tag{3}$$

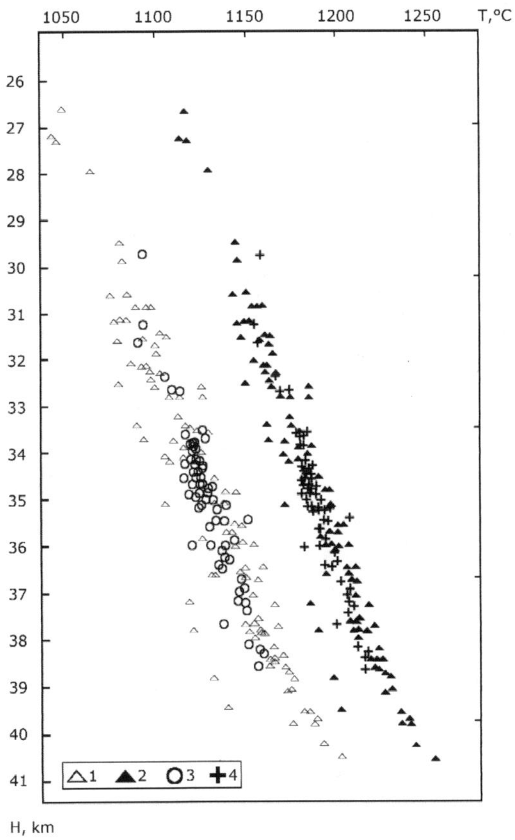

Figure 3. Temperatures and depths of formation of the eruption products in magma chambers of Tyatya volcano. 1 – products of eruptions before 1973, calculated by the formula (3); 2 – products of eruptions before 1973, calculated by the formula (2); 3 – products of the 1973 eruption, calculated by the formula (3); 4 – products of the 1973 eruption, calculated by the formula (2).

Calculations of the thermodynamic conditions for the formation of magma chambers were performed using the above equations and based on a statistically reliable amount of data on the total silicate composition of the Tyatya volcano igneous rocks. The results are shown in figure 3.

In the most of estimates of the thermodynamic state of magma chambers of ancient eruptions of Tyatya volcano, the maximum melt temperatures of 1150 and 1200 °C were at the depths of 31.0 and 36.4 km, respectively. In the magma focus of the 1973 eruption, the maximum temperatures of 1150 and 1200 °C are determined at the depths of 29.0 and 37.3 km, then during this event the magma temperatures has decreased below 34 km and increased in the earth's crust. The temperature at the mouths of the 1973 breakthroughs reached 930 °C.

In real time the temperatures of the upper mantle chambers of Tyatya volcano lower than those of the magma chambers of many previous eruptions. This conclusion confirms the assumption made in [23] that the ancient rocks of Tyatya volcano had a higher degree of melting. Nevertheless, from the results of the calculations, it follows that there was a period when the temperatures in the magma chamber were even lower than the existing ones. This is also evidenced by the presence of a small amount of andesites among the igneous rocks. It should be assumed that the dynamics of the thermobaric regime of the upper mantle magma chamber was cyclical in the Holocene.

The absence of clear boundaries between basalts and rocks of intermediate silicic acidity indicates their close relationship and unity of origin. In [24, and others], thermodynamic conditions of the liquidus-solidus area are presented at different depth intervals, in particular, at pressures up to 1210 MPa, which corresponds to a depth of about 40 km. For these pressures, the difference between liquidus and solidus is in the range of 45–50 °C. According to our calculations of the thermodynamic characteristics of the Tyatya volcano magma chambers using formulas (1) – (3), the difference between the two thermograms is within 52–60 °C. Noting the closeness of these thermograms with the diagrams in [24, et al.], we can assume that the temperatures calculated by formula (2) correspond to the liquidus boundary of the upper mantle substance, and the temperatures calculated by formula (3) correspond to the solidus boundary of the volcano's igneous rocks.

The tendency of rapid crystallization of the mineral phases of the melt in a narrow liquidus-solidus temperature range at the depths less than 40 km does not promote to the differentiation of the melt, forming a basalt and basaltic andesite composition of igneous rocks, which indicates the formation of all rocks from one permanent magma source. We emphasize that the basaltic andesites of 1973 were formed in a magma focus with temperatures of 1210–1150 °C. The main concentration was in the focus with temperatures of 1190–1170 °C, corresponding to the depths of 36–33 km.

Comparison of the temperature model of the volcano with the deep seismic section of the Kunashir Island [25, 26] shows that the region of magma chambers of Tyatya volcano coincides with the zone of absence of converted waves of earthquakes.

4. Application of statistical methods to the estimation of the tephra distribution of the 1973 eruption

Turning to a more detailed analysis of the distribution of the products of the explosive eruption in 1973, we believe that the scheme of the intensity of the fallout of stone, slag and ash material given in [2] can be clarified by applying the methods of statistical analysis of the processing of samples collected even several decades after eruption. In 2006, employees of the IMGG FEB RAS D.N. Kozlov and A.B. Belousov measured the thickness of tephra and the maximum sizes of its fragments on the northern and southeastern slopes of Tyatya volcano at 53 points [27, 5] (table). At the same time, there were 45 measurements of the tephra thickness and 34 measurements of the maximum fragment sizes.

Table 1. The results of measuring the thickness of tephra and the size of its fragments of the eruption of the Tyatya volcano in 1973. m is the thickness of the pyroclastic deposits, d is the size of the tephra fragments, a dash means the absence of data.

Point number	m, cm	d, cm	Point number	m, cm	d, cm	Point number	m, cm	d, cm	Point number	m, cm	d, cm
1	3	4	15	10	-	29	-	103	43	40	18
2	9	9	16	20	10	30	-	420	44	11	-
3	10	13	17	22	12	31	-	150	45	10	-
4	13	14	18	15	9	32	150	30	46	7	1
5	15	16	19	7	-	33	70	26	47	5	-
6	20	18	20	9	-	34	3	-	48	6	-
7	20	12	21	7	-	35	6	-	49	-	1
8	19	20	22	5	-	36	3	-	50	4	4
9	-	25	23	28	25	37	5	-	51	4	-
10	66	31	24	40	28	38	6	-	52	3	-
11	50	22	25	40	35	39	12	-	53	3	-
12	40	19	26	-	42	40	-	15			
13	30	12	27	-	44	41	20	-			
14	35	11	28	-	48	42	30	28			

In publications [27, 5], schemes of tephra distribution in isopachs and isopleths (lines of equal thickness of pyroclastic deposits and equal size of tephra fragments, respectively) are presented, as well as features of changes in the sequence of deposits. Within the framework of this study, two- and three-dimensional regression mathematical models of the distribution of tephra thickness have been built.

For the entire set of measurements, the coefficient of linear correlation between the thickness of the pyroclastic deposits and the size of tephra fragments was 65.9 %. A joint analysis of the samples taken on the southeastern slope of the volcano and in the maar eruption zone shown that the residual variance amount 56.5 %. It excludes the possibility of linear approximation of the values of one quantity based on another. Considering the difference in the mechanism and intensity of the eruption on the northern and southeastern slopes, the relationship between the thickness of the pyroclastic deposits and the size of tephra fragments was studied separately for the products of the Otvazhny crater eruption and the tephra of the Vlodavets and Radkevich maars. Two-dimensional least squares regression models are constructed. The correlation coefficient in the first case was 89.9 %, in the second 98.6 %. These indicators make it possible to use the obtained linear models for approximating the thickness of the pyroclastic deposits at the points where only the size of its fragments was measured, as well as for inverse approximation. We also constructed polynomial regression equations for the two studied factors of degrees from 2 to 6 but none of them provided an approximation quality higher than 60 %, exponential curve interpolation shown a result of 64 %. In addition, equations in polar coordinates are obtained for isopachs and isopleths constructed in [5].

Three-dimensional trend analysis allowed to construct tephra scatter models. To describe the distribution of the thickness of the pyroclastic deposits, a logarithmic regression model is effective, the equation of which is (figure 4):

$$\bar{m} = 65.5073 - 23.1728 \ln \sqrt{(x - 4.275)^2 + (y + 4.195)^2},$$

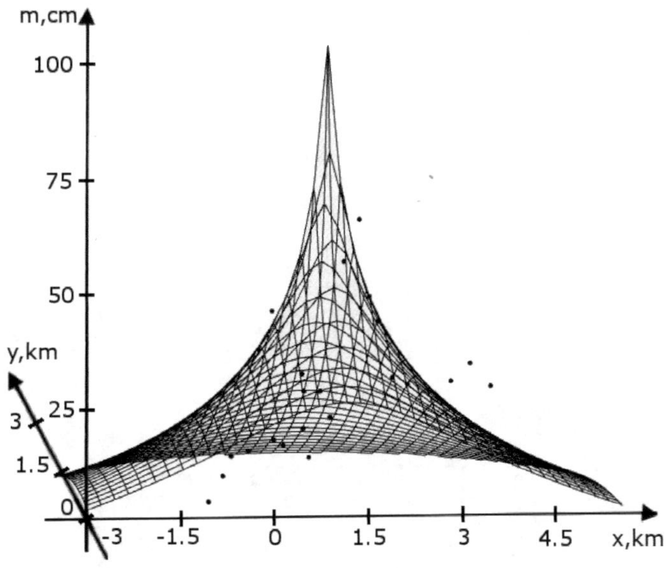

Figure 4. Logarithmic regression model of the thickness of the pyroclastic deposits erupted in 1973 (m) as of 2006. The abscissa axis is to the east, the ordinate is to the north.

Analysis of variance allowed us to evaluate the quality of approximation of the tephra distribution pattern represented by the trend surface with a multiple correlation coefficient equal to 71 %. A logarithmic regression model was also obtained for the size distribution of tephra fragments.

In [2], a model of distribution of the thickness of the igneous material from the Otvazhny crater to the southeast at up to 3 km is presented, and in [27] a curve was constructed for the same process but already up to 10 km. Over 34 years, as a result of erosion and consolidation on the southeastern slope of the volcano, the pyroclastic deposits thickness in 1973 decreased from the Otvazhny crater to the coast according to the dependence, which can be determined by the exponential relationship: $\Delta z = ax^b e^{cx}$, where Δz is the change in the pyroclastic deposits thickness,

x is the distance from the Otvazhny crater ridge to the coast; a, b, c – constants ($a = 0.5375, b = 0.61974, c = -0.001856$) (figure 5).

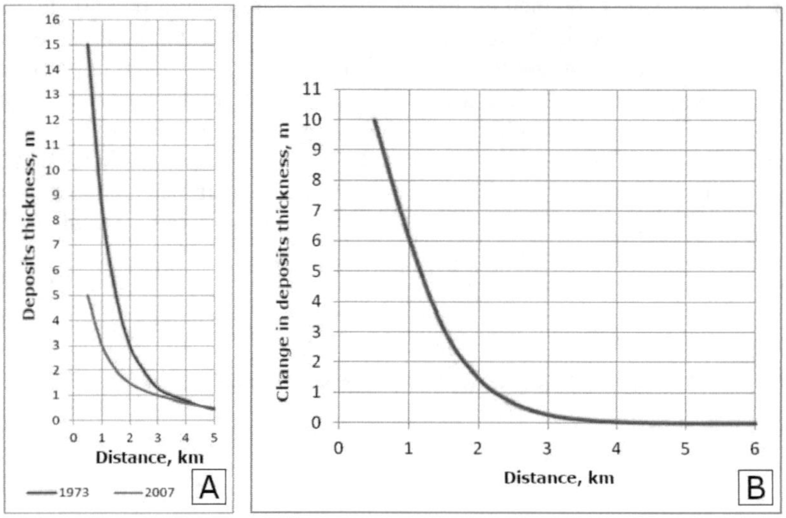

Figure 5. Estimation of the pyroclastic deposits thickness and the rate of its decrease based on the results of measurements in 1973 and 2007 on the southeastern slope of Tyatya volcano 120° in azimuth. Mark 0 corresponds to the ridge of Otvazhny crater: A – deposits thickness in 1973 and 2007; B – change in deposits thickness in 1973 over 34 years.

The resulting exponential relationship made it possible to assess the influence of erosion and consolidation of deposits on the change in their thickness. The rate of decrease in the pyroclastic deposits thickness near the Otvazhny crater in the most intense distribution of igneous material averaged 29.4 cm/year.

5. Conclusion

1. All the products of the Tyatya volcano eruptions are in the tholeiite field. The products of the 1973 eruption mainly characterized by basaltic andesite composition (75 %), are enriched in FeO, partly TiO_2, depleted in alumina, which indicates a change in the thermobaric state of the upper mantle magma chamber.
2. As a result of the evolution of the magmatic activity of Tyatya volcano, the temperature of the upper mantle part of the volcano's magma focus decreased by 1973, while in the upper crustal part of the volcano it increased. The identified zone of the upper mantle magma chamber coincides with the zone of absence of converted waves of earthquakes.
3. The use of two- and three-dimensional regression mathematical models is an effective method for describing the distribution of the pyroclastic deposits thickness erupted in 1973 from two eruption centers. These methods are recommended to be applied to assess the distribution of products of explosive volcanic eruptions in the Kuril Island arc.

References

[1] Rybin A V, Razzhigaeva N G, Guryanov V B, Degterev A V and Chibisova M V 2015 *Proc. Int. Sci. Conf. Geodynamic Processes and Natural Hazards. Lessons of Neftegorsk (Yuzhno-Sakhalinsk)* vol 2 (Vladivostok: Dalnauka) pp 271–273

[2] Markhinin E K et al. 1974 *Geology and Geophysics* **10** pp 20–31

[3] Abdurakhmanov A I and Fedorchenko V I 1976 *Proc. SakhNII* Peculiarities of mechanism of the Tyatya volcano eruption in July 1973 *Volcanism of the Kuril-Kamchatkan Region and Sakhalin Island* vol 48 (Yuzhno-Sakhalinsk) pp 44–52

[4] Maleev E F 1975 *Bull. of the Volcanological Stations* **51** pp 19–27
[5] Belousov A B, Belousova M G and Kozlov D N 2017 *Volcanology and Seismology* **4** pp 48–56
[6] Goosev G M, Zhukova E N, Nepeina L A, Pavlov A L and Slobodskoiy P M 1976 *Proc. USSR Academy of Science* vol 227 pp 1438–41
[7] Radkevich E A, Radkevich R O, Markhinin E K, Dolganova P S and Shutova G S 1977 *Geology and Geophysics* **2** pp 62–70
[8] Gushchin A V 1978 *Petrological Research of the Island Arcs Basalts* (Moscow: IPE AS USSR) pp 132–156
[9] Frolova T I, Burikova I A, Gushchin A V, Frolov V T and Syvorotkin V L 1985 *Origin of the Island Arcs Volcanic Series* (Moscow: Nedra)
[10] Fedorchenko V I, Abdurakhmanov A I and Rodionova R I 1989 *Volcanism of the Kuril Island Arc: Geology and Petrogenesis* (Moscow: Nauka)
[11] Martynov Yu A, Khanchuk A I, Kimura J-I, Rybin A V and Martynov A Yu 2010 *Petrology* **18** 5 pp 512–535
[12] Nakagawa M et al. 2002 *The Island Arc* **11** 4 pp 236–254
[13] Le Bas M J, Le Maitre R W, Streckeisen A and Zanettin B 1986 *J. Petrol.* **27** 3 pp 745-750
[14] Miyashiro A 1974 *Amer. Jour. Sci.* **274** 4 pp 321–355
[15] Irvine T N and Baragar W R A 1971 *Canadian J. Earth Sci.* **8** 5 pp 523–548
[16] Marakushev A A and Bezmen N I 1992 *Mineralogical and Petrological Criterions of Ore Content of Igneous Rocks* (Moscow: Nedra)
[17] Pecherskiy D M, Bagin V I, Brodskaya S Yu and Sharonova Z V 1975 *Magnetism and Formation Conditions of Igneous Rocks* (Moscow: Nauka)
[18] Rasskazov S V, Demonterova E I, Ivanov A V, Brandt I S and Brandt S B 2007 *Evolution of the Late Cenozoic Magmatism at the Boundary of the Tuva-Mongol Massif (East Tuva)* (Irkutsk: IEC SB RAS)
[19] Albarede F 1992 *J. Geophys. Res.* **9** B 7 pp 10997–11009
[20] Faloon T J, Green D H, Danyushevsky L V and Faul U H 1999 *J. Petrol.* **40** 9 pp 1343–75
[21] Sugawara T 2000 *J. Geophys. Res.* **105** B 4 pp 8457–72
[22] Herzberg C, Asimow P D, Arndt N, Niu Y, Lesher C M, Fitton J G, Cheadle M J and Saunders A D 2007 *Geochemistry, Geophysics, Geosystems* **8** 2 pp 1–3
[23] Martynov Yu A, Martynov A Yu, Chashchin A A and Rybin A V 2005 *Russian J. of Pacific Geology* **24** 3 pp 22–31
[24] Pugin V A and Khitarov N I 1978 *Experimental Petrology of the Deep-Seated Magnetism* (Moscow: Nauka)
[25] Zlobin T K, Fedorchenko V I, Petrov A V and Nemchenko G S 1982 *Russian J. of Pacific Geology* **1** pp 92–100
[26] Zlobin T K 1987 *Structure of the Earth's Crust and Upper Mantle of the Kuril Island Arc (According to the Seismological Data)* (Vladivostok: FEB AS USSR)
[27] Kozlov D N, Belousov A B and Belousova M G 2008 *Proc. 2nd Sakhalin Youth Sci. School Natural Hazards: Study, Monitoring, Forecast (Yuzhno-Sakhalinsk)* (Yuzhno-Sakhalinsk) pp 264–267

Temperature regime of the Daginsky thermal springs (Sakhalin Island) during their reconstruction in 2019-2021

R V Zharkov

Institute of Marine Geology and Geophysics Far Eastern Branch of Russian Academy of Science, Yuzhno-Sakhalinsk, 693022, Russia

E-mail: rafael_zharkov@mail.ru

Abstract. The article presents the results of studies of the temperature regime of the Daginsky thermal springs. Daginsky thermal springs are the most famous and visited on Sakhalin Island (Russia). In the fall of 2019, their reconstruction and arrangement of the surrounding landscape began. As a result of the reconstruction of the main thermal springs, their hydrological and temperature regime was violated, the water temperature dropped. In 2020, in the springs most popular with visitors (Aleksandrovsky, Tsentralny, Mechta), the temperature became several degrees lower, which led to the termination of their use. In 2021, after the completion of the design of the spring bath and the surrounding area, the temperatures increased. In April 2021, the temperature of the Patriot spring was 40.5 °C, the Partizan spring 38–39 °C, the Central spring 36–39 °C, the Aleksandrovsky spring 34.6–36.0 °C, the Mechta spring 34.4 °C (at the bottom up to 45.5 °C), Molodost spring 35 °C (at the bottom up to 37–41.3 °C).

1. Introduction

Daginsky deposit of thermal mineral waters is located in the north-east of Sakhalin Island, on the coast of the Nyisky Gulf of the Sea of Okhotsk. Active exploration of the Daginsky deposit began in the 1950s by V.V. Ivanov [1], N.D. Tsitenko [2], M.A. Stein [3, 4], B.T. Komissarenko [5], operational reserves of thermal mineral waters were approved in 1990–1991 [6]. From the 1990s to the early 2000s hydrothermal fluids have practically not been studied. Modern periodic field studies of the Daginsky thermal springs have begun since 2004 [7–11]. In all studies, the main role was assigned to the conditions for the formation of hydrothermal vents, the variability of physical and chemical indicators, and the assessment of the prospects for the use of hydrothermal vents in balneotherapy. The reconstruction of the springs and the improvement of the surrounding territory have begun in the fall of 2019 by the order of the Government of the Sakhalin Region. Six springs were selected for reconstruction (figure 1). The baths of the springs were made of steel (figure 2), they all have the same dimensions (193×193 cm), while the morphometric parameters and the flow rate of certain thermal springs were not taken into account. The bath is mounted on a platform with pontoons measuring 8×8 m; the top is covered with a dome about 6 m in diameter and 3.2 m high, made of an aluminum frame and plexiglass (figure 3).

During the reconstruction of the main thermal springs, their hydrological regime was changed, the water temperature decreased. The purpose of our research is to reveal the features of the change in the temperature regime of the Daginsky thermal springs during the period of their reconstruction. For this purpose, field work was carried out with spring temperature measurements (January 21–22, 2020;

May 19–21, 2020; October 5–6, 2020; April 7–8, 2021). The data obtained were compared with the water temperatures in the springs before their reconstruction.

2. Methodology

The temperature of the thermal springs was measured with electronic thermometers with a KTKhA 01.02R-T310 thermal converter, the operating temperature range of this thermocouple is from -200 °C to +1350 °C, the measurement accuracy is 0.1 °C. When determining the maximum temperature, the duration of the measurements was several minutes. The SAT SDS Hotfind-LXS thermal imager was used to survey infrared (thermal) imaging of the areas with outcrops of hydrothermal vents. The range of measured temperatures of the thermal imager is from -20 °C to + 1000 °C, the accuracy is ± 2 °C, the angular field is 24×18 °, the spatial resolution is 1.1 mrad, and the resolution of the active matrix is 384×288 points. We also used the CAT S61 (Flir) portable thermal imager.

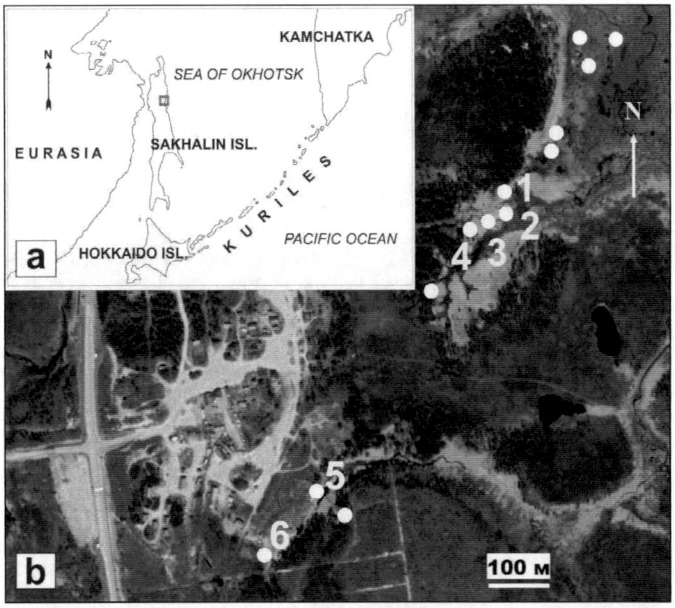

Figure 1. The geographical position of the Daginsky thermal springs (a); reconstructed thermal springs (b): 1 – Patriot, 2 – Partizan, 3 – Central, 4 – Aleksandrovsky, 5 – Mechta, 6 – Molodost (satellite image from Google Earth).

Figure 2. The bath of the spring.

Figure 3. Dome of the Central thermal spring.

3. Results

There are three areas of thermal mineral waters discharge at the field: North, Central and South. The springs of the Northern section discharge within the littoral zone of the Nyisky Bay and are flooded with sea waters and the Nelbuta River at the high tide. The springs of the Northern section were not included in the reconstruction project. Four main thermal springs have been reconstructed in the Central section of the Daginsky deposit of thermal mineral waters: Patriot, Partizan, Central, Aleksandrovsky. The hydrothermal fluids of these springs have similar physical and chemical properties. Thermal waters are methane (free gases contain 77–83 % methane), sodium chloride, pH 6.8–7.8, TDS 1.3–2.7 g/l.

The ***Patriot spring*** had an average temperature of 40 °C and was very popular with visitors before the reconstruction. After the reconstruction, the water temperature in January 2020 was within the range of 34.9–36.5 °C and the spring was practically not used. In May 2020, the average water temperature was 38.3–39.4 °C, and the spring was rarely used by visitors. At the beginning of October 2020 and in April 2021, the temperature has reached 40.5 °C, the spring is actively used by visitors. The water in the bath is relatively clean, the depth to the metal bottom in April 2021 was 88 cm.

The ***Partizan spring*** had an average water temperature of 39.0–41.5 °C before the reconstruction. After the reconstruction, the water temperature in January 2020 was 38.0–38.5 °C. The average temperature in May 2020 reached 38.5 °C, in October 39.0 °C. In April 2021, the average temperature was 38–39 °C, the water in the bath was relatively clean, the depth to the metal bottom was 104 cm. Today it is the most visited spring.

The ***Central spring*** (Dikiy, according to [6]; now it was erroneously designated as Pioneer) warmed up to 38–40 °C on average before the reconstruction. In January 2020, the temperature was 30.0–31.8 °C. In May 2020, the average temperature of the water in the bath was 25.2–27.5 °C, in October (after the work had been carried out by the contractor to reduce the flow of groundwater) it reached 39.1 °C. In April 2021, the average temperature was about 36 °C, at a depth it reached 39 °C. The spring was practically not visited, the water in the bath was relatively clean, the depth to the metal bottom was 100 cm.

The ***Aleksandrovsky spring*** was the hottest and most visited spring before the reconstruction. The average temperature in different years and seasons was within the range of 45 °C, in March 2017 it was 40.0 °C, in October 2019 it was 42.0 °C. After the reconstruction, the water temperature in January 2020 was in the range of 26.5–27.5 °C. In May 2020, the average temperature was 33.6–34.1 °C (figure 4), in October (after the work had been carried out by the contractor to reduce the flow of groundwater) it reached 39.3 °C. In April 2021, the water temperature was 34.6–36.0 °C, the depth to the metal bottom was 96 cm. The water in the bath is peaty, silty sediments are observed on horizontal surfaces, metaphyton mats and foam float on the surface. When mounting the platform of this spring, a surface turf layer of peat with a thickness of 10–50 cm was removed, there is an open peatland with cold surface water around the spring now; the temperature of the cold spring on the south side of the platform in April 2021 was 4.5 °C. Due to the low temperature and turbidity of the water, the spring is currently not used.

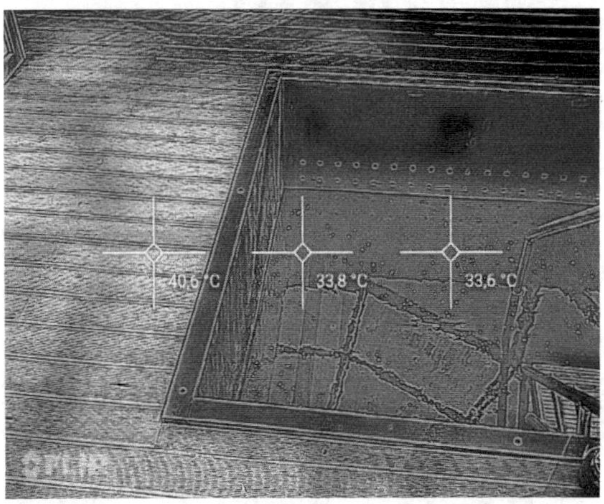

Figure 4. Infrared image of the Aleksandrovsky spring (21.05.2020).

In the southern section of the Daginsky deposit of thermal mineral waters, 2 main thermal springs were reconstructed: Mechta and Molodost. The spring fluids have similar physical and chemical properties. Thermal waters are methane (free gases contain up to 91 % methane), sodium chloride, pH 6.8–8.1, TDS 2.5–2.6 g/l.

The ***Mechta spring*** had an average water temperature in the range of 36–39 °C before the reconstruction, while temperatures of 40–45 °C were recorded in the muddy bottom of the spring. Despite the extremely unsightly appearance, the spring has always been very popular. After reconstruction in January 2020, the average water temperature in the spring was 26.5–27.7 °C, in May 2020 it reached about 32 °C in the near-surface layer, in October (after the contractor's work on changing the design of the hot tub and reducing the flow of groundwater) 35.5 °C in the near-surface layer and up to 39.5 °C at the bottom. In April 2021 (figure 5), the water temperature was 34.4 °C at a depth of 50 cm, the temperature of the sandy bottom (the metal perforated bottom of the bath was removed) reached 45.5 °C in some places, the average depth of the spring was 114 cm. There are practically no visitors at the spring.

The ***Molodost spring*** was also very popular with visitors before the reconstruction. The water temperature was within the range of 37–39 °C. After the reconstruction in January 2020, the temperature was in the range of 30.1–32.2 °C, in May the average temperature in the hot tub was 35.2–36.1 °C, in October (after the contractor's work on changing the design of the hot tub and reducing the flow of groundwater) was 35.4–37.0 °C. In April 2021, the average water temperature was 35 °C (figure 6), the temperature of the sandy bottom (the metal perforated bottom of the bath was removed) reached 37–41.3 °C in some places, the average depth of the spring was 111 cm. There are practically no visitors at the spring.

Figure 5. The Mechta spring (07.04.2021).

Figure 6. The Molodost spring (07.04.2021).

4. Conclusion

Comprehensive studies carried out in 2019–2021 showed changes in the temperature regime for several springs in comparison [7, 11] with the period before their arrangement had been started: the average water temperature became several degrees lower due to an incorrect design solution and unskilled construction work. It was found that, as a result of the arrangement, the pits of the springs were significantly expanded, the participation of cold groundwater increased, therefore, the hydrothermal fluids entering them with a certain flow rate do not heat the increased volume of water to the previous temperatures. The attempts made by the builders to block the flow of cold groundwater into the trenches of the springs gave a small temporary effect, but at the same time worsened the water exchange in the springs. Today, out of 6 equipped springs, only 2–3 springs are actively visited, depending on the season of the year. For a complete restoration of the temperature regime of the Daginsky springs, it is necessary to carry out their reconstruction based on a scientific approach and modern data. It is necessary to continue routine monitoring of the temperature of the springs. Regular observations will reveal changes and predict the further development of the situation.

References

[1] Ivanov V V 1954 *Sakhalin resort resources and the prospects for their therapeutic use: report of the integrated squad of the Sakhalin expedition* (Moscow) p 265
[2] Tsitenko N D 1961 Waters of Daginsky hot springs on Sakhalin Island *Trudy VNIGRI* **181** pp 203–213
[3] Shteyn M A 1962 Determination of parameters and depths of thermal groundwater *Trudy SakhKNII* **12** pp 162–165
[4] Shteyn M A 1967 Thermal waters of Sakhalin and issues of their use *Regional geothermy and distribution of thermal waters in the USSR* (Moscow: Nauka) pp 274–280
[5] Komissarenko B T 1964 *Mineral springs and therapeutic mud of Sakhalin and Kuriles* (Yuzhno-Sakhalinsk: Sakhalinskoye knizhnoye izdatel'stvo) p 115
[6] Zavadskiy I G 1991 Exploration work at the Daginsky deposit of thermal waters in the Nogliki District: a report for 1990-1991 (YUzhno-Sakhalinsk: Sakhalingeologiya) p 218
[7] Zharkov R V 2008 *Proc. 2nd Sakhalin Youth Sci. School Natural Hazards: Study, Monitoring, Forecast (Yuzhno-Sakhalinsk)* Daginsky deposit of thermomineral waters in the north of Sakhalin Island (Yuzhno-Sakhalinsk: IMGiG DVO RAN) pp 285–290
[8] Shakirov R B, Syrbu N S and Obzhirov A I 2012 Isotope-gasgeochemical features of methane and carbon dioxid distribution on Sakhalin Island and adjacent shelf of the Sea of Okhotsk *Bull. of Kamchatka Regional Association "Educational – Scientific Center". Earth Sciences* **20** 2 pp 100–113
[9] Chelnokov G A, Bragin I V and Kharitonova N A 2018 Geochemistry mineral waters and gases of the Sakhalin Island (Far East of Russia) *J. of Hydrology* 559 pp 942–953
[10] Pavlova V Yu, Zharkov R V 2018 GPR surveys of the discharge zone of the Daginsky hydrothermal system (Sakhalin Island) *Geosystems of Transition Zones* **2** 4 pp 323–331
[11] Zharkov R V 2018 Modern physico-chemical features of the thermomineral waters of the Daginsky deposit (Sakhalin Island) *Monitoring. Science and technology* **4** 37 pp 6–11

The shores of the contact zone of the Subarctic and moderately cold seas

V V Afanasev* **and A V Uba**

Institute of Marine Geology and Geophysics Far Eastern Branch of Russian Academy of Science, Yuzhno-Sakhalinsk, 693022, Russia

E-mail: *vvasand@mail.ru

Abstract. Longitude sectorality and latitudinal zonality of morpholithogenesis on the coast of Sakhalin is the result of the unique geographical position of the island, which is a zone of mutual influences and interactions of neighboring geosystems of the subarctic Sea of Okhotsk and the moderately cold northern part of the Sea of Japan – a contact geographical structure. Based on the digitizing of maps of morphogenetic types of shores, modern morpholithodynamic settings and lithological complexes of the coast, geomorphological and morphodynamic differences of the shores of Sakhalin Island are shown, the eastern shores of which have a subarctic appearance, and the western shores are washed by the temperate sea.

1. Introduction

The geosystems boundaries are relatively mobile zones of mutual influences and interactions of neighboring geosystems – contact geographical structures [1, 2, 3]. At the same time, most of the geomorphological boundaries can be considered as a kind of contact zones distinguished in physical geography, and the vast majority of terrestrial processes and phenomena are confined to them [4, 5]. Earlier we noted that the morphoclimatic position (geospatial parameters and related differences in morpholithodynamic parameters) of Sakhalin Island and the island-arc systems of the Northern Pacific allow us to consider them as contact zones of different types of seas, as well as seas and oceans [6].

The Sakhalin shores of the moderately cold northern part of the Sea of Japan and the subarctic Sea of Okhotsk are located 30–140 km from each other, and the ocean and sea shores of large islands in island-arc systems are separated by only a few kilometers. Thus, the morpholithodynamic characteristics of the shores of the contact zones, along with the peculiarities of hydrodynamics and thermal regime of the separated water areas, also determine the morphoclimatic parameters of the contact zone itself.

2. Methods and results

The presented work implements the methodology of morpholithodynamic studies based on both traditional and modern methods of obtaining and analyzing geospatial and geological-geomorphological information. Analysis of maps of morphogenetic types of the shores of the Northern Pacific, modern morpholithodynamic settings and lithological complexes of the coast of O. Sakhalin Island was carried out in the Quantum GIS geographic information system, the contours correspond to the accuracy of the original maps [7, 8]. To obtain the length of the coastline of each type, the Add

IOP Conf. Series: Earth and Environmental Science **946** (2021) 012033 doi:10.1088/1755-1315/946/1/012033

Geometry Attributes tool with the Ellipse Calculation option was used. Calculations were performed on the ellipse WGS84 EPSG:7030.

The main features of morpholithogenesis specificity from the marine and oceanic side of the contact zones were established by analyzing maps of morphogenetic types of the shores of the Northern Pacific (figure 1).

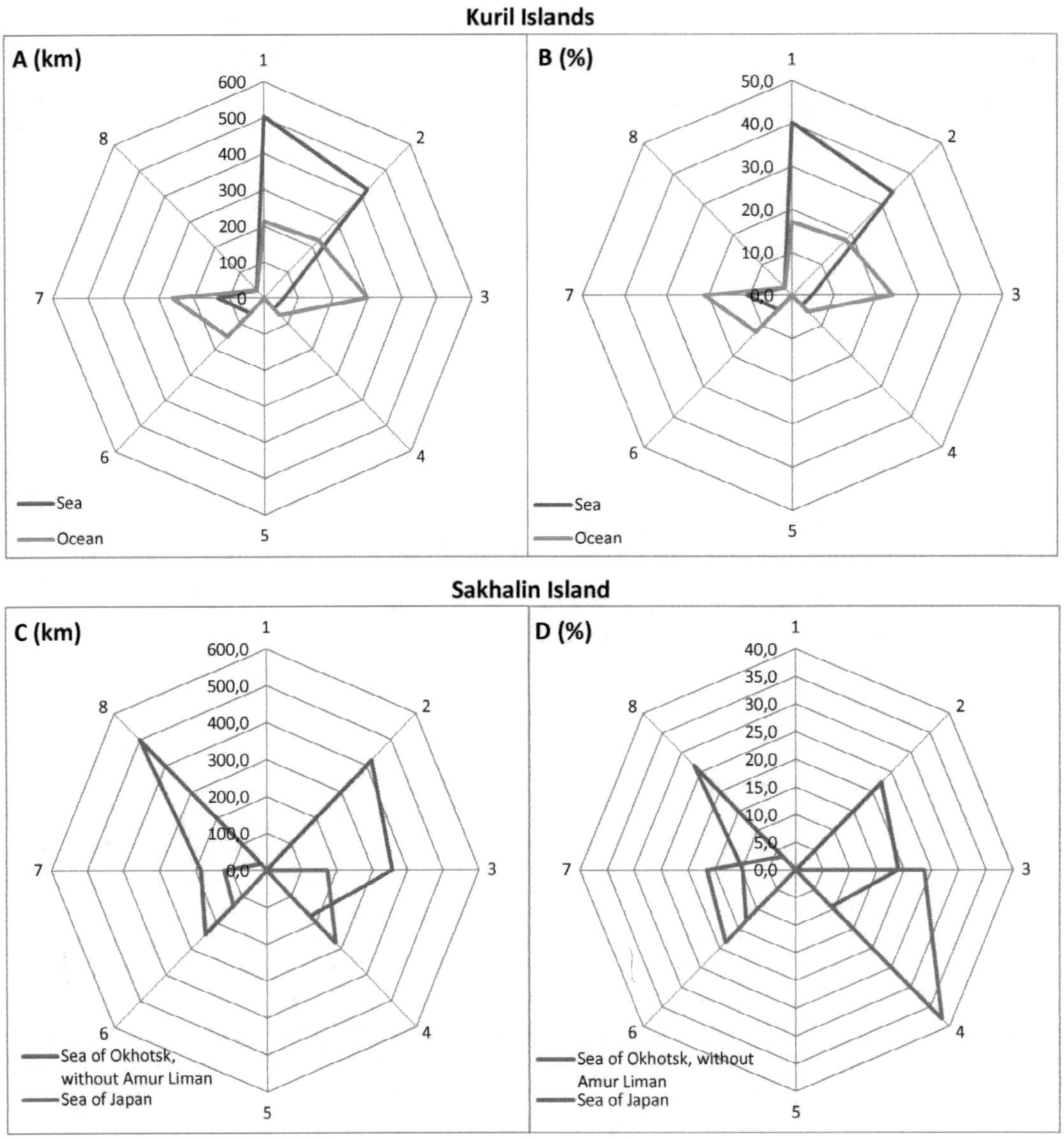

Figure 1. Distribution of morphogenetic types of coasts of the island-arc systems of the North Pacific: 1 – with glacial-tectonic dissection (fjord), tectonically determined, 2 – abrasion-denudation; 3 – abrasion (aligned and coiled); 4 – abrasion dead; 5 – thermoabrasion and ice; 6 – abrasion-accumulative (leveled and coiled); 7 – created by wave processes, beach; 8 – created by wave processes, lagoon.

As expected in accordance with the previous field studies, in percentage terms, the seashores of the 1st and 2nd types prevail over the ocean shores of these types, both the Aleutian Islands and the Kuril

Islands [6]. Abrasive shores in the Aleutians predominate on the sea side, and on the Kuril Islands, on the contrary, on the ocean side. The extent of the abrasive-accumulative shores is higher on the ocean side of the island-arc systems. The length of accumulative shores is also greater from the ocean side both in the Kuriles and in the Aleutians. The revealed features of morphogenetic types of sea and ocean shores indicate the predominant contribution of hydrodynamic and thermal regimes of the separated water areas to this distribution. Morphoclimatic parameters of the actual contact zone (island arc), unlike those on Sakhalin Island, do not affect this distribution. The morphotectonic and structural-formational conditionality of the development of the shores of island arcs is undoubtedly significant and requires further research. The obtained distributions of coast types from the sea and ocean sides of island-arc systems served as a starting point for updating the problem of the peculiarities of the development of the shores of linear contact zones separating marine basins, as well as seas and oceans.

Using the example of Iturup Island (Kuril Island Arc) and Chugunidak Island (Aleutian Island Arc), the features of the modern coastal morpholithogenesis of the contact zone and Aeolian accumulation, the so-called short-range transport, which refers to specific coastal relief-forming processes summing impacts in the Paleogeographic Middle-Late Holocene time scale were considered [9]. The most significant differences in the modern morpholithodynamics of the shores were established when analyzing the morphometric parameters of the megafestons of the beach on the island of Iturup. It turned out that on opposite sections of the Vetrovoy Isthmus, the size of ocean megafestons is more than twice the size of Okhotsk megafestons.

When analyzing the Aeolian morpholithogenesis associated with the wave processing of mainly pumice-pyroclastic material during explosive eruptions in the Holocene, Neopleistocene, it was found that the dunes on the sea side of the Vetrovoy Isthmus of Iturup Island were formed as a result of the destruction of high coastal ledges composed of pumice-pyroclastic material of the Neopleistocene, during the period of increased erosion with sea level rise about 1.5 thousand years ago. The formation of younger dunes of the ocean coast is largely due to the processing of pyroclastic material that entered the coastal zone about 1000 years ago directly during the eruption [10]. A significant proportion of ash material in pyroclastics also caused a very rapid formation of pelitic sediments in a lagoon-type basin on the ocean side of Vetrovoy Isthmus of Iturup Island.

Sakhalin Island, whose unique geographical position allows us to consider it as a kind of contact zone of the border strip of the subarctic and temperate zones, characterized by a consistent change of natural conditions and temperature regime from north to south, has become a key area for detailed studies of the features of morpholithogenesis of the shores of linear contact zones separating marine basins, as well as seas and oceans.

The northern and eastern shores of Sakhalin Island are affected by the physical and geographical conditions of the subarctic zone. The formation and development of the shores is determined by the cold Sea of Okhotsk, which belongs to the category of subarctic freezing seas. The shores of eastern Sakhalin are characterized by the presence of permafrost rocks on the coast, including at the base of lagoon embankments, the development of thermal abrasion and the existence, up to mid-July, of coastal ice.

The western and southern shores of Sakhalin Island are more affected by the physical and geographical conditions of the temperate zone. The moderately cold conditions of the northern sector of the Sea of Japan provide a relatively frequent temperature transition through zero, the absence or short-term formation of ice and soldered ice.

Detailed zoning of morpholithodynamic conditions of the Sakhalin coast generally confirms the results of the analysis of the map of morphogenetic types of the shores of the Northern Pacific (figure 2).

There are about 2 times more abrasive shores from the Sea of Okhotsk than from the Japanese side. There are 2.5 times more dead abrasive ledges on the shores of the Sea of Japan, there are significantly more abrasive-accumulative and terraced shores. Interestingly, in general, the length of the shores with a dead abrasive ledge is higher in percentage terms on the Sea of Okhotsk than on the Sea of Japan. Such a distribution of coast types, along with the structural and geological features of the structure of the coasts, is probably due to a change in the intensity of destruction of the shores of the Okhotsk and

Japan Seas, during the warm periods of the Holocene, characterized with a longer period of "open sea" at negative air temperatures [11].

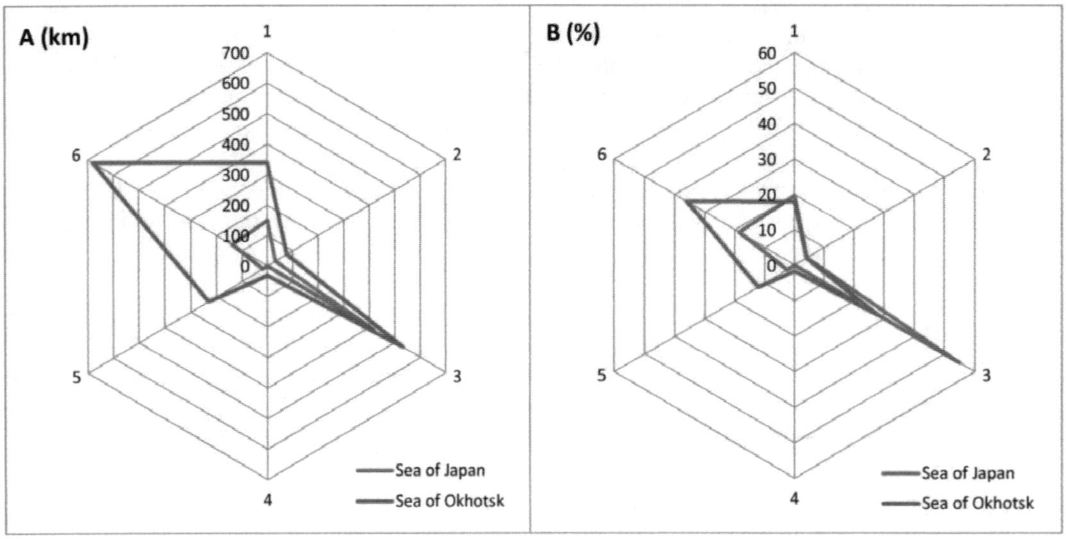

Figure 2. Distribution of morpholithodynamic conditions along the shores of Sakhalin Island. 1 – active cliff in bedrock, 2 – active coastal ledge in weakly consolidated and loose formations, 3 – dead cliff in bedrock, 4 – dead cliff with a low terrace, 5 – dead coastal ledge in weakly consolidated and loose formations, 6 – accumulative Holocene coastal forms.

It should be noted that the Sakhalin shores of the Japanese and Okhotsk Seas practically do not differ in terms of resistance to destruction (figure 3).

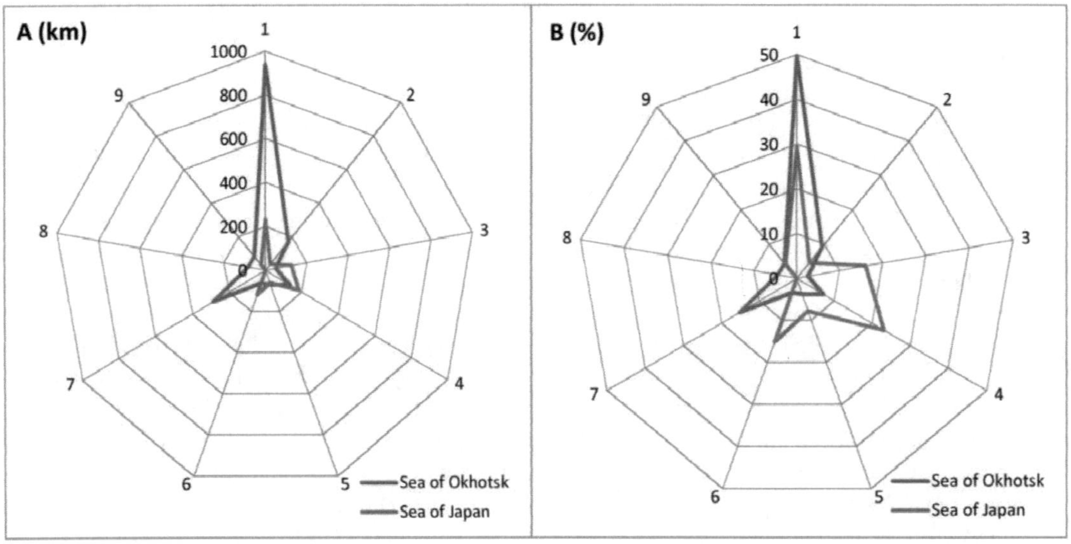

Figure 3. Distribution of lithological complexes of the coast of Sakhalin Island (using [8]). Rock groups: 1 – loose; 2 – weakly lithified (Maruyam Formation); 3 – weakly lithified (Okobykai Formation); 4 – strongly lithified (Kholm Formation); 5 – strongly lithified (Pila Formation); 6 – strongly lithified (Arakai Formation); 7 – strongly lithified (Ostrin Formation); 8 – strongly lithified (Langer Formation); 9 - strongly lithified intrusive rocks form promontories protruding into the sea.

3. Conclusion

For the first time, a quantitative characteristic of the results of the main relief-forming processes was obtained and the originality of the modern morpholithodynamics of the shores of the Northern Pacific was clearly demonstrated. An equally important result of the analysis is the actualization of the problem of the development pecularities of the shores of the linear contact zones of the Northern Pacific, separating the sea basins, as well as the seas and oceans. Complex geological and geomorphological studies confirm the morpholithodynamic and geomorphological specifics of the shores of the contact zones.

References

[1] Likhacheva E A, Makkaveev A N and Lokshin G P 2010 Physical essence of geomorphological boundaries *Geomorphology* **2** pp 3–10
[2] Baklanov P Ya 2014 Geographical and geopolitical factors in regional development *Regional studies* **2** 44 pp 4–10
[3] Skrylnik G P 2017 Basic levels of stability in the general organization of geosystems of the Earth *Successes of modern natural science* **11** pp 101–106
[4] Likhacheva E A and Timofeev D A 2007 To the hierarchy of geomorphological systems from the standpoint of their organization *Geomorphology* **4** pp 3–8
[5] Likhacheva E A and Timofeev D A 2008 Analysis of geomorphological systems: basic concepts *Geomorphology* **2** pp 14–21
[6] Afanasyev V V, Ignatov E I and Uba A V 2020 Morphogenetic differentiation of the shores of the contact zones of the Northern Pacific *Hydrosphere. Dangerous processes and phenomena* **2** 2 pp 158–172
[7] *National Atlas of Russia* 2007 ed A V Borodko et al (Moscow: PKO “Cartography”) **2**
[8] Polunin G V 1992 *Map of lithological complexes and exogenous processes of the Sakhalin land and shelf on a scale of 1:500000* (Khabarovsk, Yuzhno-Sakhalinsk)
[9] Afanasyev V V 2019 About a new type of Aeolian morphogenesis on volcanogenic shores (Iturup Island, Bolshaya Kuril ridge) *Geosystems of transition zones* **3** 4 pp 423–427
[10] Afanasiev V V, Gorbunov A O, Uba A V and Dunaev N N 2020 The manifestation of caldera-forming volcanism in the formation of the coast (on example of Iturup island of the Great Kuril Ridge) *Springer Geology* **1** pp 51–61
[11] Afanasyev V V, Romanov A O and Uba A V 2017 Dynamics of shores in the cold period *Geosystems of transition zones* **1** pp 23–29

IOP Conf. Series: Earth and Environmental Science **946** (2021) 012034 doi:10.1088/1755-1315/946/1/012034

Recreational resources of the coasts of the southern part of Sakhalin Island and the southern part of the Crimean Peninsula (problems and solutions)

V V Afanasev[1,*], A V Uba [1], A B Faustova[2] and N A Vaseikina [1]

[1] Institute of Marine Geology and Geophysics Far Eastern Branch of Russian Academy of Science, Yuzhno-Sakhalinsk, 693022, Russia
[2]LLC "Dalmorberegotechnika", Yuzhno-Sakhalinsk, 693022, Russia

E-mail: *vvasand@mail.ru

Abstract. The possibilities of recreational-oriented use of the coast are studied on the basis of the trend analysis in the development of the coasts and regional features of the exploitation of the coasts of Sakhalin Island and the Crimean Peninsula. It is shown that the loss of recreational capacity of coastal territories is the result of both man-made and natural components of development. The conclusion is made about the degradation of a significant part of the recreational resources of the coasts. The concept of a model of integrated use and sustainable development of natural mineral resources, hydrogeological and recreational resources is proposed.

1. Introduction
Recognizing that among the natural resources which determine the recreational potential of the coasts, climatic ones play a special role, one should not underestimate the importance of the state of natural complexes of recreational territories themselves [1]. Degradation of recreational resources of the coast of Southern Sakhalin is a process that, alas, is already coming to an end before our eyes. Not only does the loss of low sea terraces and dune systems as a result of erosion significantly reduce the recreational capacity of coastal areas, but also the extraction of common minerals on the coast has turned previously picturesque places into a desert. The total area of flooded quarries on sea terraces unsuitable for any kind of use is already about 45 unattended hectares. There are no urban beaches in any Sakhalin city located on the coast. The only embankment, which moreover exists in the system of coastal protection solutions only in Nevelsk But it is the embankment that is the face of the seaside city. The need to protect transport communications and residential areas also caused the loss of recreational capacity of coastal areas. Despite the fact that today 30 beaches have been officially opened in Sevastopol, morpholithodynamic features of the shores of the southern Crimea significantly complicate beach holidays. The purpose of the presented work is to substantiate the increase in the recreational capacity of the coasts in the conditions of accelerating evolutionary changes and increasing anthropogenic load.

2. Methods and results
The presented work implements the methodology of geomorphological research based on both traditional and modern methods of obtaining and analyzing geospatial and geological-

geomorphological information. The satellite digital terrain model SRTM (Shuttle Radar Topography Mission) was used to calculate the volumes [2]. The initial relief model for quarries in the vicinity of Sevastopol was reprojected in WGS 84 UTM Zone 36N. The volume is calculated for a figure bounded by two surfaces: by the SRTM model from the bottom, and by a surface constructed inside the calculated polygon from the top, taking into account the relief surrounding the quarry, using triangulation method.

Until recently, estimates of the anthropogenic load on the shores of Sakhalin were significantly underestimated [3]. When assessing the anthropogenic load on the shores of southern Sakhalin, it was found that the concrete, sheet pile and ridge walls have a length of about 100 km [4]. It is known that port hydraulic structures and shore-protecting concrete structures are localized mainly in areas developing according to the abrasive type. On accumulative sites, coast protection is not needed, port facilities are practically impossible to build, and portability is a significant obstacle to the functioning of the port. The need to protect transport communications and residential areas also caused the loss of recreational capacity of coastal areas and a decrease in the incoming part of the budget of sediments of the coastal zone.

Irretrievable removal of sediments from the coastal zone has a no less significant impact on the development of coastal morpholithystems than the limited stabilization of active coastal ledges by protected areas of residential territories and transport communications [5]. Only in the south of Sakhalin, for example, this is about 10 million m^3 of sediment seized from the beach and avandune zones [6]. Given the morpholithodynamic mechanisms of coastal development, this is equivalent to the loss of about 3 km^2 of low accumulative coastal plain as a result of erosion, which, in fact, has already happened. Given the almost ubiquitous distribution of peatlands on low coastal terraces, the process of their erosion from the point of view, for example, of the carbon balance, can be characterized as a man-made carbon release into the atmosphere-ocean system.

The difficult sanitary and hygienic situation caused by the discharge of untreated wastewater into the Aniva Bay by several cities and the danger of contamination of the water area by a new LNG plant is the reason for the limited use of recreational resources of this southernmost water area of Sakhalin island. At the same time, the highest concentrations of harmful substances are found in coastal-marine sediments and during severe storms they are released into coastal waters [7, 8].

There are no official beaches on the entire coast of the island. Most of the places where people swim and rest are spontaneous places. Those, that are officially registered, are the places of rest near the water, where only sunbathing is allowed. Therefore, the development of beach infrastructure is the most relevant and necessary type of recreation. Only two beaches – in Yablochny and Pionery (specialized children's beach "Aquamarine" near the camp "Chaika") Kholmsky district – on the eve of the season were accredited and received the status of safe. Two more places: Lake Verkhnee in the Yuzhno-Sakhalinsk Park and the Sunny beach on the Tunaycha, are suitable only for relaxing near the water. It is impossible to swim there according to Russian norms [9].

The most interesting for the development of water recreation are the shores of lagoon lakes (Lake Busse, Lake Izmenchivoye, Lake Tunaycha, etc.) and the south-west of Sakhalin, washed by the warm branch of the Tsushima current of the Sea of Japan. The favorable period for bathing here lasts up to two months from July to August [10, 11].

The climate in most of the territory of the Sevastopol region belongs to the subtropical oceanic with hot summers and without a pronounced dry season [12, 13].

The tourist season begins with the May holidays until the end of September, the peak is reached in July-August. The possibilities of beach recreation are limited by the abrasive-topple-landslide type of shores and a high degree of technogenic development of the coast. Thus, despite the approximately 2.5 times longer duration of favorable climatic conditions, the possibilities of beach holidays in the Sevastopol region are quite limited. There is no infrastructure to provide winter recreation here.

Considering that the increased intensity and activity of coastal erosion in the southern part of Sakhalin Island and the Crimean Peninsula is an objective reality, the strategy for reducing economic and environmental damage in coastal areas is thus determined by the forecast of vulnerability of the coast to expected changes and the possibility of implementing a system of measures to reduce it.

One of the solutions to the problem may be the development of tourist and recreational complexes for year-round use, which can literally become high-tech factories for the production of recreational services. We are talking about the reclamation of spent quarries in the form of equipped artificial reservoirs (figure 1, 2). Until recently, a similar project, in its most advanced form, functioned in Japan. Located in Miyazaki, on the Island of Kyushu, the world's largest indoor water park was called the Seagaia Ocean Dome [14].

Figure 1. A sand pit on the coast of southern Sakhalin.

Figure 2. Swan Lakes Recreation Center – on the site of old swampy sand pits.

Based on the developed principles, the possibilities of recultivation of quarries of flux raw materials in the vicinity of Sevastopol are very promising (figure 3).

The project Swan Lakes Recreational Zone" is based on three artificial lakes with an area of 7.9 hectares each, with ground feeding and one with an area of 2.6 hectares without hydraulic connection with groundwater (pools covered with domes) with which all architectural solutions are connected [6, 15]. It is around the lakes that tourist facilities and a beach will be built. It is assumed that 250 thousand people will be able to rest under the dome for a year with an average rest duration of six days and up to 3 thousand people at a time.

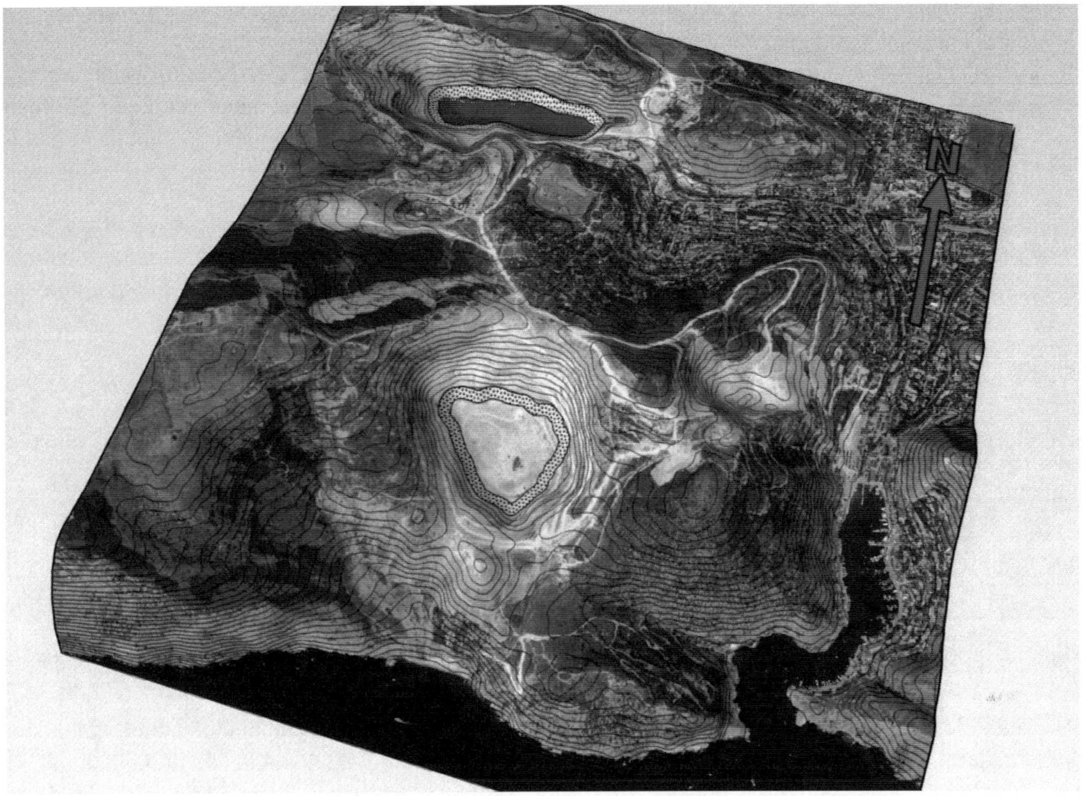

Figure 3. Perspective isometric projection of the relief model of the quarries of flux raw materials in the vicinity of Sevastopol.

Great potential for use is the Balaklava quarries – Kadykovsky (flooded with groundwater, now used as a backup source for the water supply of Sevastopol, the volume of excavation is 53 143 870 m^3) and Psilerakhsky (flux limestone is being mined, the volume of excavation is 46 156 827 m^3).

Previously, options for recreation-oriented recultivation of quarries have been repeatedly proposed [16]. However, in our case we are talking about structures with artificial climate [15]. The length of the coastline of the artificial reservoir of the Kadykovsky quarry is 2 km. When creating a 40 m wide beach, the capacity of the recreational area will be 16 000 people (a unit of 5 m^2/person). The perimeter of the bottom of the Psilerakhsky quarry is about 1300 meters and when creating an artificial reservoir, its recreational capacity will exceed 10 000 people.

3. Conclusion

The development of recreational areas with an artificial climate will allow:

- to increase the contribution of the southern part of Sakhalin Island and the Crimean Peninsula to the development of the tourist and recreational complex in the Russian Federation;

- to create conditions for attracting domestic and foreign investments aimed at creating modern, meeting international standards, tourist and sanatorium-resort complexes;
- radically and for the better change the image of the Sakhalin Region and the Sevastopol region not only in Russia, but also in the countries of the Asia-Pacific region and the EU;
- to promote the positive dynamics of economic development by creating a favorable investment climate and not only in the field of tourism;
- to use the scientific and technical potential, market infrastructure, transport, geographical and historical advantages of the Sakhalin Region and the Republic of Crimea,
- to create several tens of thousands of new jobs in the highly profitable sector of the economy;
- to compensate for the loss of recreational capacity of coastal areas as a result of both man-made and natural components of development.

References

[1] Ignatov E I, Orlova M S and Sanin A Yu 2014 *Coastal morphosystems of the Crimea* (Sevastopol: SPC "EKOSI-Hydrophysics")

[2] Rodriguez E, Morris C S, Belz J E, Chapin E C, Martin J M, Daffer W and Hensley S 2005 *An assessment of the SRTM topographic products, Technical Report JPL D-31639* (Jet Propulsion Laboratory, Pasadena, California)

[3] Aibulatov N A 2005 *Russia's Activity in the coastal zone of the sea and environmental problems* ed V I Osipov (Moscow: Nauka)

[4] Afanasev V V and Ignatov E I 2009 Geoecology of the coastal zone of Sakhalin Island *Problems of regional ecology* **6** pp 275–280

[5] Ignatov E I 2005 Coastal morphosystem as an object of geomorphological research *Geomorphology* **2** pp 3–11

[6] Afanasev V V 2020 *Morpholithodynamic processes and development of the shores of the contact zone of the subarctic and temperate seas of the Northern Pacific* (Yuzhno-Sakhalinsk: IMGG FEB RAS) p 234

[7] Report of the Ministry of Natural Resources and Environmental Protection of the Sakhalin Region "On the state and environmental protection of the Sakhalin region in 2010"

[8] ASTV news (*Electronic materials*) https://astv.ru/news/criminal/2019-07-23-kupat-sya-v-anivskom-zalive-opasno-dlya-zdorov-ya

[9] Russian Federation GOST R 55698-2013 (Electronic materials) https://docs.cntd.ru/document/1200107030

[10] *Agro-climatic resources of the Sakhalin region* 1973 (Leningrad: Hydrometeoizdat)

[11] *Hydrology and hydrochemistry of the seas* 2003 **IX** (St. Petersburg: Hydrometeoizdat) Issue 1

[12] Climate Sevastopol (*Electronic materials*) https://en.climate-data.org/asia/russian-federation/sevastopol/sevastopol-3011

[13] Kottek M, Grieser J, Beck C, Rudolf B and Rubel F 2006 World Map of the Köppen-Geiger climate classification updated *J. Meteorol. Z.* **15** 3 pp 259–263

[14] Gadgetreview (*Electronic materials*) https://www.gadgetreview.com/the-seagaia-ocean-dome-was-the-worlds-largest-indoor-beach

[15] Eden Project (*Electronic materials*) https://www.visitcornwall.com/things-to-do/attractions/south-coast/ st-austell/eden-project

[16] Parfenova I A, Repetskaya A I and Kladchenko E S 2016 The concept of recultivation of the Kadykovsky quarry with the organization of a tourist and recreational facility (Sevastopol, Balaklava) *Ecosystems* **6** 36 pp 51–65

Coastal dynamics of the Gulf of Kalamita (Western Crimea)

I V Agarkova-Lyakh

Institute of natural and technical systems, Sevastopol, 299011, Russia

E-mail: iva_crimea@mail.ru

Abstract. The performed research showed that since the first third of the 20th century until today the beach width at various sections of the Gulf of Kalamita has decreased two- to fourfold, and at some of them beaches have totally disappeared. The average coast erosion rate for the last 100 years has been 1.3 m/yr, with the maximum being 7.8 m/yr. These processes are accompanied with depletion of debris on the beach and underwater slope, erosion of bench deposits, and destruction of coastal structures. This is due to deficiency of beach-forming material resulted from intense economic management in the coastal area. Implementation of scientifically and technically sound coastal protection projects is essential to preserve the beaches of the Gulf of Kalamita.

1. Introduction

Beaches of the Gulf of Kalamita are the main recreational resource and tourist attraction in the Western Crimea. From the 1930s, the beach width in some sections of the bay began to decrease gradually followed by almost complete degradation in the 1980s. Now, these processes are a serious regional problem, the main cause being active economic management in the coastal area, the scale of which on the world ocean shores is increasing [1, 2].

The paper describes the beach and coastal dynamics of the Gulf of Kalamita based on long-term observations of the 20th to early 21st cc.

2. Methods and results

The paper uses file materials from the leading research organizations of Crimea [3-8] and results of our own field research for 1998–2017. The main information processing methods were comparative-geographical, comparative-historical and cartographic ones. During the fieldwork, we used methods of beach and coast observation at key locations, instrumental and semi-instrumental measurements of coast and beach morphometric features, and collection of photographic evidence of the coastal situation. The key monitoring area was the section of the Sakskaya Bay-Bar coast between G. Titov Zvezdny Children's Recreation Camp (CRC) and the village of Pribrezhnoye.

The Gulf of Kalamita is situated off the western coast of Crimea and is bounded by the Cape Evpatoriyskiy in the north and the Cape Lukull in the south (figure 1). The total length of the gulf's shores is about 66 km (according to a 1:200 000 scale map). Previously, the gulf coasts were an example of an erosive-accumulative pair consisting of two different genesis areas: the northern accumulative (from the Cape Evpatoriyskiy to Lake Kyzyl-Yar) and the southern erosion (from Krasnaya Gorka to the Cape Lukull) ones merged by a material exchange into a single lithodynamic system.

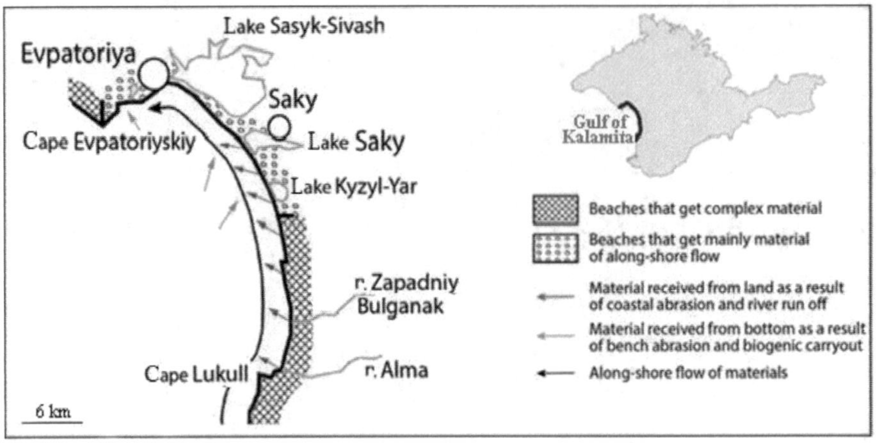

Figure 1. Beach feeding sources and genetic types [9].

The beaches of the southern erosive section are formed due to sediment runoff material of the Zapadniy Bulganak and Alma rivers, products of coast and underwater slope erosion, biogenic emissions and longshore drift material (figure 1). The beaches are attached to coast slopes and composed of sandy, boulder-sandy and gravel-pebble sediments.

The beaches of the northern accumulative section are formed mainly by longshore drift material transported from the south. The longshore drift is directed from the mouth of the Belbek River to Evpatoriya and includes sediment runoff material of the Kacha and Belbek rivers, material received from the bottom, coast and bench erosion products (figure 1). These beaches are classified as full-profile (free) beaches comprising sandy, gravel-sandy and gravel fractions.

Further, let us consider changes in the beach widths at various sections of the Gulf of Kalamita in the 20th to early 21st cc. from the Cape Lukull to the Cape Evpatoriyskiy.

In the mid-1980s, there was a natural beach 8.0–10.0 m wide at the mouth of the Alma River and an artificial beach of the same width near the village of Peschanoye [8]. The beach shrinkage observed back then led to construction of coastal protection structures. Consequently, by the 1990s, there appeared 15 groins 50.0–85.0 m and long standing 50.0–500.0 m apart [10]. The space between the groins was filled with rock debris. The cliff to the north of the groin group was cut and terraced. By 1994, a nearly 850-meter long breakwater seafront was built. The constructed complex did not exist for long. In 2005, the coastal protection structures on the beaches of the village council were partially or completely destroyed.

In the mid-1980s, the beaches north of Peschanoye and near Beregovoye were 5.0 to 8.0 m wide and north of Beregovoye they were 10.0 to 15.0 m wide [8]. At that time, a sloping terraced breakwater seafront was constructed at the mouth of the Zapadniy Bulganak River (Beregovoye) [10]. In 20 years, the distance from the cliff edge to the nearest recreation buildings in Beregovoye was only 2.5 to 5.0 m [8]. The restoration of the coastal protection structures in Beregovoye, which began in 2013, has not been completed yet [10].

Near the urban-type settlement of Nikolaevka, a critical situation of beach degradation arose in the late 1970s threatening the functioning of a number of recreational facilities. In response to this, in 1981–1982, construction of a sloping terraced waterbreak seafront started in the southern and northern parts of Nikolaevka. The seafront occupying 1.5 km of the coastline was finished by 1990.

In the southern part of Nikolaevka, in the mid-1980s, the beach was about 30.0 m wide [8]. In ten years after the seafront construction, the beaches shrank dramatically, and somewhere they disappeared completely. In 2005, the sloping terraced seafront footing was eroded and concrete slabs collapsed. During 2006–2008, six transverse groins were built in this area to form artificial beaches and protect the remaining seafront [11]. This resulted in an increase of beach widths south of the groins and their rapid degradation to the north accompanied by destruction of the coastal protection structures.

In the 1990s, in the centre of Nikolaevka, the width of the beach protected by the sloping terraced seafront was 23.0–34.0 m. By 2005, it had decreased to 10.0–18.0 m near the lifeguard station [8].

To the north of Nikolaevka, in the late 1970s through mid-1980s, there was an unstable 10-meter wide beach. The beach got stable only after construction of two groins on the cape near the settlement in 1992. As a result, since the late 1990s, at the northern section in front of the cape, beach material was accumulating. At a distance of about 1.5 km south of the cape, the beach width was 25.0–35.0 m; further to the south, that was 18.0–20.0 m. As of 2005, this section generally remained stable, with the beach width being 22.0–30.0 m and its seasonal variation ranging within 5.0–7.0 m [8].

In the mid-1980s, to the south of the Bogaily Lake bay-bar, the beach was 18.0 m wide. Between Bogayly and Kyzyl-Yar lakes, the beach width was 7.0–12.0 m, and near Krasnaya Gorka it was 10.0 m. In the same years, the width of the southern Kyzyl-Yar bay-bar was 200.0 m. After construction of a hydraulic structure on the northern bay-bar in 1979–1982, the beach in front of it started accumulating at an average rate of 7.5 m/year [12]. In the early 2000s, the southern Kyzyl-Yar bay-bar shrank 1.5 times, and the coastline of its northern part protruded into the sea by 120.0 m, to the length of the hydraulic structure.

Similar processes were observed on the Sakskaya Bay-Bar. According to our research, in the early 1980s, near Titov Zvezdnyy CRC, there were beaches up to 60.0 m wide, of which today there are only sheer fragments left of 15.0–20.0 m width. The scarp forming here has a height of 0.1 to 1.0 m and is constantly shifting towards the land. The beach in the northern part of the Sakskaya Bay-Bar is relatively stable. In the mid-1980s, the width of various parts of the beach ranged within –3.0 to +5.0 m. During the study period, the width of the beaches in the northern part of the Sakskaya Bay-Bar remained almost unchanged and exceeded 50.0 m.

On the Lake Sasyk-Sivash bay-bar, the average beach width in the mid-1980s ranged within 30.0–50.0 m. During this period, maximum values of 50.0–100.0 m were observed for the full-profile (free) beaches opposite Pribrezhnoye, where the coastline changes its direction and longshore drift material accumulates.

On the northern Sasyk-Sivash Bay-Bar, the coastal curve turns to the west to form the Evpatoriyskaya Bay, which extends to the Cape Karantinniy.

In the late 1950s and early 1980s, the width of the Evpatoriya's beaches at the eastern outskirts of the city in the area of Simferopolskaya Street was 30.0–50.0 m. After this territory had been flooded during a storm in November 1981, the sea began to destroy a part of this street together with the seafront. Today, attempts of individual owners to protect their houses from destruction have led to formation of rubbles of shell stones, broken stairs, concrete slabs, etc. in the seaside area.

In 1968–1972, Tereshkova Seafront was built in the historic district of the city. The seafront was a 1.8 km long concrete wall with a shaped breakwater and protruded for 30.0–50.0 m into the sea. After this reconstruction, the beaches have totally disappeared there [11].

Until the 1950s, at the Cape Karantinniy and adjacent areas, there was a broad full-profile (free) beach narrowing to 10.0–20.0 m to the east. In the late 1970s, at the Cape Karantinniy, a 200.0 m long mole of the Evpatoriya Commercial Seaport was built, whereafter a wide bank formed on the cape's east side, and on its west side (area of Gorkiy Seafront), the coast began to retreat intensely. This construction resulted in the loss of up to 50.0 m of the city's beach width [11]. After the mole erection, the beaches between the Capes Karantinniy and Yevpatoriyskiy have been constantly degrading.

In the early 20th c., the beach width at the today's M. Gorkiy Seafront was 21.0–85.0 m, but in 2005, it was only 15.0–35.0 m. Further west, at the today's Frunze Park through *Zolotoy Bereg* Resort, the beach width was 50.0–80.0 m in the early 20th c., 30.0–40.0 m in the mid-1980s, and in 2005, it was only 6.0–15.0 m. Over 90 years, between 1915 and 2006, the beach width in the today's *Orlyonok* Resort House area decreased more than twofold [8]. According to estimations [11], there are no beaches within a three-kilometer zone in the central part of Evpatoriya; within another 1.5-kilometer section of the coastal area, they have significantly shrunk.

The width of the Moynaki Lake bay-bar decreased from 125.0–150.0 m in 1989 to 90.0–100.0 m in 2005. Intense erosion and coastline retreat are supported by erosion scarps up to 0.8 m high. On some beaches, the sediments are almost completely washed away and the bedrock loam soils are exposed.

Publications concerning sediment deposits on the beaches of the Gulf of Kalamita in the late 1980s indicate that their largest amounts were concentrated on accumulative bay-bars of saltwater lakes and in the zone of longshore drift discharge (apex of the gulf) [7]. Their average specific volume was increasing from south to north (figure 2). Thus, in the southern part of the gulf, between Peschanoye and Nikolaevka, the average specific volume of beach sediments was 7.4–12.0 m^3/m increasing to 33.9–42.0 m^3/m towards the Lake Bogaily bay-bar and the settlement of Novofedorovka. At the Sasyk-Sivash bay-bar, the average specific volume of sediments reached its maximum of 85.0 m^3/m. On the Evpatoriya beaches, the sediment amount decreased by a factor of 3 to 5, varying from 17.6 m^3/m near the Cape Karantinniy to 28.0 m^3/m near the Cape Evpatoriyskiy [7]. The comparison of variation coefficients between the two coastal areas had shown that the beaches of the Cape Karantinniy and Nikolaevka (36.4 % and 25.7 %, respectively) were the most dynamic while the beach at the Sasyk-Sivash Bay-Bar was the most stable. Five out of seven testing areas showed an insignificant average annual decrease of the beach width: from 0.7 to 3.6 m (figure 2).

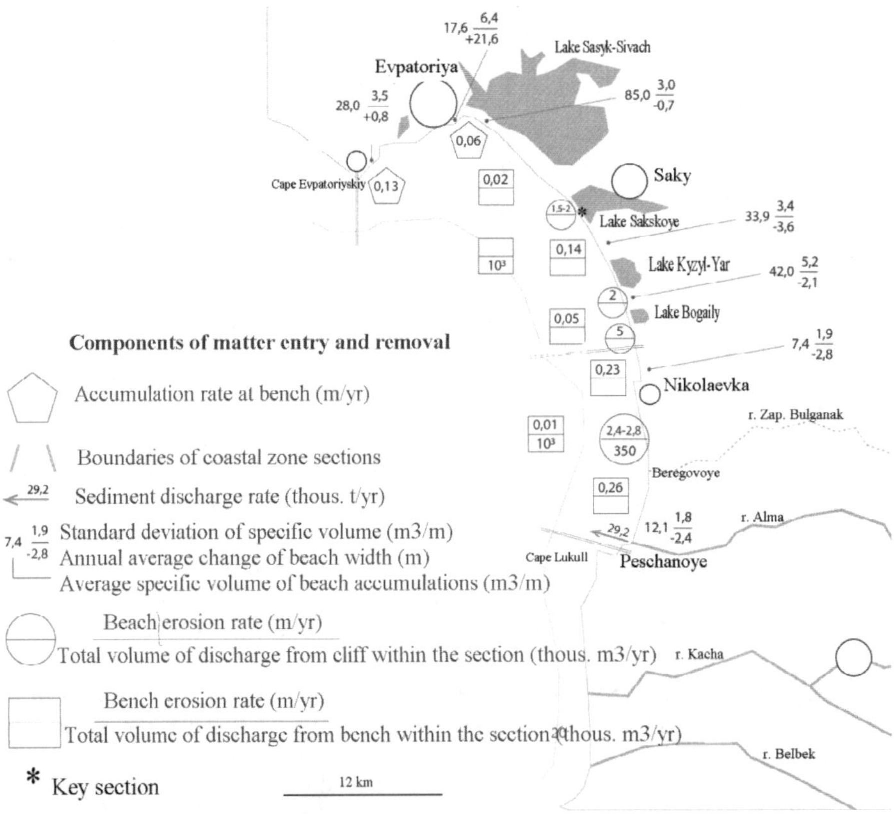

Figure 2. Quantitative indicators of coastal zone dynamics.

Thus, it should be stated that since the 1980s until now, the width of beaches in the Gulf of Kalamita has been dramatically decreasing along their entire length from the Cape Lukull to Evpatoriya, except in stable sections and areas of hydroengineering construction with a disturbed natural lithodynamical regime. As a result, the average beach width of 10.0–15.0 m is extremely insufficient to dampen the energy of sea waves causing their coasts to erode and retreat.

Today, erosion is the dominant coastal process and the main source of debris material. According to the data [8], over the last 100 years, the erosion average rate in the Gulf of Kalamita has been 1.3 m/yr, with a maximum of 7.8 m/yr. The rate of retreat varies from year to year and in the seasonal cycle it is associated with the period from November to March when the wind wave activity increases.

An objective assessment of the gulf's coast erosion rates requires comparing the long-term observation series presented in the table. The highest erosion rates are noted at the southern erosion section. At the accumulative section, they are two or more times lower.

Table 1. Coast retreat rate in the Gulf of Kalamita.

Coast section	Observation period	Average retreat rate, m/yr	Sourse
Evpatoriya	1970s	3.75	[13]
	no data	1.0–2.0	[8]
Southern Sakskaya Bay-Bar	1941–1963	2.1	[6]
Saky section	1970s	1.0	[13
Sakskaya Bay-Bar	1979–1982	2.6	[4]
Sakskaya Bay-Bar near G. Titov Zvezdny CRC	1983–2005	1.5–2.0; 3.76 (maximal)	[9, 12]
Southern Kyzyl-Yar Lake bay-bar	no data	5.0	[7]
Krasnaya Gorka section	no data	2.0–2.8	[7]
Northern bay-bar Bogaily Lake	1994–2005	1.36	[8]
Southern bay-bar Bogaily Lake	no data	5.0	[7]
Section between Nikolaevka and Kyzyl-Yar Lake	1940–1973	1.46	[6]
North of Nikolaevka	1976–1982	from 0.6–1.0 to 6.0	[3]
Alma River mouth, near Peschanoye and Beregovoye	1976–1982	3.0	[3]
Between the Cape Lukull and the end of the bay-bar Bogaily Lake	no data	2.0–2.8	[7]
Entire gulf	1905–2005	1.3; 7.8 (maximal)	[8]

As the coastline retreats, erosion of the seabed and sediments occurs on the underwater slope. Its rate decreases from the south to north of the gulf: from 0.23–0.26 m/yr to 0.02 m/yr [7] (figure 2). Only at the bottom near the Capes Karantinniy and Evpatoriyskiy, there is low accumulation with the rates of 0.06 and 0.13 m/yr, respectively, associated with the interception of the material near the Evpatoriya Seaport breakwater and discharge of longshore drifts.

The current coastal conditions in the Gulf of Kalamita are attributable to an extremely low amount of incoming debris material. With the dramatic reduction of the sediment runoff and low shell production, the main source of beach-forming material in the coastal zone is erosion of coastal slopes and the underwater slope, which have a specific volume of 2.7 $m^3/m \cdot yr$ and 1.7 $m^3/m \cdot yr$ of sediments, respectively [14]. Altogether, they yield about 4.4 $m^3/m \cdot yr$ sediments on average over a long-term period but are not sufficient to completely saturate the Kalamita longshore drift. The unsaturated flow of sediment, which is formed here, with a capacity of 65.0 thous. m^3/yr discharges completely to the bottom of the Evpatoriyskaya Bay leaving the beaches of the northern Gulf of Kalamita without any refeeding.

As a result, the lack of material in the coastal zone is offset by reduction of beaches and coastline retreat, erosion of the bottom, and sediments of the underwater slope. The current deficit of sediments is caused by intense economic activities in the coastal zone, as discussed in [6, 12]. The problem of the beach shrinkage and coastal retreat can be solved by implementing a system of coastal protection measures.

3. Conclusion

The formerly accumulative coasts of the Gulf of Kalamita are now actively retreating along almost their entire length, excluding areas of hydrotechnical construction with a disrupted lithodynamic regime. According to observations since the beginning of the 20th c., the width of the beaches at various sections has decreased by a factor of 2–4; the average erosion rate has been 1.3 m/yr.

Today, the main source of beach-forming material is the products of the coastline and underwater slope erosion, but their amounts are insufficient to saturate the Kalamita longshore drift completely. This situation is due to a drastic decrease in volume of the debris material reaching the coastal zone because of an increasing pressure of economic activity there. Implementation of scientifically and technically sound coastal protection projects is essential to preserve the beaches of the Gulf of Kalamita.

Acknowledgements

The study was supported by state assignment of Institute of Natural and Technical Systems (Project Reg. No. AAAA-A19-119031490078-9).

References

[1] *Coastal erosion and protection in Europe* 2013 (Taylor & Francis Ltd.) ed by E Pranzini and AWilliams 488 p

[2] Rangel-Buitrago N G, Anfuso G, Williams A T 2015 Coastal erosion along the Caribbean coast of Colombia: Magnitudes, causes and management *Ocean and Coastal Management* **114** pp 129–144

[3] *Report on the study of landslides Crimea region in 1976–1980 and 1981–1982* 1983 (Crimean State Hydrogeological Expedition) ed I F Erysh

[4] 1986–1988 Evpatoria and Saki resort and recreation subarea. Territorial complex scheme of nature protection resort of national importance (Ukrainian State Institute of Design Cities «Giprograd») Annex 14 (Yalta) Annex 1 (Kiev)

[5] Compose inventory of above-water parts of the Crimean coast at a scale 1:200 000 1988 (Crimean State Hydrogeological Expedition; Institute of Mineral Resources) ed O S Romanyuk

[6] Romanyuk O S, Lushchik A V and Morozov V I 1992 *The conditions of formation and dynamics of the sea coast in the region of the Saky resort area* (Simferopol Institute of Mineral Resources)

[7] *A study of conditions for the development of exogenous geological processes in the coastal zone of Crimea peninsula* 1993 (Simferopol', GGP «Krymgeologiya») ed Yu P Luk'yanov

[8] State of the West Coast of the Crimea 2006 *Crimean Republican landslide management* (Yalta)

[9] Rubtsova S, Agarkova-Lyakh I, Lyamina N and Lyamin A 2016 *Proc. of Int. Conf. August 22–27 EMECS'11-SeaCoasts XXVI, St Petersburg, Russia* Modern state and dynamic of the beaches of Kalamitskiy gulf in the Western Crimea. Managing risks to coastal regions and communities in a changing world pp 644–652

[10] Goryachkin Yu N 2016 Coastal protection structures of the Crimea: West coast *Gidrotekhnika* Part 2, **2** pp 38–43

[11] Goryachkin Yu N 2016 Coastal protection structures of the Crimea: West coast *Gidrotekhnika* Part 1, **1** pp 49–54

[12] Agarkova I V 1999 Influence of economic activity on the dynamics of the Saky coast *Scientific notes of Taurida National University* **12** (51) 1 pp 15–19

[13] Shuyskiy Yu D 1979 Food detrital material of northwest and Crimean districts of the shelf of the Black Sea *Issledovanie dinamiki rel'efa morskikh poberezhiy* (M.: Nauka) pp 89–97

[14] Shuyskiy Yu D 2005 Main regularities of morphology and dynamics of the Western coast of the Crimean Peninsula *Ecological safety of coastal and shelf zones and complex use of shelf resources* **13** pp 62–72

Integrity of risk indicators of hazardous natural processes (on the example of the Krasnodar Krai)

T V Lyubimova* **and N A Bondarenko**

Kuban State University, Krasnodar, 350040, Russia

E-mail: *tv-luy@yandex.ru

Abstract. The problem of assessing risk indicators of hazardous natural processes as the basis for sustainable development of the region arises due to the peculiarities of the geological structure of the territory of Krasnodar Krai. The work is based on a comprehensive assessment of the geological conditions of the territory of Krasnodar Krai with the use of a systematic approach. A list of characteristic natural processes that pose the greatest danger to capital constructions was determined: in the mountainous part – erosion processes, on the plain – soil subsidence and associated landforms. Then, schematic maps were compiled and the geological component of the risk (its spatial-areal indicator) was determined. Subsequently, to compare these maps, overlay operations were performed, as a result of which the spatial characteristics of ARC/INFO layers were combined into a new layer and relational join of their attribute data was carried out. Adding the risks of each parameter made it possible to rank the territory into 4 zones according to the degree of integral risk of manifestation of natural processes.

1. Introduction

The beginning of systematic research aimed at assessing the danger of natural processes and risk analysis falls on the middle of the 20th century. A review of modern views on the problem of assessing environmental risks both in foreign and Russian literature [1–5] shows that until now there is no unified methodology for their assessment. So, the terms "risk" and "risk assessment" are still ambiguous. There is domestic experience in assessing risk indicators of endogenous and exogenous processes, experience in developing methodological foundations for assessing the risk of exogenous geological processes when considering regional features, some approaches to assessing natural and anthropogenic factors in relation to urbanized areas were proposed [6–8].

Krasnodar Krai is one of the main economic, recreational and tourist centers in the south of Russia. The expansion of economic activity is inevitably associated with the impact on the environment, but the natural environment itself can become a source of risk and pose a threat to life and health of the population and economic entities. In this regard, finding the solution of issues of sustainable and effective development of society, protection of territories, buildings and structures from hazardous geological processes come to the fore in the region.

Assessment of the integral seismogenic risk for the Black Sea coast is scarcely developed. This risk includes hazard factors caused by both direct seismic effects and subsequent "slow" threats associated with the structural features of the upper part of the geological profile, geomorphological and geological-engineering conditions [9].

One of the ways to solve natural risk management issues is a comprehensive assessment of the geological conditions of the territory of Krasnodar Krai on the basis of geological-engineering zoning. Over the past decade, big volume of survey work has been carried out in Krasnodar Krai, and a large amount of new engineering and geological information has been accumulated. Using these data, potential risk of such exogenous geological processes as landslides, erosion processes, flooding, etc. has been assessed [10].

The purpose of this work is an integral assessment of the risk indicators of natural processes for the territory of Krasnodar Krai. The main tasks of the study come down to assessing the degree of activity of certain geological processes, compilation of a set of maps that have theoretical and applied value, as well as ranking the territory according to the degree of risk using modern geographic information systems.

2. Materials and research methods

The spatial position of the territory of Krasnodar Krai is associated with such tectonic structures as the North-West Caucasus and the Scythian plate, which determine the nature of manifestation of endogenous and exogenous geological processes. Endogenous processes are due to the crustal stress, active tectonic disturbances, and high seismicity. The diversity of types of exogenous geological processes is determined by the geological and geomorphological structure and climatic factors. The most common are processes driven by surface waters. They are flooding, rainwash and soil subsidence in the northern part of Krasnodar Krai, river erosion, rainwash and landslides – in the central part, landslides, mudflows, river, ravine and sheet erosion – in the southern part.

To assess the territory according to the degree of potential risk, the scoring method was used within the framework of evaluative engineering-geological zoning [11] carried out using the ArgGis programs where each estimation parameter is a separate GIS layer of the digital map. By expert analysis, weights were assigned to the established criterion factors depending on their importance and an ordered scoring scale was created according to the increase (decrease) of the properties. The final assessment is the average weighted risk index of the manifestation of the process based on the integral index (in points) and the values of the area of each process development. This method makes it possible to overcome the complexity of assessing non-uniform parameters.

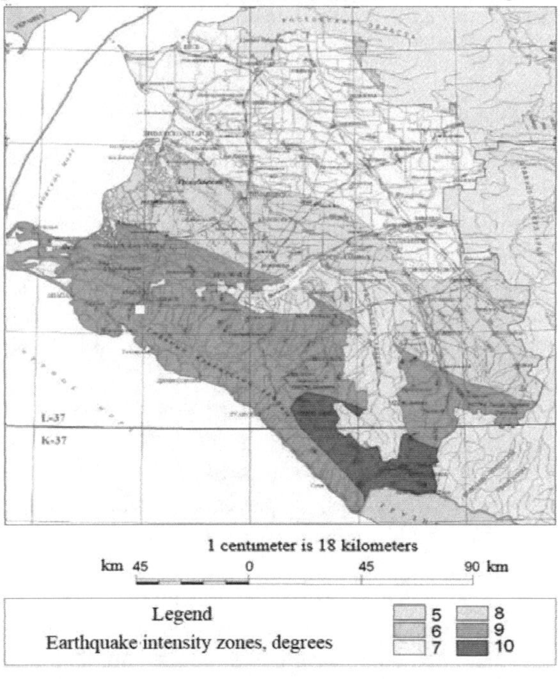

Figure 1. Map of general seismic zoning with a degree of seismic hazard (1%).

When assessing the seismic hazard of the territory of Krasnodar Krai various methods were used: compilation of seismotectonic diagrams, historical materials on paleoseismicity, micro-seismic zoning, seismogeodynamic modeling based on the dependence of the maximum magnitude of possible earthquakes in the area of the seismogenic block of the Earth's crust. Figure 1 shows a map of the general seismic zoning of Krasnodar Krai used in this work.

In the course of studying the complex of hazardous exogenous processes, the most significant factors were identified: distribution of specific soils, extent to which the territory is affected by subsiding landforms, ruggedness of relief and affection of the territory by gully erosion. To reflect the extent to which the territory is affected by each process, coefficient K was calculated, expressed as a percentage based on the total area of the affected territories. In this work, a pairwise compilation of schematic maps (M 1: 2500000) of the density of the gully network and degree of erosion of temporary streams as well as of distribution of loess-type rocks and location of subsiding landforms was carried out. As a result, the geological component of the risk (its spatial-areal index) was determined.

For example, in figure 2, which shows the degree of erosion of temporary streams the territory of Krasnodar Krai, 12 regions are identified. A schematic map (figure 3) of the density of the gully network shows 6 areas with difference in ruggedness index. In figure 4, according to the total area of subsiding landforms, expressed as a percentage of the area of a unit square, 9 regions were identified. Figure 5 shows the zoning distribution of subsidence rocks.

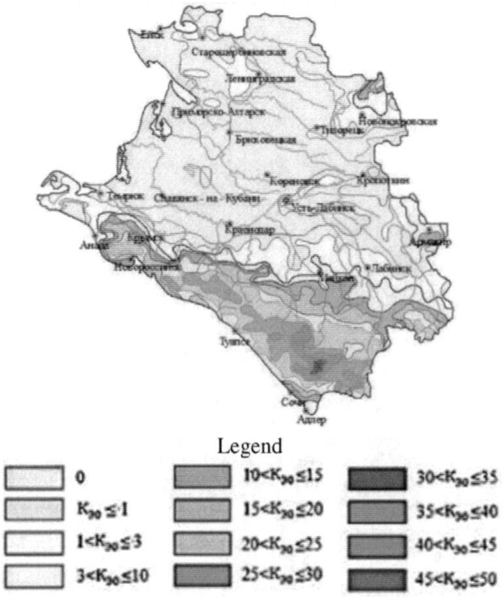

Figure 2. Schematic map indicating the degree of erosion of temporary streams.

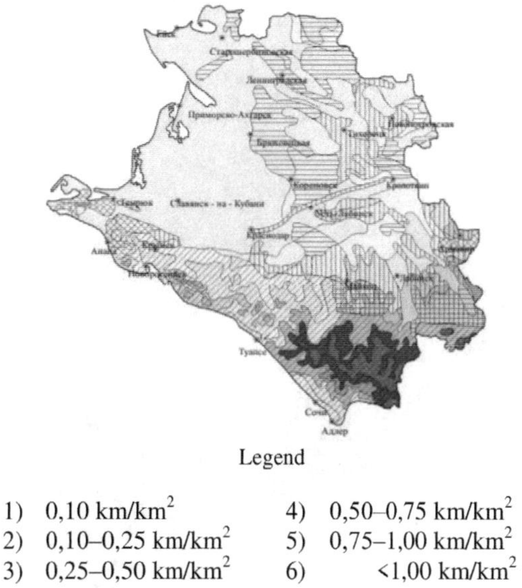

Figure 3. Schematic map is a diagram of the density of the gully-beam network.

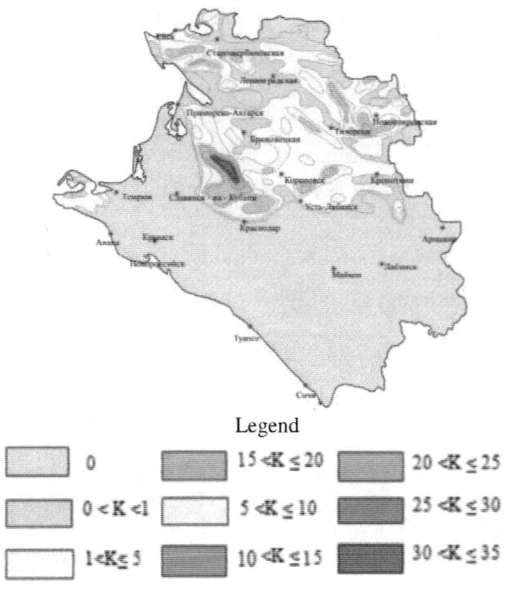

Figure 4. Schematic map is a diagram of the distribution of subsidence landforms.

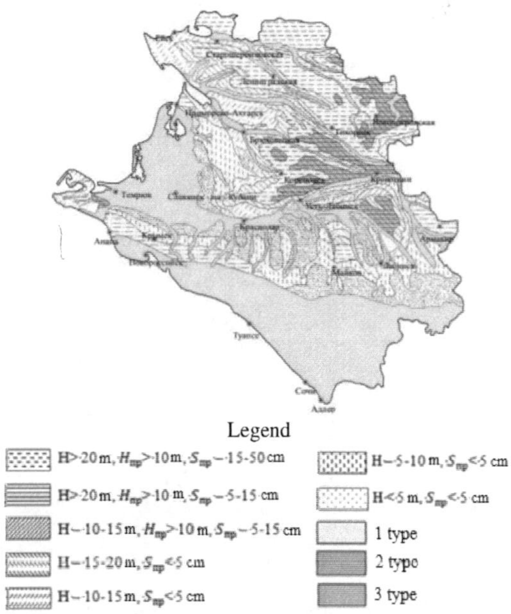

Figure 5. Schematic map of the distribution of loess-like rocks.

IOP Conf. Series: Earth and Environmental Science **946** (2021) 012036 doi:10.1088/1755-1315/946/1/012036

Subsequently, to compare these maps, overlay operations were performed, as a result of which the spatial characteristics of ARC/INFO layers were combined into a new layer and relational join of their attribute data was carried out.

3. Results

Figure 6 shows a map that provides information on the intensity of erosion forms and the density of the gully network distribution. Similarly, a map of loess-type rocks and the location of subsiding landforms in an integral form was created (figure 7).

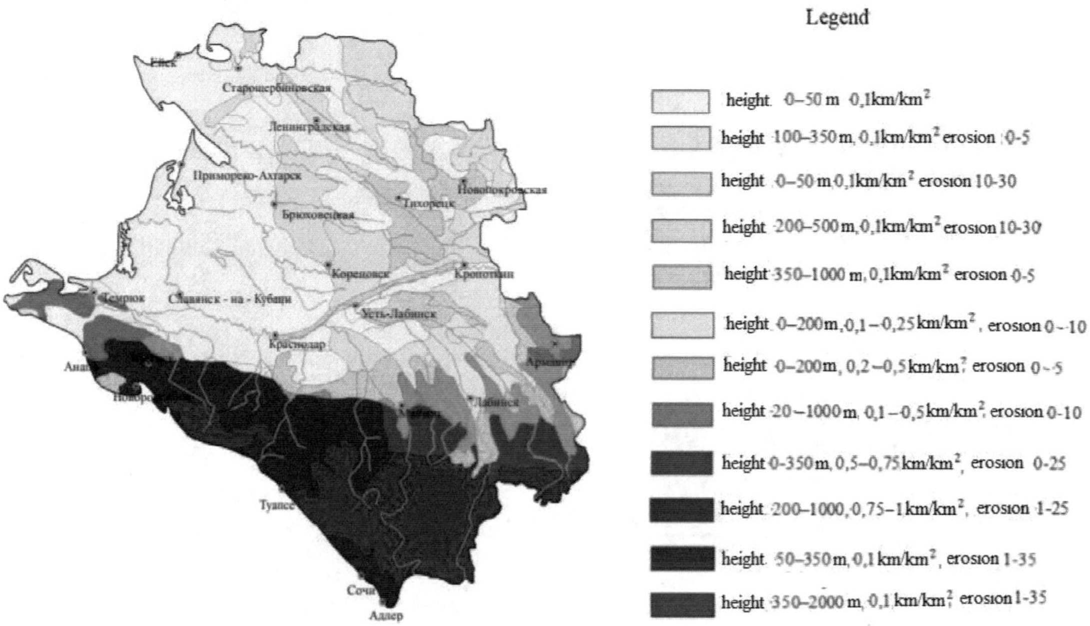

Figure 6. Combined schematic map is a diagram of the intensity of erosion forms and the density of the gully-beam network.

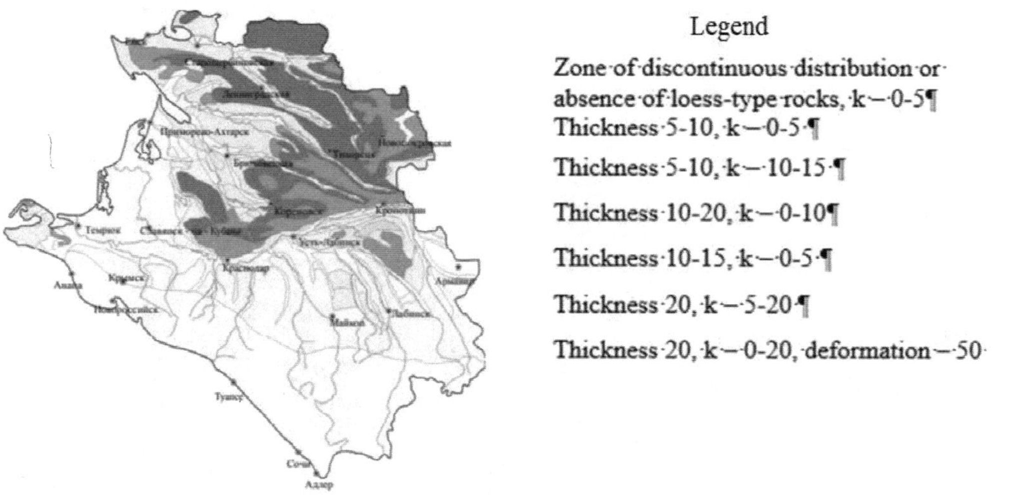

Figure 7. Combined schematic map of the distribution of loess-type rocks and the location of subsiding landforms.

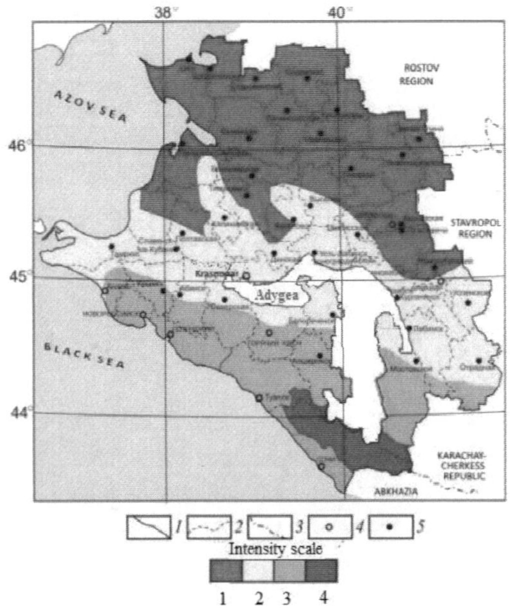

Adding the risks of each of the considered parameters by summing the above maps made it possible to rank the territory according to the degree of integral risk of manifestation of endogenous and exogenous processes. Figure 8 shows the results of zoning the study area according to the risk indicators with the allocation of relatively homogeneous parts and division of the criteria into classes. Areas with similar natural conditions were combined into four regions (zones) according to the level of potential risk of hazardous endogenous and exogenous processes: low, medium, increased and high.

Figure 8. Map of the integral risk of endogenous and exogenous processes in Krasnodar Krai. The intensity scale of the levels of potential risk of hazardous natural processes: 1 – low, 2 – medium, 3 – increased, 4 – high.

4. Conclusion

The performed integral assessment of hazardous endogenous and exogenous geological processes makes it possible to assess potential natural risks at the regional level. The lowest potential risk of natural hazards is characteristic of the plain territories the territory of Krasnodar Krai, the highest potential risks are typical for mountainous areas, for which the development of risk management measures will be required, since their consequences may require significant material expenses and time to be eliminated.

References

[1] Kosinova I and Kustova N 2008 Theory and methodology of geoecological risks *Vestnik VSU, Series Geology* **2** pp 189–197

[2] Osipov V, Eremina O and Kozlyakova I 2017 Assessment of exogenous hazards and geological risk in urbanized areas (review of foreign experience) *Geoecology, Engineering geology, Hydrogeology, Geocryology* **3** pp 3–15

[3] Cascini L and Ferlisi S 2014 Introduction to the thematic set of paper on the quantitative analysis of landslide risk *Bull. of Engineering Geology and Environment* **73** 2 pp 207–208

[4] Sharma V 2006 *Engineering Geology for Tomorrow's Cities, IAEG 2006, 6–10 Sept.* Zonation of landslide hazard for urban planning: case study of Nainital town, Kumaon Himalaya, India pp 3–191

[5] Vranken L, Vantilt G, Van Den Elckhaut M, Vanderckhole L and Poesen J 2019 Landslide risk assessment in densely populated hilly area *Landslides, 2015* **12** 4 pp 787–798

[6] Burova V N Features of zoning of urbanized areas for risk assessment from hazardous natural processes *Geoecology, Engineering geology, Hydrogeology, Geocryology* **6** pp 106–111

[7] Larionov V, Frolova N and Ugarov A 2017 Seismic risk assessment *Geoecology, Engineering geology, Hydrogeology, Geocryology* **2** pp 22–37

[8] Osipov V, Larionov V, Sushchev S, Frolova N, Ugarov A, Kozharinov S and Barskaya T 2015 Assessment of the seismic risk in the territory of B. Sochi *Geoecology, Engineering geology, Hydrogeology, Geocryology* **1** pp 3–19

[9] Stognii G and Stognii V 2019 *Proc. VII Int. Sci. and Pract. Conf.* The problem of assessing the seismogenic risk of the Black Sea coast of the Krasnodar Territory (Krasnodar, Kuban: State University) pp 226–229

[10] Lyubimova T, Bondarenko N, Stognii V and Pogorelov A 2017 Map of the integral risk for manifestation of exogenous geological processes as a basis for solving various theoretical and applied problems in the Krasnodar Territory *Natural and technical sciences* **2** 114 pp 140–145

[11] Bondarik G and Pendin V 1982 Methodology of quantitative assessment of engineering-geological conditions and special engineering-geological zoning *Engineering geology* **4** pp 82–89

On the issue of the impact of adverse and dangerous natural processes on the use of coastal recreational areas of the Russian Federation

A Yu Sanin[1,2,*] **and V Kulakovskaya**[2]

[1]Zubov State Oceanological Institute, Roshydromet, Moscow, 119034, Russia
[2]Lomonosov Moscow State University, Moscow, 119447, Russia

E-mail: [*]eather86@mail.ru

Abstract. Adverse and dangerous natural processes are a deterrent factor in the economic, in particular, recreational use of the coastal territories of Russia. They significantly reduce their tourist potential. At the same time, a decrease in the recreational value of territories is one of the consequences of human impact on the natural landscapes of the coastal zone. It seems important to consider the impact of adverse and dangerous natural processes specifically for coastal recreational areas due to their exceptional importance for domestic tourism. It, in turn, plays a huge role both for the economy of Russia and certain regions of our country, and for Russians themselves: tourists and local residents.

As a result of the research, for each of the seaside recreational areas available in Russia, the study revealed the most characteristic adverse and dangerous natural processes. Some of them are more or less manifested everywhere, for example, abrasion processes or storms. Others, in particular, seismic threat, are characteristic only for certain coastal territories. Measures to counteract dangerous natural processes are also somewhat similar for all the regions considered. However, when developing measures to counteract adverse and dangerous natural processes, the peculiarities of coastal regions should also be taken into account. Such features include climatic and relief characteristics, the landscapes of the region, the existing structure of human activity and the intensity of economic development of the region, the types of tourism that are characteristic of it, and the level of popularity of the region among recreants, as well as its recreational potential.

1. Introduction

Mass tourism is characteristic for only a small percentage of the sea coasts and coasts of inland waters of the Russian Federation, but it is one of the key areas of domestic tourism, the role of which is great for both the country's economy and its citizens. Recently, the role of domestic tourism has been noticeably increasing due to the COVID-19 pandemic and its consequences, in particular, the falling incomes of many Russians and the complication of tourist trips abroad. Recreational activity should be recognized as a priority for those coastal territories for which, due to climatic conditions and other factors (the level of development of recreational infrastructure, preferences of tourists, transport accessibility, etc.), mass recreation is most characteristic, as it happens in many cases abroad [1]. The main purpose of the research is to study adverse and dangerous natural processes in coastal recreational areas. Due to the significant importance of such territories (they play a key role for domestic tourism) the research focused on them.

Tourism in the coastal zone is subject to the negative impact of adverse and dangerous natural processes, which leads to material damage and human casualties. A number of researchers have studied the influence of such processes for the territory of the Russian Federation as a whole, including for coastal territories [2, 3]. In general works on nature management in the coastal zones of the Russian Federation, the influence of adverse and dangerous natural processes is somehow taken into account [1,4–9], as well as in regional works. In some of them consideration of various adverse and dangerous natural processes is a key or one of the key issues [6, 10]. The authors identified and characterized the adverse and dangerous natural processes for the coastal recreational territories of the Russian Federation and in general terms for those that relates to inland water bodies, and offered recommendations to reduce their negative impact on human economic activity. The available scientific literature contains recommendations on counteraction adverse and dangerous natural processes, however, for the entire territory of Russia as a whole or for certain regions. The authors in this article have prepared and summarized them specifically for all seaside recreational areas, taking into account their peculiarities.

Different regions of the Russian Federation are characterized with a various adverse and dangerous natural processes. Some of them manifest themselves exclusively or to the greatest extent in the coastal zone (for example, the processes of abrasion and erosion of the coasts), others cover a significant area of territory, including coastal zones. They include abnormal air temperatures, strong winds, heavy precipitation and floods as a consequence, earthquakes, etc. Dangerous processes characteristic primarily to mountainous territories, in particular, slope processes (landslides etc.), are provoked by abrasion and are found on many parts of the coast [2]. Abrasive processes pose a significant threat to engineering structures, including future ones, which will appear on the areas of coastal zones currently being developed for recreational purposes. At present, they are characteristic of 41 % of the coasts of the seas and 36 % of the coasts of water reservoirs [11]. For some water bodies, this part is noticeably higher. For example, up to 70 % of the coasts of the Sea of Azov are subject to erosion [12].

Adverse and dangerous natural processes in coastal areas are also currently characterized with activation due to human impact. So, the removal of sand and pebbles within the coastal zone enhances abrasion processes, deforestation on slopes intensifies mudslides, as well as floods, there are some other examples. Consequently, adverse and dangerous natural processes can be considered as a geoecological problem along with a decrease in the recreational potential of landscapes under human influence, environmental pollution and some others.

However, the nature-human system is characterized by two-way connections. If adverse and dangerous natural processes are the most noticeable negative consequence of the human impact on nature, it is also necessary to take into account the human impact on nature, which leads to a significant change in natural landscapes. The negative consequences of human impact on both the recreational potential of coastal territories and their natural landscapes as a whole are described both in Russian [4, 5, 7, 9] and foreign [13–16] literature.

The purpose of the research is to characterize adverse and dangerous natural processes characteristic of coastal recreational areas of Russia and to propose measures to counteract them. The objectives of the research include at least four:

1) List the areas of the coast on which dangerous natural processes is manifested quite acutely and for which it is recommended to minimize their economic use.
2) Show the importance of informing the public society about dangerous natural processes.
3) Outline ways to combat dangerous natural processes on the most valuable parts of the coast.
4) Justify the necessity to ban capital construction in beach areas.

2. Research methods and materials used

The following methods were used to conduct the research and prepare the current article:

- Comparative descriptive, which made it possible to compare the coastal territories and those related to inland water bodies, as well as Russian and foreign coastal territories, to identify their geoecological risks and their features.

- The method of scientific synthesis, which made it possible to generalize and compare with each other the information previously obtained from large-scale topographic maps, navigation maps for coastal waters, thematic maps, in particular, maps of the spread of adverse and dangerous natural phenomena, published literature and information taken from other sources.
- The method of keys, which, in turn, allows applying the method of analogues, of course, taking into account the individual characteristics of coastal territories. It was used to interpolate characteristics of adverse and dangerous natural processes in the territory for which there was a lack of information.

3. Research results

Adverse and dangerous natural processes characteristic for all coastal regions of Russia, which are characterized by mass recreation, as well as for coastal zones of inland water bodies, are shown in Table 1 below. The table is compiled based on author's research, the above and other literary sources, available maps and remote sensing data of the Earth. Here natural dangerous processes are typified into atmospheric, hydrological and hydrogeological (first group) and geological.

Table 1. Adverse and dangerous natural processes typical of coastal recreational areas and possible measures to counteract them.

Coastal region of Russia	The most typical atmospheric, hydrological and hydrogeological natural dangerous processes peculiar to the region	The most typical geological natural dangerous processes peculiar to the region	Possible countermeasures against natural dangerous processes minimization the damage from them.
The southern coast of Crimea	Upwelling of cold waters in the coastal zone, thunderstorms, storms, sudden jumps in atmospheric pressure and humidity, droughts, hail.	Rockfalls, landslides (including man-made ones) and landslides, mudslides (especially in the South-Eastern Crimea), abrasion processes), earthquakes (which, in turn, are the "trigger" for rockfalls) and landslides), karst processes (as a rule, in areas adjacent to the coastal zone).	Minimization of economic use of the coastal areas with the most active manifestation of natural dangerous processes, creation of coast protection structures (and maintenance of existing ones) on the most valuable sites. Consideration of seismic hazard in the construction of any engineering structures. Prohibition of capital construction in the area of beaches and in river valleys with increased mudflow hazard.
The West Coast and all the other coasts of Crimea	Storms, extreme heat, dry winds and dust storms.	Landslides, abrasion processes (the rate of retreat of the coasts in some areas reaches 1 meter per year or more [7]), water erosion.	Minimization of the economic use of coastal areas with the most active manifestation of these processes, the creation of artificial beaches, the prohibition of capital construction within the beaches.
Black Sea coast of The Caucasus	Heavy rainfall and thunderstorms, tornadoes, floods, surges in the mouths of large rivers, extremely high temperatures, droughts, sudden jumps in	Mudslides, avalanches, landfalls and rockfalls, landslides, abrasive processes, seismic threat. Landslides are especially active in the southern part of Large Sochi, where they	Restriction of construction in the valleys of gullies, for which mudslides are most characteristic, informing the population about safety measures in case of avalanches and stopping them in those areas where they threaten the

	atmospheric pressure and humidity, sea storms, bora in the surroundings of Novorossiysk	affect the slopes for 50–80 %. The capacity of mudslides is up to 100 thousand m^3 or more [17]. Landslides and mudslides are also provoked by a technogenic factor (deforestation, pruning of slopes during road construction, etc.). The repeatability of mudslides is 10–20 years.	life and health of people, buildings and structures. Defense of coasts of coastal areas where slope processes and abrasions threaten the health and life of recreants. Monitoring of railways and highways in areas subject to landslide processes. Prohibition of capital construction in the area of beaches and in mudslide-prone river valleys. Consideration of the seismic threat.
Azov sea coast within the territory of Russian Federation	Driving-surging processes. Storms, strong winds (including wringing wing, which can lead to deaths among recreational swimmers) extreme temperatures, droughts, dry winds, thunderstorms, ice hummocks.	Abrasive processes. The rate of coastal retreat in some areas is higher here than on the coasts of all other recreational coastal territories, up to several meters per year [12]. Landslides within the coastal zone.	Compliance with safety measures to avoid human casualties during storms. Avoiding the placement of engineering structures near the edge of the sea on the abrasive areas of the shores and near them. Coastal protection works, in particular, the creation of artificial beaches on the most valuable sites.
Caspian sea coast	Strong winds, storms, dust storms, dry winds, extremely hot temperatures, droughts. Driving-surging processes, storms, fluctuations in sea level (only in 1975–1995 he rose two meters [9]).	Abrasion, slope processes on the shores and in mountainous areas, seismic threat, erosion processes in the coastal zone.	Economic development, with consideration to the significant amplitude of both short-term and long-term fluctuations in sea level, protection of the most important areas from them. Positioning the region as an "autumn" resort to avoid trips during hot temperatures.
The Baltic coast within the Russian Federation	Storms, blizzards, snowfalls, fogs, icing, ice hummocks. The superimposition of driving-surging processes on storms leads to a rise in the level by 3–4 m [6].	Abrasion processes, slope processes on the coasts, water erosion.	Creation of protective dams for the most important sections of the coast to protect from driving-surging processes, shore protection in certain areas, avoiding the use of areas with the greatest manifestation of natural dangerous processes.

The coast of the Sea of Japan near Vladivostok	Storms, thunderstorms, heavy rains and floods, ice hummocks, tsunamis. Tsunamis have been repeatedly recorded over the past 100 years (up to several meters), theoretically higher tsunamis are also possible [18].	Abrasion processes, mudslides in some areas, slope processes on the coasts and adjacent territories, water erosion.	Maintenance of the warning system of the population about the tsunami hazard and informing about the procedure in this case, shore protection work, if necessary, avoiding economic development of areas with the most pronounced natural dangerous processes.
Coasts of inland bodies of water actively used for recreation	It depends on the region: all of the above adverse and dangerous natural processes may manifest themselves. Flooding, waterlogging and overgrowth of coastal waters with hydrophilic vegetation. In the coastal zone of some waterbodies, in particular, lakes of Altai and Baikal.	Landslide processes on the coasts, abrasion and erosion of the coasts.	Avoiding the use of sites with the most pronounced landslide processes, preventing the construction of engineering objects in the immediate vicinity of the water's edge in areas with strong processes of abrasion and coastal erosion. Compliance with restrictions on the economic use of water protection zones and coastal protection areas.

4. Conclusion

As follows from Table 1, for most recreational coastal areas, many measures to counteract adverse and dangerous processes are also similar. It is possible to distinguish such common measures for all studied areas:

1) Minimization (if possible) of the economic use of coastal areas where adverse and dangerous natural processes are most acute. These include, in particular, the bottoms of beams, especially their estuaries facing the seashore, areas of the coast with active abrasion processes, landslide slopes, etc.
2) Informing recreants and local residents (residents, however, as a rule, have a much better idea of the threats from dangerous natural processes than recreants) about the risks and dangers caused by adverse and dangerous natural processes. So, tornadoes approaching the shore can become the object of photography, and storm waves encourage some recreationists to swim in the sea, despite the risk.
3) The struggle for the most attractive areas of the coast for economic activity: the implementation of coasts protection works (methods of so-called soft and hard shore protection can be applied), anti-landslide measures, etc.
4) Prohibition of capital construction in the area of beaches in connection with the threat from storm waves, and in some cases, driving-surging processes for such engineering structures on the seashores and on the coasts of large inland bodies of water.

Without taking into account the negative impact of adverse and dangerous natural processes, effective environmental management is impossible in coastal tourist areas, as they threaten the life and health of local residents and existing engineering objects. Without due regard to adverse and dangerous natural processes both recreational and other types of economic use of territories in the present and its planning for the future are impossible.

References

[1] Dolotov Yu S 1996 *Problems of Rational Use and Protection of Coastal Areas of the World Ocean* Moscow (Scientific world) p 304
[2] Kutepov V M and Sheko A I 2002 *Natural Hazards of Russia* **3** (Moscow: Publishing house of the firm "KRUK") p 348
[3] Kuzmin S B 2009 *Dangerous Geomorphological Processes and the Risk of Environmental Management* (Novosibirsk: Publishing house "GEO") p 195
[4] Aybulatov N A 2005 *Russia's Activity in the Coastal Zone of the Sea and Environmental Problems* (Moscow: Nauka) p 364
[5] Ignatov E I 2006 *Morphosystem Analysis of the Coasts* Moscow (Smolensk: Magenta) p 406
[6] Gordeeva S M and Malinin V N 2014 *Variability of the Sea Level of the Gulf of Finland* (St. Petersburg: Publishing House of RGGMU) p 179
[7] Goryachkin Yu N and Ivanov V A 2006 *The Level of the Black Sea: Past, Present and Future* (Sevastopol: Publishing House of the Marine Hydrophysical Institute) p 210
[8] Malinin V N 2012 *Ocean Level: Present and Future* (St. Petersburg: Publishing House of RGGMU) p 260
[9] Safyanov G A 1996 *Geomorphology of Sea Coasts* (Moscow: Publishing House of the Geographical Faculty of Moscow State University) p 400
[10] Klyukin A A 2007 *Exogeodynamics of the Crimea* Simferopol (Tavria) p 320
[11] Ragozin A L 1993 Ranking of Dangerous Natural and Technological Processes by Socio-Economic Damage From Their Manifestation on the Territory of Russia *Problems of safety in emergency situations* **2** pp 50–61
[12] Selivanov A O 2001 *Coastal Disaster on the Sea of Azov. Myth or Real Threat*? (Moscow: GEOS) p 84
[13] Barkin D 2002 Eco Tourism for Sustainable Regional Development *Current Issues in Tourism* **5** pp 245–253
[14] Carter R W G 1987 *Coastal Environments* (London-Toronto) p 617
[15] Cicin-Sain B and Knecht R W 1998 *Integrated Coastal and Ocean Management: Concepts and Practices* (Washington D C: Island Press) p 517
[16] K Robert and J Alder 1999 *Coastal Planning and Management* (London and New York: ESF13) p 414
[17] Malyneva I V Activity of Mudslides and Landslides on the Black Sea Coast of the Caucasus and Adjacent Mountainous Areas at the Beginning of the 21st Century (*Electronic materials*) https://docplayer.ru/31661558-Aktivnost-seley-i-opolzney-na-chernomorskom-poberezhe-kavkaza-i-prilegayushchih-gornyh-rayonah-v-nachale-xxi-veka.html
[18] Ignatov E I, Frol V V, Lokhin M Yu and Nikiforov A V 2008 *Geomorphological Problems of Tsunami Hazard (on the Example of the Sea of Japan)* (Smolensk: Magenta) p 128

IOP Conf. Series: Earth and Environmental Science **946** (2021) 012038 doi:10.1088/1755-1315/946/1/012038

Toxic element concentrations in lichen *Platismatia interrupta* around the geothermal power station on Kunashir Island

A K Ezhkin* and A V Kordyukov

Institute of Marine Geology and Geophysics, Far Eastern Branch of the Russian Academy of Science, Yuzhno-Sakhalinsk, 693022, Russia

E-mail: *ezhkin@yandex.ru

Abstract. The results of the study of the gross content of 14 chemical elements in lichen Platismatia interrupta collected around the geothermal power station "Mendeleevskaya" on Kunashir Island are presented. The highest exceedance was noted for As in the research area. The gross content of As varies from 0.8–2.6 mg/kg in control areas and up to 5–9 mg/kg in the impact zone, i.e. the maximum excess in lichen is noted by 11.6 times. For other elements, there were no strong exceedances in the impact zone compared to the control areas. According to the results of regression analysis, a statistically reliable inverse exponential relationship between the content of As and the distance to the power station (adjusted R2: 0.86, p-value: 0.001) is traced. The average value of the enrichment coefficient for As exceeds 100, which indicates the non-substrate origin of this element.

1. Introduction

Industrial development of geothermal fields is one of the most economical and promising methods of power generation. However, operations carried out on geothermal fields can destroy local ecosystems, including mechanical damage to natural landscapes, water and air pollution [1, 3]. Geothermal steam often contains a large amount of salts and among them are very often ions of F^-, B^{3+}, $As^{3+, 5+}$ and heavy metals. Geothermal steam also contains from 0.5 to 6 % by weight non-condensable gas, mostly CO_2, H_2S, CH_4, NH_3 [7]. Concentrations of H_2S in natural steam range significantly from 30 mg/l and up to 500-1000 mg/l as per the particular deposit.

The content of hydrogen sulfide in the air above 50 mg/l paralyzes a human, and death occurs at higher concentrations. Such accumulations of hydrogen sulfide occur on geothermal fields [4]. A toxic substance – ammonium in high concentrations – 88 mg/l was found in the condensate of the steam discharged into the environment by the geothermal power plant on the Iturup Island [5]. The release of carbon dioxide into the atmosphere together with the evaporation of a large amount of water can lead to thermal pollution of the environment and local changes in climatic conditions, especially in the areas with big power stations [7]. In the areas of modern volcanism of the Kuril region, all uses of thermal waters are possible. There are high regional average values of deep heat flow – up to 92 mW/m^2 and there are many thermic anomalies – natural hydrothermal outputs of a wide variety of composition and temperature [7].

The purpose of this work is to assess the gross content of a number of toxic elements in thalluses of the lichen *Platismatia interrupta* W.L. Culb. et C.F. Culb. and to determine the spatial distribution of toxic chemical elements around the geothermal power station "Mendeleevskaya" on Kunashir Island. Research on lichen accumulation of pollutants in the area of operation of the geothermal power station "Mendeleevskaya" on Kunashir Island is performed for the first time. There is not much research on the impact of geothermal power plants on the environment by lichenoindication methods, due to the specifics of production and the difficult access to facilities. There are two such geothermal power stations on Kuril Islands. The first researches of impact of geothermal power stations and geothermal wells on lichens were carried out in Italy and Mexico [16, 8, 10]. Laboratory studies of lichens collected on geothermal fields in Italy showed an increased level of toxic elements such as S, B, As, Hg and others [2, 6, 9].

2. Research area

Research was performed around the geothermal power station "Mendeleevskaya" in the vicinity of the town of Yuzhno-Kurilsk at the foot of the Mendeleev volcano on Kunashir Island (figure 1). Mendeleev volcano shows intense fumarolic activity. The "Mendeleevskaya" geothermal power station operates on the basis of the geothermal field "Goryachiy Plyazh" since 2002 year and provides an electricity to Yuzhno-Kurilsk town. The power of the station at the moment is 3.6 MW. The steam-water mixture is discharged from the pipes in the vertical and horizontal directions, in windless weather, the steam tail of the vertical pipe can reach a height of more than 10 m and extends tens of meters around the power plant. The vegetation of the research area is typical for the island of Kunashir and is represented by coniferous forests formed by *Picea glehnii* (F.Schmidt) Mast., *P. jezoensis* (Siebold & Zucc.) Carrière and *Abies sachalinensis* (F.Schmidt) Mast. with the dominance of the first. The grass layer is formed by representatives of the genus *Sasa* Makino & Shibata. The average age of coniferous trees in the surveyed forest areas is 100–150 (200) years.

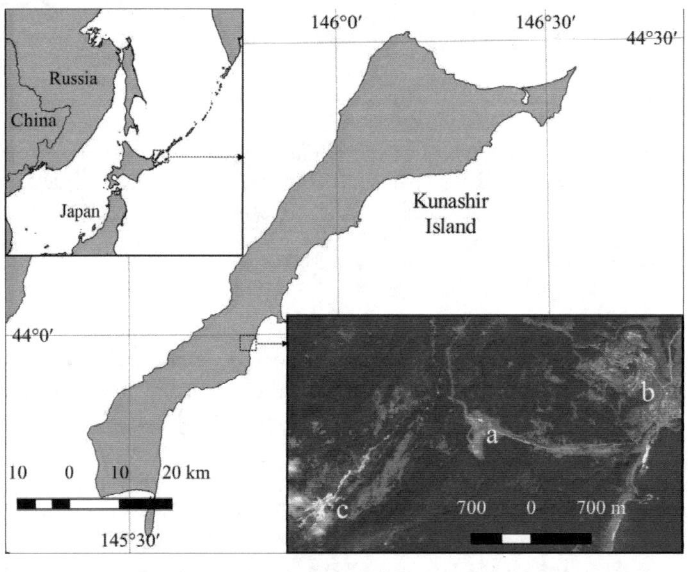

Figure 1. Research area: a – geothermal power station "Mendeleevskaya", b – Goryachiy Plyazh village, c – north-eastern solphataric field of the Mendeleev volcano.

3. Materials and Methods

The widespread lichen *Platismatia interrupta* from the bark of *Picea glehnii* was used for chemical analyses. Samples of the lichen were collected in August 2015 in the vicinity of the geothermal power station "Mendeleevskaya" from 9 sites remoted from the power plant at different distances in all directions. Thalluses of the lichen were investigated on the content of 14 chemical elements – Al, S, Cr, Mn, Fe, Co, Ni, Cu, Zn, As, Mo, Cd, Pb, Hg by a mass spectrometry method with inductively connected plasma on the Agilent 7500a device by the technique FR.1.31.2009.06787. The required amount of test substance is 1 g of dry weight of lichen material. Samples were taken from at least 6 tree trunks per site in each study area on 1.2–1.8 m above the level of soil cover, as described in the "Monitoring with Lichens" manual [13].

To assess the effect of the composition of the Earth's crust on the elemental composition of samples [17, 12, 15], widely used calculated enrichment factors were used for each element (X) with respect to the content of this element of the reference material and the content in the sample and reference material of the reference element by the formula:

$$EF = \frac{(X/reference\ element)\ in\ lichen}{(X/reference\ element)\ in\ reference\ material}$$

For reasons described [17, 11], the upper continental crust [18] was chosen as the reference material, and aluminum is chosen as the reference element (Fe, Ti and Sc are also often used as the reference element).

4. Results and discussions

Gross content (mg/kg) of chemical elements in the lichen *Platismatia interrupta* (except Cd, the contents of which were obviously smaller than the detectable values) are shown in the Table 1.

Table 1. Gross content (mg/kg) of chemical elements in the lichen *Platismatia interrupta* at 9 sampling points.

№	Al	S	Cr	Mn	Fe	Co	Ni	Cu	Zn	As	Mo	Hg	Pb
1	752.0	763.5	1.5	204.2	577.4	0.5	1.4	4.0	32.4	8.6	0.3	0.0	3.5
2	401.6	517.1	3.3	240.6	407.4	0.4	0.8	5.1	28.7	5.0	0.3	0.0	3.8
3	507.3	806.8	3.5	99.7	473.4	0.5	2.0	4.8	35.1	9.0	0.3	0.0	6.1
4	1397.1	797.4	3.4	160.6	984.1	0.5	1.5	5.0	40.5	7.5	0.3	0.0	5.8
5	547.2	674.9	1.8	76.3	536.4	0.5	2.1	4.9	29.0	1.3	0.4	0.0	5.1
6	627.8	933.7	5.4	99.9	593.4	0.6	1.6	5.0	37.5	2.6	0.4	0.1	5.9
7	690.8	659.6	3.5	162.7	620.0	0.4	1.1	3.7	33.4	1.5	0.2	0.1	5.2
8	605.7	577.0	1.4	133.7	542.3	0.5	1.4	3.9	28.6	0.8	0.3	0.0	4.5
9	660.8	757.5	3.5	168.2	524.7	0.4	1.3	4.7	34.0	1.1	0.2	0.0	5.3

Geothermal power plant steam is discharged through two pipes located in the northern and southern parts of the station. Plots 1–4 are located in forest areas most exposed to steam. Their distance from the pipes is 122–242 m. They are in the impact zone. Damaged lichens were noted on trees in the impact zone. The trees themselves on the edge of the forest are strongly damaged or dead (figure 2).

Figure 2. Damaged fir trees in impact area of the geothermal power station "Mendeleevskaya" on the Kunashir Island (left picture), geothermal steam discharge (right picture).

Sampling plots 5–9 are remoted from pipes by distance 468–1028 m; they are less exposed to emitted vapors. They represent control areas. For most of the elements, the gross content in the impact and control zones do not have statistically significant differences. Data for all sites are uniform – the coefficient of variation does not exceed 33.5 % for all elements except Al (39 %) and Cr (40 %). The main exception is As. In control areas, its content is 0.8–2.6 mg/kg, and in the impact zone – 5–9 mg/kg, the coefficient of variation is 78 %. The maximum excess relative to the minimum is 11.6 times.

Based on the results of regression analysis for As only, a statistically reliable inverse exponential relationship between its content and the distance to power station (adjusted R^2 0.86, p-value: 0.001) is traced. Mapping of As content in the lichen *Platismatia inerrupta* by inverse distance weighting method was also performed. The scattering diagram and spatial distribution map are shown in figure 3.

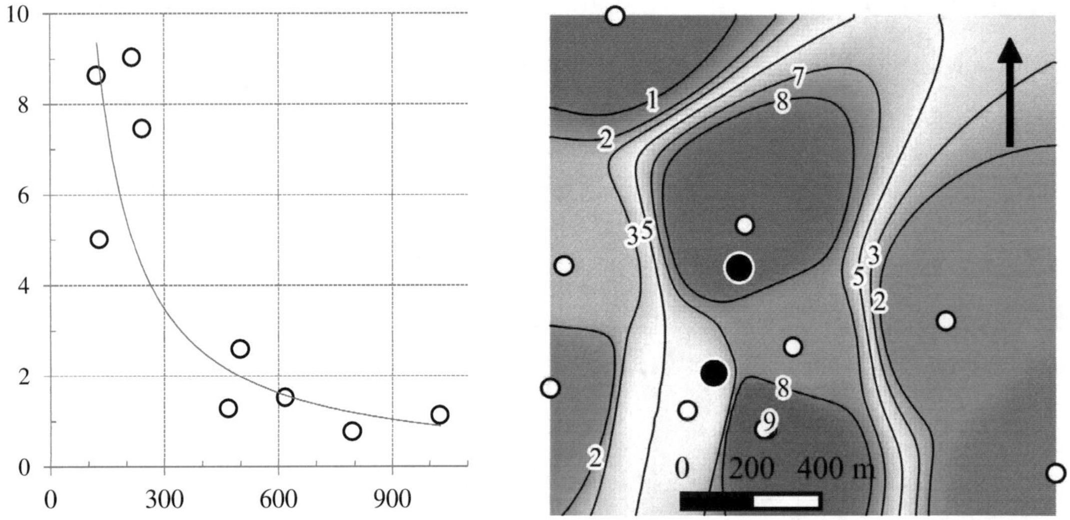

Figure 3. Scatter plot of gross content of As in the lichen *Platismatia interrupta* (y axis, mg/kg) at various distances from power station discharge pipes (x axis, m) and spatial distribution of elements (IDW interpolation). Full circles indicate the places of discharge of vapors (pipes), hollow circles – sampling areas, isolines on the map show the content of As in lichen thalluses in mg/kg.

For the remaining elements, the "zero" hypothesis was confirmed – no statistically significant correlations between the gross content in the lichen and the distance to the used predictors were noted. As predictors, the distance to pipes, access road and power station border were used. The spatial distribution of the gross content of these elements does not show patterns and general trends.

An enrichment factor (EF – enrichment factor) was calculated to estimate the ratio of anthropogenic to natural lithogenic sources of chemical elements entering the environment. The EF values are shown in figure 4.

It is generally accepted that EF values up to 5 indicate that the accumulation of elements in lichen thalluses comes from a reference material – the Earth's crust [17, 12, 11]. Thus, it can be concluded that Fe, Co, Ni and Cr (as well as the reference material Al) have a generally lithogenic occurrence. The next group of elements – Cu, Mn, Mo, Pb, Zn – have average EF values from 21.6 to 63.9 with minimum values from 10.4 to 35.3. Moreover, their spatial distribution is relatively uniform – the coefficient of variation is 11–17 % for all these elements, except Mn (33.4 %).

The gross content of S is characterized by average homogeneity (coefficient of variation 16.7 %). At the same time, the average EF is 149.4 (74.9–208.8), which indicates the unsubstantiated nature of its origin. We can assume that the high S content is associated with the activity of the Mendeleev volcano – 2 km southwest of the power plant is the northeast solfataric field, on which sulfur was mined in significant quantities in the past. In addition, the Kislyi stream flows from the solfataric field. Its channel approaches the power station up to 450 m. Along the stream there are three groups of thermal springs. Their waters also contain H_2S [19].

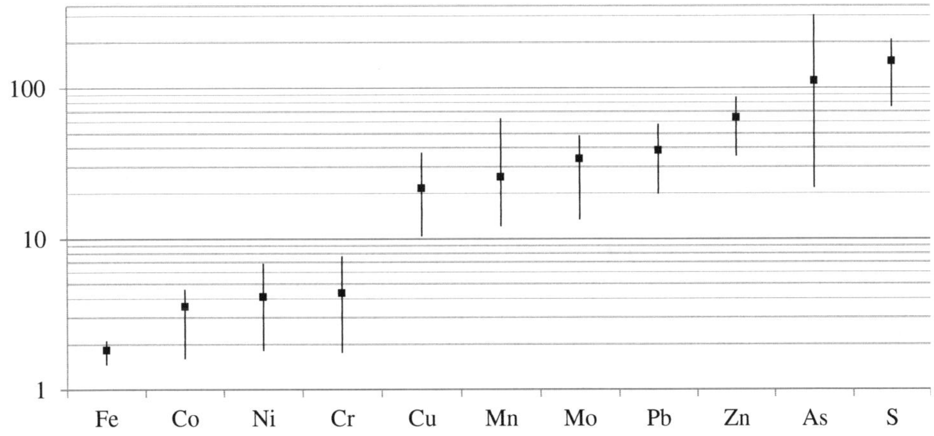

Figure 4. Values of enrichment factors (EF – enrichment factor) for the examined elements.

As is the only element characterized by high EF (average value is 111, ranging from 21 to 302) and high sample heterogeneity – a coefficient of variation of 77.5 %. In addition, as mentioned earlier, the gross content of As in lichen thalluses correlates closely and statistically reliably with the distance to the pipes.

As is a toxic substance and has an adverse effect on plants. The increased content of As in the soil leads to the dumping of needles and the death of thin roots of coniferous trees, reduces the growth of plants and development of mycorrhiza [14]. This is probably one of the reasons for the damaged trees and necrosis of lichens in the research area.

5. Conclusion

The results of the analysis of the content of toxic elements in the thalluses of the lichen *Platismatia interrupta* in the vicinity of the geothermal power station "Mendeleevskaya" show a mostly homogeneous distribution. In accordance with the values of the accumulation coefficient, Cu, Mn, Pb, Zn, As and S have a non-substrate origin.

For most elements, the relationship between their content and the predictors considered was not confirmed. The spatial distribution of their gross content does not show patterns and general trends and requires further study.

The high value of EF for S in combination with the average uniformity of distribution suggests that the source of S compounds is the solphataric activity of the Mendeleev volcano.

High EF value for As, non-uniformity of distribution and high correlation coefficient of gross content of As with distance to power station discharge pipes are noted. Due to usually increased content of As in geothermal products, it is very likely that the source of As supply to lichen thalluses is a steam discharged by a geothermal power plant. Apparently, As is one of the reasons for the noted necrosis of lichen thalluses and the damaged trees in the impact zone. The impact of power station on the surrounding ecosystems was noted at a distance of up to 400 m from steam discharge pipes.

References

[1] Ármannsson H and Kristmannsdóttir H 1992 Geothermal environmental impact *Geothermics* **21** 56 pp 869–880
[2] Bargagli R and Barghigiani C 1991 Lichen biomonitoring of mercury emission and deposition in mining, geothermal and volcanic areas of Italy *Environmental Monitoring and Assessment* **16** 3 pp 265–275
[3] Belousov V I and Belousova S P 2002 *Natural disasters and environmental risks (on the example of the development of geothermal energy)* (Petropavlovsk-Kamchatsky: Publishing House of KSPU)

[4] Ellis A J 1977 Geothermal fluid chemistry and human health *Geothermics* **6** 3 pp 175–182

[5] Ezhkin A K, Zharkov R V and Kordyukov A V 2015 Assessment of the impact of the Okeanskaya geothermal power plant (Baransky volcano, Iturup Island) on the environment by the lichenoindication method *Bulletin of the FEB RAS* **2** pp 109–117

[6] Ferrara R, Maserti B E, Andersson M, Edner H, Ragnarson P, Svanberg S and Hernandez A 1998 Atmospheric mercury concentrations and fluxes in the Almadén district (Spain) *Atmospheric Environment* pp 3897–3904

[7] Kononov V I 1983 *Geochemistry of thermal waters of the regions of modern volcanism (rift zones and island arcs)* (Publishing House "Science) p 379

[8] Loppi S 1996 Lichens as bioindicators of geothermal air pollution in central Italy *Bryologist* **99** 1 pp 41–48

[9] Loppi S and Bonini I 2000 Lichens and mosses as biomonitors of trace elements in areas with thermal springs and fumarole activity (Mt. Amiata, central Italy) *Chemosphere* **41** 9 pp 1333–36

[10] Loppi S and Nascimbene J 1998 Lichen bioindication of air quality in the Mt. Amiata geothermal area (Tuscany, Italy) *Geothermics* **27** 3 pp 295–304

[11] Loppi S, Pirintsos S A and De Dominicis V 1999 Soil contribution to the elemental composition of epiphytic lichens (Tuscany, central Italy) *Environmental Monitoring and Assessment* **58** 2 pp 121–131

[12] Nash T H and Gries C. Lichens as indicators of air pollution in *The Handbook of Environmental Chemistry* (New York: Springer-Verlag) Ed O Hutzinger **4** Part C pp 1–29

[13] Nimis P L, Scheidegger C and Wolseley P A 2002 *Monitoring with lichens—monitoring lichens* in Monitoring with Lichens—Monitoring Lichens (Springer Netherlands) pp 1–4

[14] Ormrod D P 1978 *Pollution in Horticulture* (Elsevier, Amsterdam, The Netherlands) p 260

[15] Paoli L, Munzi S, Guttová A, Senko D, Sardella G and Loppi S 2015 Lichens as suitable indicators of the biological effects of atmospheric pollutants around a municipal solid waste incinerator (S Italy) *Ecological Indicators* **52** pp 362–370

[16] Peralta M G and Carmona A C 1995 Líquenes como indicadores biológicos en el campo geotérmico Los Azufres, Michoacácan, México *Geotermia* **11** pp 137–143

[17] Puckett K J and Finegan E J 1980 An analysis of the element content of lichens from the Northwest Territories, Canada *Canadian Journal of Botany* **58** 19 pp 2073–89

[18] Rudnick R L and Gao S 2003 Composition of the continental crust *Treatise on geochemistry* **3** pp 1–64

[19] Zharkov R V 2014 *Thermal springs of the South Kuril Islands* (Vladivostok: Dalnauka) p 378

Epiphytic mosses in urban sites and the possibility of their use in biomonitoring (Yuzhno-Sakhalinsk)

V V Kaganov[1,2]

[1] Institute of Marine Geology and Geophysics FEB RAS, Yuzhno-Sakhalinsk, 693022, Russia

[2] Sakhalin Branch of the Botanical Garden-Institute FEB RAS, Yuzhno-Sakhalinsk, 693023, Russia

E-mail: vladimirkaganov@mail.ru

Abstract The article presents the study results on epiphytic mosses. The research area is Yuzhno-Sakhalinsk and its surroundings. To assess the epiphytic cover the method of occurrence measuring on the tree trunks with the help of a frame 10×40 cm (400 cm^2) was used at 15 locations within the study region. 15 species of epiphyte moss and their distribution characteristics were identified. *Populus maximowiczii* A. Henri was selected as a substrate because it predominates in urban vegetation and grows naturally in floodplain areas where control plots were set up.

1. Introduction

Populus maximowiczii was selected as a substrate as it is most widely represented in the plantings of the city [1]. This species was massively introduced into landscaping in the 1960's and 1970's of the last century. To date, those trees are about 70 years. Poplars were mainly planted in courtyards, city squares and along roadside edges. This species was chosen because of its availability. In river valleys, wild poplar forms pure poplar, willow-poplar and other mixed forests. Poplar is of low maintenance, takes root when planted in the city, and grows very fast. Taking into account all the above factors, it was decided to select *Populus maximowiczii* as the most affordable and successful substrate for bioindication.

Recently, the study of urban flora has attracted considerable interest due to the issue of preserving biological diversity. For the first time, high sensitivity of epiphytic bryophytes to contamination with sulfur oxides was noted at the beginning of the 20th century [2, 3]. With moss as indicators, the following characteristics are often defined by researchers: habitat, species composition, life forms, plant cover, abundance, spore-bearing specimens, etc. One of the main conditions for biomonitoring is assessing and comparing indicators within a single control plot [4].

The study was aimed at assessing the species structure of epiphytic bryophytes inhabiting *Populus maximowiczii* bark.

2. Study Region

Geographically, the study took place in Yuzhno-Sakhalinsk, its surroundings and the valley of Belaya river (Sokol settlement of Dolinsky district). Yuzhno-Sakhalinsk is located in the south-eastern part of Sakhalin Island in the central part of Susunaiskaya lowland elongated longitudinally between two ridges: the Mitsulsky in the west, and the Susunaisky in the east.

IOP Conf. Series: Earth and Environmental Science **946** (2021) 012039 doi:10.1088/1755-1315/946/1/012039

The area belongs to Yuzhno-Sakhalinsk climatic region with humid monsoon climate, relatively significant rainfall of about 860 mm per year, warm southern winds in summer and cold northern and north-westerly winds in winter. The average annual temperature is 2–3°C, the average minimum of -13,5°C being observed in January and the average maximum of 17,2°C being recorded in August. Frequent fogs and significant clouds characterize the climate of Yuzhno-Sakhalinsk as chilly and humid [5].

Specific geographical location and climatic conditions prevent the area from dispersal and removal of air pollutants from various sources, both stationary and mobile (rail and road transport, solid fuel boilers, etc.). As a result, a dense long-lasting smog layer is formed in the lower atmosphere. This phenomenon is often observed in cool seasons.

3. Materials and Methods

The study touches upon some distribution peculiarities of leaf-stemmed epiphytic moss in various conditions of anthropogenic impact. The research area was roughly divided into the following plots: strong impact zones (public gardens), a buffer park zone, and woodlands including control plots located at a distance from urban buildings. This distribution is due to the remoteness from the dense road network [1]. In urban areas, five plots were set up in public gardens, which are characterized by close proximity to public roads and are located in vicinity to intersections of major city roads: Sakhalinskaya Street – Mira Avenue, Lenina Street – Pogranichnaya Street, Lenin Street – Pobedy Avenue, Komsomolskaya Street – Sakhalinskaya Street. Another four plots were organized in the park zone in the Yuri Gagarin City Park of Culture and Recreation. All in all, five plots remote from major public roads were arranged on the territory of the Institute of Marine Geology and Geophysics of the Far Eastern Branch of the Russian Academy of Sciences, floodplain areas along Krasnoselskaya, Rogatka and Belaya rivers. Control plots are shown in figure 1.

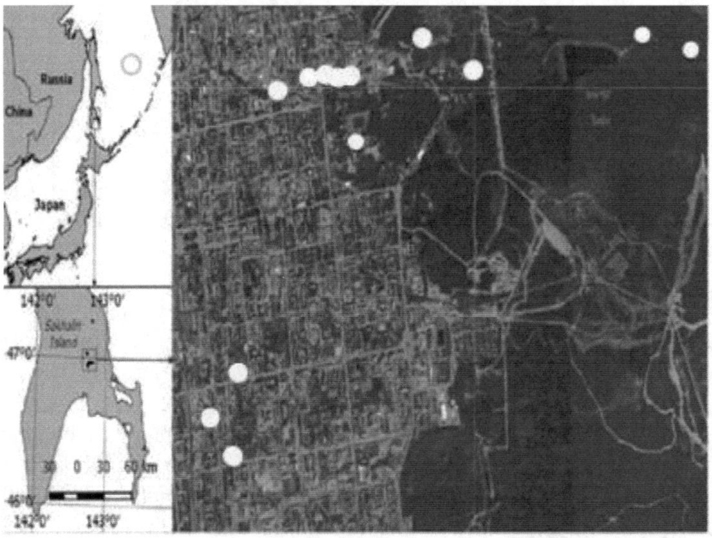

Figure 1. Outline map of the study region (Yuzhno-Sakhalinsk).

Moss cover studies were conducted in the summer of 2020–2021 in two stages:

1. reconnaissance operations including test areas (plots) arrangement and herbarium materials collection for subsequent identification;
2. measurement of the cover characteristics.

Sample trees were selected to be of the same diameter, approximately the same age, with at least ten trees in the plot, without inclination and/or with inclination not exceeding 10° relative to the ground. To assess the moss cover, the number of species and species occurrence frequency were measured. To determine the frequency, a 10×40 cm (400 cm^2) frame was used divided into 4 micro-sites, 10×10 cm each. The frame was applied from four sides of the tree trunk at height from 1,1 to 1,5 m. The frame image is shown in figure 2. The total of more than 2 400 micro-sites were set up in the research area. Based on these measurements, the occurrence frequency was calculated.

Figure 2. Frame image 10×40 (400 cm^2) on a tree trunk.

Samples were collected with location coordinates fixed. Garmin eTrex 30 GPS navigator system was used. The altitude above sea level was recorded and the exposure was marked. Samples were gathered in paper envelopes made of wrapping paper. 10x and 15x magnifying lenses were used. Particular attention was paid to samples with sporophytes for reliable subsequent samples identification.

Samples were dried by means of generally accepted methods described in modern literature. The collected species were placed in paper envelopes with a field label indicating: collection area (site point number), substrate, collection date, cardinal direction and other data. Identification was carried out using conventional bryophytes techniques [6, 7, 8]. When identifying the samples, the following microscopes were used: Biolam, Micromed-2, MBS-10, preparation needles, tweezers, slides and cover glasses for microscopy. Materials were processed and identified in Phyto- and Geo-Ecology Laboratory of the Institute of Marine Geology and Geophysics of the Far Eastern Branch of the Russian Academy of Sciences.

4. Findings of the Study

15 species with different participation share in poplar epiphytic cover formation were identified.

The species identified were assigned to two orders, 12 families and 13 genera. Most species belong to the order Hypnales Dumort. (14 species), only one species was assigned to the second order Orthotrichales Dixon (*Orthotrichum sordidum* cf. *obtusifolium* Hedw.).

Among 12 identified families, the dominant ones are *Pylaisiaceae* Schimp. (3 species), two species in each of the families *Anomodontaceae* Kindb., *Brachytheciaceae* Schimp., *Neckeraceae* Schimp. The families *Entodontaceae* Kindb., *Leskeaceae* Schimp., *Leucodontaceae* Schimp., *Orthotrichaceae* Arn., *Hylocomiaceae* (Broth.) M.Fleisch., *Scorpidiaceae* Ignatov & Ignatova include one species each.

The dominant genera are presented by *Anomodon* Hook. & Taylor и *Pylaisia* Bruch et al. (two species each). The other genera have one species each: *Brachythecium* Bruch et al., *Entodon* Müll.Hal., *Homalia* Brid., *Leskea* Hedw., *Leucodon* Schwägr., *Myuroclada* Besch., *Neckera* Hedw., *Orthotrichum* Hedw., *Pleurozium* Mitt., *Sanionia* Loeske, *Stereodon* (Brid.) Mitt.

Typical boreal leaf-stemmed moss species are dominant for the entire study area, such as *Leskea polycarpa* Hedw. (more than 60 %), *Orthotrichum sordidum* cf. *obtusifolium* Hedw. (about 32 %), *Brachythecium salebrosum* (F.Weber & D.Mohr) Bruch et al. (23,55 %) и *Anomodon thaustus* Mull.Hal. (19,13 %). The following species are the least common: less than 1 % – *Entodon* sp., *Myuroclada maximowiczii* (G.G. Borshch.) Steere & W.B.Schofield; from 1 to 3 % – *Neckera pennata* Hedw., *Sanionia uncinata* (Hedw.) Loeske, *Leucodon sciuroides* (Hedw.) Schwagr., *Pylasia* sp., *Pleurozium schreberi* (Brid.) Mitt., *Stereodon vaucheri* (Lesq.) Lindb. ex Broth., *Anomodon longifolius* (Brid.) Hartm.

The table 1 below shows identified species with an average total occurrence for the entire study area and by anthropogenic impact zones. A dash indicates a species absent in the zone. For convenience, the types of bryophytes are placed in alphabetical order.

Table 1. List of mosses in the study area.

Mosses species	Occurrence, %	zone type		
	Average mount	Public gardens	Park	Woodlands
Anomodon longifolius (Brid.) Hartm.	2,71	5,63	–	2,08
Anomodon thaustus Mull.Hal.	19,13	29,38	22,03	8,65
Brachythecium salebrosum (F.Weber & D.Mohr) Bruch et al.	23,56	–	78,91	6,30
Entodon sp.	0,37	–	–	0,93
Homalia trichomanoides (Hedw.) Bruch et al.	3,54	–	1,14	8,10
Leskea polycarpa Hedw.	60,88	14,38	74,22	90,74
Leucodon sciuroides (Hedw.) Schwagr.	1,55	–	2,08	2,49
Myuroclada maximowiczii (G.G. Borshch.) Steere & W.B.Schofield	0,38	–	–	0,94
Neckera pennata Hedw.	1,04	–	–	2,60
Orthotrichum sordidum cf. *obtusifolium* Hedw.	31,88	22,5	23,44	45,31
Pleurozium schreberi (Brid.) Mitt.	2,22	–	6,25	1,39
Pylaisia polyantha (Hedw.) Bruch et al.	8,18	–	5,47	16,81
Pylaisia sp.	2,08	0,63	6,25	0,52
Sanionia uncinata (Hedw.) Loeske	1,25	–	–	3,13
Stereodon vaucheri (Lesq.) Lindb. ex Broth.	2,50	–	1,00	5,59

5 species of epiphytic mosses were identified in public gardens. The highest occurrence is in *Anomodon thaustus* (29,38 %) and *Orthotrichum sordidum* cf. *obtusifolium* (22,5 %). The lowest occurrence is in *Pylaisia* sp. (0,63 %). Samples layers are of small size and with oppression. In some zones, sporophytes were absent. Most likely, the species reproduction was mainly vegetative in such conditions. The occurrence analysis in this zone showed that urban settlements of mosses have not been completed yet. This can be influenced by a number of factors. However, one of the key factors is the degree of moisture content in epiphyte habitats.

In parks, the number of mosses increased and amounted to 10 species. Here, the dominant species are *Brachythecium salebrosum* and *Leskea polycarpa*. The species listed are highly vital. Low occurrence (no more than 3 %) was observed in the following species: *Stereodon vaucheri*, *Homalia trichomanoides*, *Leucodon sciuroides*. The species have an average vitality, with sporogones being present abundantly.

15 species were recorded in woodlands. The most common species are *Leskea polycarpa* and *Orthotrichum sordidum* cf. *obtusifolium* build an epiphytic moss cover. The lowest occurrence in this zone was noted in such species as *Pylaisia* sp., *Myuroclada maximowiczii*, *Entodon* sp., and *Pleurozium schreberi*. Their occurrence is not more than 2 % of each. Since the control plots were located in places of natural poplar growth (such as valley forests and riverine areas), mosses are much more vital here. Poplar trunks are abundantly covered both at the upper level of bryophyte settlement and on flanges (butt ends).

For the total study area, *Leskea polycarpa* species was singled out enabling to conduct short-term monitoring and having small area coverage. This species is present in all zones. Its vitality is average and it can be recommended as a subject item for regular monitoring studies.

5. Conclusion

Populus maximowiczii was selected as a phorophyte prevailing in public gardens and growing naturally in floodplains areas, where study plots were set up as control ones. 15 species with different participation share in poplar epiphytic cover formation were identified on poplars. The species identified were assigned to two orders, 12 families and 13 genera. Among the 12 identified families, the dominant ones are *Pylaisiaceae* (3 species) and 2 species each in the family *Anomontaceae, Brachytheciaceae, Neckeraceae*.

The research area was roughly divided into three zones (public gardens, parks and woodlands), where different numbers of epiphytic mosses were identified on poplars. 5 species were recorded in public gardens, 10 species were noted for parks, and 15 species were identified in woodlands. In each zone, a characteristic set of species was identified that prevail in the upper distribution of epiphytic cover.

At this stage, they cannot be considered resistant to anthropogenic pollution for a number of reasons, such as short-term monitoring, or poor bryophyte species study for poplars. Moss oppression was noted for all control areas. Bryophytes remote from water sources seem to directly affect their occurrence, especially in urban environment. Therefore, moss studies on the banks of open reservoirs, such as rivers, lakes, water basins and other objects, are common in the literature. Research should be continued in this direction.

References

[1] Kaganov V V, Kordukov A V and Ezhkin A K 2019 Monitoring of recreational areas of Yuzhno-Sakhalinsk and its surroundings *IOP Conf. Ser.: Earth Environ. Sci.* **324** 012034

[2] Gilbert O L 1968 Bryophytes as indicators of air pollution in the Tune Valley *New Phytologist* **67** pp 15–30

[3] LeBlanc F and DeSloover J 1970 Relation between industrialization and the distribution and growth of epiphytic lichens and mosses in Montreal *Canadian J. of Botany* **48** pp 1485–96

[4] Byazrov L G 2002 *Lichens in environmental monitoring* (Moscow: Scientific World) ed D A Krivolutsky p 336

[5] Zemtsova A I 1968 *Sakhalin climate* (Leningrad: Gidrometizdat) p 197

[6] Ignatov M S and Ignatova E A 2003 *Moss flora the Middle European Russia* vol 1 (Moscow: KMK Scientific Press) p 608

[7] Ignatov M S and Ignatova E A 2004 *Moss flora the Middle European Russia* vol 2 (Moscow: KMK Scientific Press) p 352

[8] Bakalin V A 2012 *Bryophytes of Sakhalin* (Vladivostok: MSUN) p 310

IOP Conf. Series: Earth and Environmental Science **946** (2021) 012040 doi:10.1088/1755-1315/946/1/012040

Possibility of using satellite-based monitoring for large-scale mapping and research of dynamics of mud volcanic landscapes

A V Kopanina[1] and K A Shvidskaya[1,*]

[1] Institute of Marine Geology and Geophysics Far Eastern Branch of Russian Academy of Science, Yuzhno-Sakhalinsk, 693022, Russia

E-mail: *kristina66689@mail.ru

Abstract. Currently Earth remote probing to study vegetation dynamics and monitor volcanic activity is of great scientific interest. The purpose of this study is to create a large-scale outline map of Yuzhno-Sakhalinsk mud volcano which will include the topography objects, mud fields of eruptions of various years and gryphons, and to perform semi-automatic classification of Yuzhno-Sakhalinsk mud volcano. Work was performed with QGIS software using the following modules: «QuickMapServices», «Freehandrastergeoreference», «LatLanTools», and «Semi-AutomaticClassificationPlugin». We developed an outline map of Yuzhno-Sakhalinsk mud volcano on a scale of 1:10000, which shows how the mud flows have changed directions over the last 70 years, as well as mud fields have been formed over the last 20 years. Using semi-automatic classification of satellite images from Sentinel-2A satellite in various color channel sets, we obtained two premaps of Yuzhno-Sakhalinsk mud volcano vegetation on a scale of 1:50 000. Satellite monitoring of YuSMV activity allows us to track the eruptive activity of the volcano, and assess its impact on vegetation.

1. Introduction

Study of vegetation dynamics and assessment of biophysical parameters of vegetation by methods of Earth remote probing (hereinafter ERP) is of interest for a range of scientific and applied tasks [1]. Analysis of current satellite images and data from multiyear field observations of plants allows us to assess changes of volcanic landscapes and vegetation recovery speed. ERP was used to study the activity of a number of active magma and mud volcanos: volcanos of the Kuril Islands [2–4], volcanos such as Shinmoedake and Sakurajima (Japan) [4], Kilauea (Hawaii) [4], Saint Helen's (USA) [5], Ol Doinyo Lengai (Tanzania) [6], etc.

Most scientific works on satellite monitoring of volcanic activity [2, 4], studying of vegetation restoration dynamics and calculation of vegetation index NDVI [3, 5–6], were made for areas up to 50 hectares. The main purpose of our research is to study potential use of ERP data for areas less than 1 hectare.

The object of this study is Yuzhno-Sakhalinsk mud volcano (hereinafter YuSMV) located 20 kilometers north-west of Yuzhno-Sakhalinsk city (N 47°04′53′′E 142°34′43′′); the volcano is listed as a natural monument [7–8]. Data on mapping of this object is very scarce. Existing maps and outline maps of YuSMV contain information on its geological structure [9], location of gryphon [9–10], and types of vegetation [7, 9].

YuSMV eruptions are quite frequent. Strong eruptions were recorded in 1959, 1979 and 2001, smaller ones between 1994 and 1996, then in 2011 and 2020. During an eruption, the volcano throws out a large amount of mud volcanic breccia (about 10^3–10^5 cubic meters) and underground gas (often burning) over a short period of time. This is why mud volcanism is a dangerous geological process which must be taken into account during construction and operation of engineering structures. At gryphon stage, the volcano continually emits relatively small amounts of mud volcanic breccia and gases from gryphons (eruptive instruments on the body of a mud volcano). The total number of gryphons on a volcano remains practically unchanged from year to year and is usually from 50 to 70 [10]. Changes in the temperature mode of the gryphons are caused mostly by variation in the speed of water and mud mixture moving through the gryphon channels [11].

Mapping of YuSMV area and vegetation is necessary for assessing the multiyear dynamics of its activity and identifying additional indicators of its activity.

The purpose of this study is to develop a large-scale (1:10000) outline map of YuSMV which shows the dynamics of the mud volcano landscape, and to look for adequate methods of semi-automatic classification of YuSMV vegetation. In our understanding, the area where YuSMV is located is the current eruptive center and the adjacent medium-altitude mountainous forests with total area up to 8–10 square kilometers [12].

2. Materials and methods

Outline map of YuSMV was developed with indication of topographic objects, boundaries of mud fields of eruptions of various years and gryphons, using QuantumGIS 3.16 (hereinafter QGIS) information system. The following was used as the basis for the YuSMV outline map: multispectral image of south-east part of Sakhalin as of October 19, 2020 from satellite Sentinel-2A [13], satellite image of the area of research (4x4 square kilometers) from SASPlanet software (Yandex-Sputnik section), «OpentopomapRussia» and «Topomap (marshruty.ru)» topographic maps from «QuickMapServices» module installed in QGIS. QGIS is an open-source geological information system [14]. We also used «Freehandrastergeoreference» modules to simplify loading and adjustment of bitmap images, and «LatLanTools» module to show locations on the outline map using known coordinates.

We used «Semi-AutomaticClassificationPlugin» (hereinafter «SACP») module installed in QGIS for semi-automatic classification of YuSMV vegetation (hereinafter classification).

3. Results and discussion

Topographic objects shown on the outline map of YuSMV include the railroad, a forest road, a path, rivers (the Alat and the Puta) with their outlets and tributaries, primary heights (295 m and 354 m). Using the outline provided by O. Melnikov [9], we updated our outline map of YuSMV with boundaries of mud fields which appeared in 1959, 1966, 1979, 1996, and 2001, a large gryphon mound 6–7 meters in diameter at the base and about 1 meter tall, as well as smaller gryphons (73 ea). Boundaries of the 2011 mud field were formed using the results of a field survey by V. Yershov [10]. Scopes of YuSMV emissions and their boundaries from the 2020 eruption were studied in detail by teams from the plant ecology and geoecology lab and geospheres material composition research lab of the Resource Sharing Center of the Institute of Marine Geology and Geophysics of the Far East Branch of the Russian Academy of Science when the teams performed field work at the volcano in 2020.

YuSMV mapping using satellite images allowed us to identify landscape changes in its territory. In the last 60–70 years, YuSMV eruption fields have changed quite significantly since the 2001 eruption. Before 2001, mud emissions from large eruptions of 1959, 1966, and 1979 went south-east towards the outlet of the Puta River. In 2001 a large mud flow went 600 meters north-west from the supposed eruption center. It reached and blocked the outlet of the Alat River. Therefore, mud fields from the more recent eruptions have a different shape. Judging by the image from SASPlanet software (Yandex-Sputnik section), the latest YuSMV eruption of 2020 is relatively large, but not as powerful as the 2001 one. Direction of the 2020 primary mud flow is the same as in 2001 (figure 1). Mud flows of the 2020 eruption have completely destroyed all vegetation along the 2001 flow and its boundaries, which is clearly seen on the satellite images.

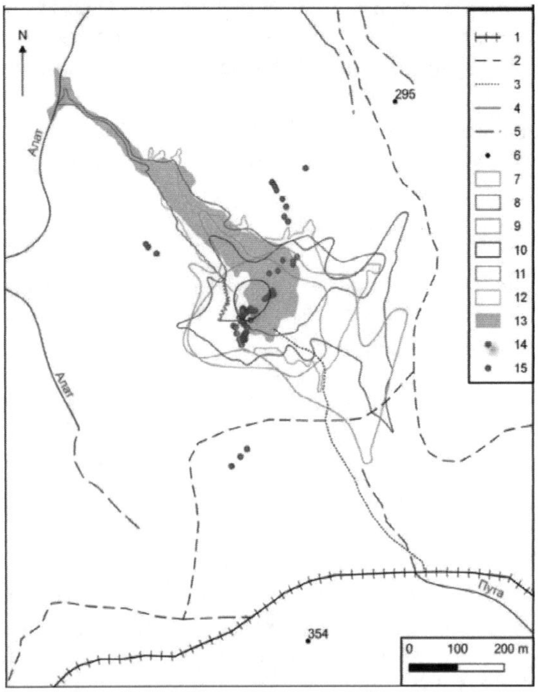

Figure 1. Outline map of YuSMV: 1 – railroad; 2 – forest road; 3 – path; 4 – rivers; 5 – rivers outflows; 6 – primary heights; 7 – mud fields after the 1959 eruption; 8 – mud fields after the 1966 eruption; 9 – mud fields after the 1979 eruption; 10 – mud fields after the 1996 eruption; 11 – mud fields after the 2001 eruption; 12 – mud fields after the 2011 eruption; 13 – mud fields after the 2020 eruption; 14 – a large gryphon; 15 – smaller gryphons and salsas.

Semi-automatic classification, or controlled classification (classification with a learning curve), is placing each of the image pixels into a certain class of objects matching a certain area in the spectral properties. This type of classification is used when a learning curve can be set on the image. A learning curve is a set of sample areas which were placed into a certain class of objects based on the data of on-site inspections, detailed images or maps, and which are clearly identifiable on the image [15].

According to scientific and practical recommendations of A. Karpachev [16–17], Landsat and Sentinel-2A satellite images can be processed and classified successfully. A. Karpachev classified the above satellite images in QGIS using «SACP» module to identify areas of dry and disturbed forest in Orlovskoye Polesye national park (Oryol Region, Russia). He found that Sentinel-2A images were clearer than Landsat images, and had a better match to the geographic data, because the parameters of visible satellite image channels for Landsat are 30 meters, while for Sentinel-2A they are 10 meters. Therefore, using Sentinel-2A satellite images seems more appropriate for small areas. Sentinel-2A satellite images allow to show the difference in the vegetation condition, including its changes over time. In our case, mud volcano monitoring covers a small area, so Sentinel-2A satellite images are more relevant for us. Visual and automatic interpretation of satellite images of YuSMV with use of on-site observations allowed us to obtain the boundaries of eruptions of 2020, 2011, 2001 (partial), the boundaries of grassland vegetation and various types of mixed and coniferous forests (figure 1, 5). We also classified the boundaries without vegetation, locations of fresh breccia and active salsas in the forest.

When we interpreted and selected channels of multispectral image of YuSMV and mapped the new mud field of the 2020 eruption, we used recommendations of D. Kurbatsky [18] and the information table for channels interpretation prepared by D. Dolgopolov [19], showing comparison of channels of multispectral images from different satellites (MODISMCD43A4, Landsat 8, Sentinel-2A).

The «artificial colors» combination (8–4–3 for Sentinel 2-A, 4–3–2 for Landsat 5, 7) is widely used for studying vegetation, monitoring drainage and soil mosaic, and also for studying crop species

[20–24]. Vegetation is shown in shades of red, soil color varies from dark brown to light brown, clouds look white of light blue. Overall, deep shades of red are indicative of healthy vegetation of all kinds, while the lighter colors mean grass and shrubs or sparce forest [18]. In our opinion, this set of channels is not suitable, because yellow leaves are shown in blue, just like a mud field or clouds, which makes classification more difficult (figure 2).

Figure 2. Image of YuSMV (Sentinel-2A, 19.10.2020) in the 8–4–3 color set. The red contour shows that the clouds, mud field, and yellow vegetation are all shown in the same color.

Figure 3. Image of YuSMV (Sentinel-2A, 19.10.2020) in the 4–6–12 color set.

Figure 4. Image of YuSMV (Sentinel-2A, 19.10.2020) in the 4–3–2 color set.

When we performed classification, we tried using two sets of channels: 4–6–12, as proposed by A. Karpachev [17], and 4–3–2 («natural colors»). In the 4–6–12 channels set the colors look more natural than in the «natural colors» set. The 4–6–12 color set has a wide range of colors, areas without vegetation are shown in bright colors; the «healthy» vegetation, dry and sparce vegetation are also visualized quite well. In the 4–3–2 color set the objects on the ground look the way a human eye sees them, with less contrast, but visually telling one kind of vegetation from another is more difficult than in the 4–6–12 set. Also, the image in the 4–3–2 color set is sharper, because channels 2, 3, and 4 have image resolution of 10 m, while for channels 6 and 12 it's 20 m (figure 3–4).

As a result of classification performed for YuSMV (1:50000 scale), areas without vegetation are clearly visualized. We can see various versions of forest vegetation: sparce forest with bamboo, synusia of tall grass in forest communities, or forest communities (figure 5). As a result of classification in the 4–6–12 color set we mostly visualized sparce forests with bamboo and synusia of high grass in small-leaved forest communities, and classification in the 4–3–2 color set shows small-leaved forest and synusia of tall grass with bamboo in mixed small-leaved and dark coniferous forest communities. Areas shaded due to area landscape peculiarities are hard to classify. We can tell which result is more accurate and which set of colors is more suitable for further studies of YuSMV only after we compare these results to field studies results.

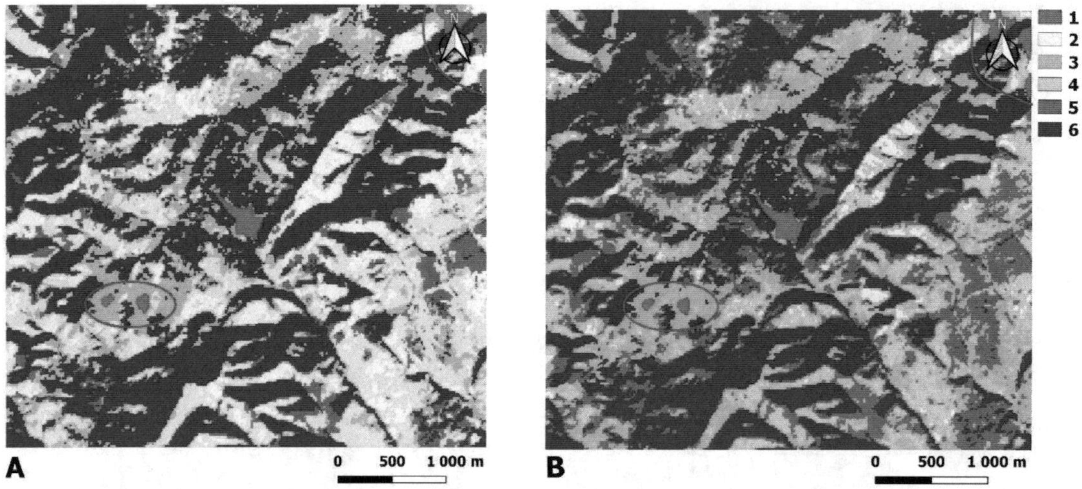

Figure 5. Results of classification in 4–6–12 (a) and 4–3–2 (b) color sets. 1 – Areas without vegetation. 2 – Sparce forest with bamboo. 3 – Small-leaved forest. 4 – Synusia of tall grass in small-leaved forest communities. 5 – Synusia of tall grass and bamboo in mixed small-leaved and dark coniferous forest communities. 6 – Mixed small-leaved and fir forest. The red contour shows clouds.

4. Conclusion

As a result of our work on creating an outline map of YuSMV using visual and automatic interpretation of satellite images of YuSMV and using on-site observations, we have created eruption boundaries for the last 20 years, and showed how large eruptions of a mud volcano change its landscape.

Using semi-automatic classification with «SACP» module in QGIS, we found that this method is suitable for large-scale mapping and study of dynamics of woody plants in volcanic landscapes. Further work on creating YuSMV vegetation map requires a comparison of this study results with data which will be obtained during field work.

Satellite monitoring of YuSMV activity is very important. It allows us to track the eruptive activity of the volcano, assess its impact on vegetation. Satellite monitoring also allows us to perform a more detailed analysis of the vegetation, both close to the volcano and several kilometers from the eruption center, trace the succession process, and determine the exact location of potential field work.

References

[1] Shabanov N V, Bartalev S A, Eroshenko F V and Plotnikov D E 2018 *Sovremennye Problemy Distancionnogo Zondirovaniya Zemli iz Kosmosa* **15** 4 pp 166–178

[2] Rybin A V, Chibisova M V and Degterev A V 2018 *Geosistemy Perehodnikh Zon* **2** 3 pp 259–266
[3] Verkhoturov A A 2020 *Vestnik of SSUGT* **25** 3 pp 139–150
[4] Melkiy V A and Verkhoturov A A 2021 *Materials of the 3d National Sci. and Practical Conf. with Int. Participation: Oil and Gas Complex: Problems and Solutions* (Yuzhno-Sakhalinsk: IMGG FEB RAS) pp 79–88
[5] Teltscher K and Fassnacht F E 2018 *J. Mt. Sci.* **15** pp 1851–67
[6] Schutter A, Kervyn M, Canters F, Bosshard-Stadlin S, Songo M and Mattsson H 2015 *Journal of Applied Volcanology* **4** 15 pp 1–18
[7] Korznikov K A 2014 *Tomsk State University Journal of Biology* **1** 25 pp 56–65
[8] *State Cadaster of Specially Protected Natural Areas of Regional Significance of the Sakhalin Region* 2021 (Yuzhno-Sakhalinsk: Agency of Forestry and Hunting of the Sakhalin Region) p 945
[9] Mel'nikov O A 2002 *The Yuzhno-Sakhalin gas-water-lithoclastic («mud») volcano is a unique object of Nature in the Far East of Russia (a guide to the excursion to the volcano for the participants of the international scientific symposium on September 24–28, 2002)* (Yuzhno-Sakhalinsk: IMGG FEB RAS) p 46
[10] Ershov V V and Nikitenko O A 2017 *Izvestiya Vysshikh Uchebnikh Zavedeniy. Severo-Kavkazskiy Region. Seriya Yestestvenniyye Nauki* **4-1** (196-1) pp 110–120
[11] Mel'nikov O A and Ershov V V 2010 *Vestnik of the FEB RAS* **6** pp 87–93
[12] Kopanina A V 2019 *IOP Conf. Series: Earth and Environmental Science* **324** 012032
[13] Website of US Geological Survey (*Electronic materials*) http://earthexplorer.usgs.gov
[14] Rua L and Bright P 2020 *QGIS for Digital Cartography in Census and Surveys* (Oceania: UNFPA, SPC) ed S Juran p 163
[15] Malysheva N V 2018 *Fundamentals of Automated Interpretation of Aerospace Images of Forests Using GIS: A Tutorial* (Moscow: MESKH) p 136
[16] Karpachev A P 2016 *Experience of Landsat Satellite Image Classification Using Semi-AutomaticClassificationPlugin in QGIS* (*Electronic materials*) https://gis-lab.info/qa/landsat_qgis_scp.html
[17] Karpachev A P 2017 *Experience of Sentinel 2-A Satellite Image Classification Using Semi-AutomaticClassificationPlugin in QGIS* (*Electronic materials*) https://gis-lab.info/qa/qgis-sacp-sentinel2a.html
[18] Kurbatsky D V 2005 *Interpretation LandsatTM/ETM+ Data Channel Combinations* https://gis-lab.info/qa/landsat-bandcomb.html
[19] Dolgopolov D V 2021 *Monitoring. Nauka i Tekhnologii* **1** 47 pp 75–83
[20] Komarov A A, Sukhanov P A and Kirsanov A D 2018 *Materials of the 2nd All-Russian Sci. Conf. with Int. Participation: Application of Remote Means Earth Sensing in Agriculture* (St. Peterburg: ARI) pp 139–145
[21] Shikhov A N, Gerasimov A P, Ponomarchuk A I and Perminova E S 2020 *Thematic Decoding and Interpretation of Space Images of Medium and High Spatial Resolution: A Tutorial* (Permian: PSU) p 191 (*Electronic materials*) http://www.psu.ru/files/docs/science/books/uchebnie-posobiya/shikhov-gerasimov-ponomarchukperminova-tematicheskoe-deshifrovanie-i-interpretaciyakosmicheskih-snimkov.pdf.
[22] Shikhov A and Marakulin Ya 2012 Assessment of the Consequence of Forest Fires in 2010 in the Perm Region https://gis-lab.info/qa/fires-perm.html
[23] Martynova M V, Sultanova R R, Gabdelkhakov A K, Rakhmatullin Z Z and Odintsov G E 2020 *Materials of the IX Int. Sci. Conf.: Forest Ecosystems in the Context of Climate Change: Biological Productivity and Remote Monitoring* (Yoshkar-Ola: VSUT) **6** pp 128–134
[24] Chashchin A N 2018 *Basics of Processing Satellite Images in QGIS: A Tutorial* (Permian: PPC Procrost) p 47

Adventive component flora of the southern part of Sakhalin Island

O O Lozhnikova

Institute of Marine Geology and Geophysics Far Eastern Branch of Russian Academy of Science, Yuzhno-Sakhalinsk, 693022, Russia

E-mail: o.lozhnikova@imgg.ru

Abstract. The article is devoted to the adventive flora of the southern part of Sakhalin Island. Nowadays, 274 species of adventive plants, belonging to 192 genera and 57 families, have been identified here. Five families (Amaranthaceae, Cannabaceae, Commelinaceae, Linaceae, and Malvaceae) are represented by invasive species only. The structure of the leading families and genera is analyzed, the participation of adventive species information of plant communities is indicated. Adventive species are classified into life forms, method of entering and degree of naturalization.

1. Introduction

The adventive component of flora is a dynamic system constantly being in the stage of active formation, both due to the deliberate, targeted bringing of new plant species during the introduction or formation of modern agricultural crops, and their accidental entering. In the past half century, a sharp increase in the rate and extent of invasion of alien species into natural landscapes has been observed all over the world. The expansion of adventive elements occurs beyond any boundaries and, as a result, leads to the transformation of local plant communities, contamination of the genetic fund of populations of native plant species and poses a threat to the conservation of biological diversity in general [1, 2].

Invasive plants, being part of the adventive component, are distinguished by their ability to quick settling and development a wide range of ecotopes. Therefore, the identification of adventive plant species, as well as monitoring of their distribution and abundance, is extremely important and relevant, since they can not only disrupt the development and balance of natural plant communities and lead to pollution of the genetic fund of local populations, but also become malicious weeds over time. The study of alien species, including invasive or potentially invasive ones, in recent decades has become one of the important directions of fundamental and applied works [3].

The number of regions of our country has the lists of stranger plants with the greatest danger to the health of people and domestic animals. For the European part of Russia, "The Black Book" was published including the characteristics of widespread and aggressive alien plant species [3]. Along with this, in some cities and regions of the Far East, adventive taxa have already been made known [4–8]. However, very little attention is still paid to the study of the adventive component of the flora on Sakhalin Island, and it is of a fragmentary nature, despite the fact that certain efforts are being made in this direction [9–13]. Regarding this fact, the purpose of this work is to analyze adventive plant species in the southern part of Sakhalin Island, the natural landscapes of which have been most transformed over the past century.

2. Materials and methods

Studies of the adventive flora of the southern part of Sakhalin were carried out in 5 geobotanical regions of the island [14], from its southern tip to the Poyasok isthmus. Administratively, this part of Sakhalin covers Tomarinsky, Dolinsky, Kholmsky, Anivsky, Yuzhno-Sakhalinsky, Korsakovsky and Nevelsky regions, which are the most economically developed.

Field studies of the adventive flora of the southern part of Sakhalin Island were carried out during the spring, summer and autumn seasons of 2018–2021, using the method of route-floristic description of the territory. They were accompanied by the collection of herbarium specimens for further identification. Studies of the most economically developed settlements were carried out repeatedly at different times of the growing season: Korsakov, Prigorodnoye, Okhotskoye, Aniva, Nevelsk, Kholmsk, Yuzhno-Sakhalinsk, Novo-Aleksandrovsk, Lugovoye, Dolinsk, Sokol, Vzmorye, Tomari, Ilyinskoye, etc. Particular attention was paid to the areas of the most intensive entering and concentration of adventive plants – railway stations, seaports, man-made territories, wastelands, landfills for solid waste, roadsides and railways, parks and squares, summer cottages, vegetable gardens and agricultural fields.

To analyze the adventive fraction of the flora, published works on the diversity and distribution of vascular plants on Sakhalin Island [15, 16], materials from the herbarium of the IMGG FEB RAS were critically worked out, and our own field studies were carried out.

The selection of adventive species by the method of entering and the degree of naturalization was carried out according to the system of F.G. Schroeder [17]. Xenophytes, ergaziophytes, and xenoergaziophytes were distinguished according to the method of entering of adventive plants, and ephemerophytes, colonophytes, epecophytes and agriophytes were distinguished according to the degree of naturalization. The analysis of life forms of plants is given according to the classifications of C. Raunkiaer [18].

3. Results of researches

Nowadays, as a result of researches, it has been established that the adventive fraction of the flora of South Sakhalin includes 274 species of vascular plants from 192 genera belonging to 57 families. Among them, the largest family is Asteraceae, which includes 36 genera and 51 species, it is 18.6 % of the total number of species of adventive flora in the southern part of the island. The most common representatives of this family are *Achillea millefolium* L., *Arctium tomentosum* Mill., *Artemisia vulgaris* L., *Conyza canadensis* (L.) Cronq., *Gnaphalium sylvaticum* L., *Lepidotheca suaveolens* (Pursh) Nutt., *Leucanthemum vulgare* Lam., *Pilosella aurantiaca* (L.) F.W. Schultz et Sch. Bip., *Senecio vulgaris* L., *Taraxacum officinale* Wigg. and others. After Asteraceae there are such families as Poaceae – 14.6 %; Fabaceae – 9.5 %; Brassicaceae – 6.9 %; Polygonaceae – 5.1 %; Caryophyllaceae – 4.4 %, Scrophulariaceae – 4.4 %, etc. The taxonomic characteristic of 10 leading families of the adventive flora of the studied area is shown in table 1.

Table 1. Spectrum of 10 leading families of adventive flora of South Sakhalin.

№	Family	Quantity of species	Share of participation, %	Quantity of genera	Share of participation, %
1	Asteraceae	51	18.6	36	18.8
2	Poaceae	40	14.6	26	13.5
3	Fabaceae	26	9.5	13	6.8
4	Brassicaceae	19	6.9	16	8.3
5	Polygonaceae	14	5.1	10	5.2
6	Scrophulariaceae	12	4.4	10	5.2
7	Caryophyllaceae	12	4.4	7	3.6
8	Rosaceae	8	2.9	7	3.6
9	Lamiaceae	6	2.2	5	2.6
10	Pinaceae	6	2.2	3	1.6
	Total:	194	70.6	133	69.3

The first triad of families includes 117 species of adventive plants, and the leading 10 families cover 194 species, or 70.6 % of the total number of taxa. As a whole the adventive flora of the southern part of Sakhalin is represented by 57 families, and 5 families, such as Amaranthaceae, Cannabaceae, Commelinaceae, Linaceae and Malvaceae, contain exclusively strangers. In this case, the multi-species genera are *Trifolium* (include 7 species), *Poa* (6), *Veronica* (5), *Vicia* (4). The rest of the genera include 3 or fewer species.

Analysis of the structure of adventive plant species due to life expectancy showed that perennial plants include 148 species, and juveniles (one- and biennials) – 126 species. All the main life forms are represented in the adventive flora of South Sakhalin, their ratio is shown in figure 1.

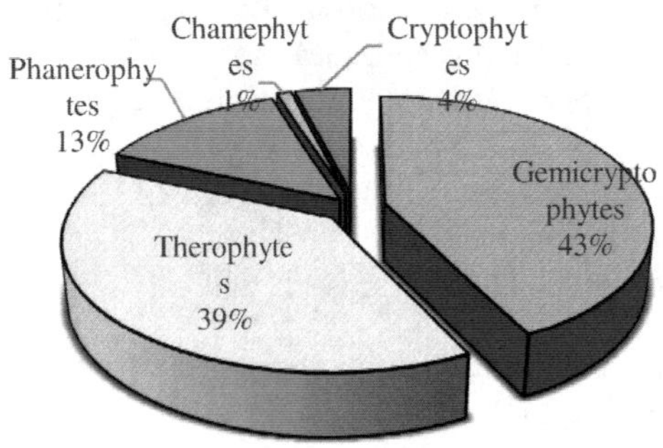

Figure 1. The ratio of life forms of the adventive flora of the southern part of Sakhalin.

In the adventive flora of the southern Sakhalin, hemicryptophytes predominate and cover 117 species (43 %); therophytes – 106 (39 %); phanerophytes – 37 (13 %); cryptophytes – 11 (4 %) and hamefits – 3 types (1 %).

Introduced species of trees and shrubs (phanerophytes) are of great environmental and aesthetic importance in the urban environment. In the first half of the last century, *Juglans mandshurica* Maxim., *Pinus koraiensis* Siebold et Zucc., *Populus nigra* L., *Salix koriyanagi* Kimura ex Goerz, were planted on the streets of Yuzhno-Sakhalinsk, many of which have survived to ourtime. In the Soviet period, such out-of-district trees and shrubs species as *Acer platanoides* L., *A. negundo* L., *Caragana arborescens* Lam., *Fraxinus pennsylvanica* Marsh., *Padus maackii* (Rupr.) Kom., *Physocarpus opulifolius* (L.) Maxim., *Pinus sylvestris* L., *Populus alba* L., *Robinia pseudoacacia* L. and etc. were widely used in the greening of the regional capital [19]. The modern dendroflora of Sakhalin is represented by 37 strangers, which is 13 % of the adventive component of the flora of South Sakhalin.

The overwhelming majority of adventive species have successfully established themselves in communities on anthropogenically disturbed landscapes. According to the degree of naturalization, four groups are distinguished, and according to the method of entering – three groups (table 2). We excluded phanerophytes from the group according to the degree of naturalization because of their use only in local areas, mainly for greening urban areas, therefore, rarely spreading beyond their original habitat, with the exception of some species, for example, *Acer negundo*.

Table 2. The ratio of adventive species in the flora of South Sakhalin according to the degree of naturalization and the method of entering.

Group of species on a way	Quantity of species	% of the total number of adventive species
Naturalizations:		
Epecophytes	140	59.1
Colonophytes	66	27.8
Ephemerophytes	11	4.7
Agriophytes	20	8.4
Total:	**237**	
Immigrations:		
Xenophytes	162	59.1
Ergaziophytes	97	35.4
Xenoergaziophytes	15	5.5
Total:	**274**	

IOP Conf. Series: Earth and Environmental Science **946** (2021) 012041 doi:10.1088/1755-1315/946/1/012041

Ergaziophytes were introduced into cultivation in different periods of urbanization as decorative, used for greening cities, or food and forage plants (*Acer negundo*, *Fraxinus pennsylvanica*, *Heracleum sosnowskyi* Manden., *Lupinus nootkatensis* Donn ex Sims, *Solidago canadensis* L. and others). Here, the *Heracleum sosnowskyi* should be especially noted, which was brought to Sakhalin in 1962 to create a highly productive silage culture and improve the fodder base of the livestock complex [20]. Possessing strong grow power, abundant seeding and high competitive ability, it quickly captured many disturbed areas, wastelands and formed single-species cenopopulations, and at the same time, actively displacing native species. Some species of cereals and legumes were brought to Sakhalin by Japanese (from 1905 to 1945) and Russian (starting from the second half of the 20th century) settlers. First of all, these were forage grasses, *Anthoxanthum odoratum* L., *Dactylis glomerata* L., *Phleum pratense* L., *Poa pratensis* L, *Trifolium repens* L., *T. pratense* L., etc. [19].

Among the adventive plants of South Sakhalin, more than 20 species have acquired an invasive status, which have successfully adapted to local natural and climatic conditions, quickly spread around the island and develop new types of habitats, have a high reproductive potential, actively compete not only with synanthropic, but also with meadow, floodplain and forest native species in anthropogenically transformed communities [16].

The adventive species in the studied area occupy different ecotopes. The most active formation of local cenopopulations of adventive plants occurs in ruderal habitats and disturbed meadows (*Cerastium holosteoides* Fr., *Galium spurium* L., *Leucanthemum vulgare*, *Lupinus nootkatensis*, *Pilosella aurantiaca*, *Rhinanthus minor* L., *Trifolium pratense*, etc.); on ravines and ditches (*Berteroa incana* (L.) DC., *Impatiens glandulifera* Royle (figure 2), *Heracleum sosnowskyi*); in vegetable gardens and arable lands (*Brassica campestris* L., *Galinsoga parviflora* Cav. (figure 2), *Spergula arvensis* L., *Stellaria media* (L.) Vill.), as well as on the sides of roads and railways (*Conyza canadensis*, *Heracleum sosnowskyi*, *Hordeum jubatum* L., *Linaria vulgaris* Mill., *Reynoutria japonica* Houtt., etc.). They grow not only in anthropogenically transformed habitats, but often appear in weakly disturbed plant communities, forming monodominant groups.

Figure 2. *Galinsoga parviflora* (on the left) is often found in agricultural fields, *Impatiens glandulifera* (on the right) – on roadsides and ditches.

At the first stages of dispersal, plants are found only singly, then their local groups appear, which, gradually expanding their occupied areas, form already closed thickets. For example, *Galega orientalis* Lam. was noticed in 1992 on the territory of the Botanical Garden in Yuzhno-Sakhalinsk with two small groups, and then the population increased approximately threefold. In

2006, individual samples were found in the northwestern part of the island on the line of the main oil and gas pipeline. Apparently, the seeds were introduced under the biological recultivation by grass mixtures [21]. Nowadays several more new localities of *Galega orientalis* have been noted on the road section Bereznyaki-Yuzhno-Sakhalinsk, as well as on the outskirts of agricultural fields. Along with this, further active spread of *Pastinaca sylvestris* Mill. and *Inula helenium* L. should be noted outside their primary habitats. Of course, the process of plant immigration is ongoing, and the flora is replenished with new taxa. This is evidenced by the identification of new adventive species in the flora of Sakhalin Island.

4. Conclusion

1) The adventive component, as the most dynamic part of the flora, is constantly being replenished with new species. Nowadays, the adventive flora of the southern part of Sakhalin Island includes 274 species of higher vascular plants belonging to 192 genera and 57 families. Five families are represented only by invasive species – Amaranthaceae, Cannabaceae, Commelinaceae, Linaceae, and Malvaceae.
2) Analysis of the species structure of the adventive flora of the southern Sakhalin showed that the Asteraceae families cover 51 species (18.6 %) and 36 genera, Poaceae – 40 species (14.6 %) and 26 genera, Fabaceae – 26 species (9.5 %) and 23 genera, Brassicaceae – 19 species (6.9 %) and 16 genera and Polygonaceae – 4 species (5.1 %) and 10 genera.
3) According to the ratio of life forms in the adventive flora of the southern part of Sakhalin, hemicryptophytes (117 species) predominate and therophytes (106 species). The dominant group due to the way of immigration occurred xenophytes (162 species), and according to the degree of naturalization, epecophytes predominate (140 species).
4) Adventive species are mainly confined to ruderal habitats, they are often found along roadsides, in settlements, in the place of abandoned settlements, and some of them have penetrated into natural cenoses and become invasive.

References

[1] McNeely J A, Mooney H A, Neville L E, Schei P and Waage J K 2001 *IUCN Gland. A Global Strategy on Invasive Alien Species* (Switzerland; Cambridge, UK) p 50
[2] Richardson D M, Pysek P, Rejmanek M, Barbour M G, Panetta F D and West C J 2000 Naturalization and invasion of alien plants: concepts and difinitions *Diversity Distrib.* **6** pp 93–107
[3] Vinogradova Yu K, Mayorov S R and Horun L V 2010 Black book of flora of Central Russia: alien plant species in ecosystems of Central Russia (Moscow: GEOS) p 512
[4] Aistova E V 2009 Check-list of adventive flora of Amur region *Turczaninowia* **12** 1–2 pp 17–40
[5] Antonova L A 2011 Adventive component of flora of the city of Khabarovsk *Vestnik IrGSHA* **44** 2 pp 42–46
[6] Antonova L A 2012 An invasive component of flora of Khabarovsky krai *Russian J. of Biol. Invasions* **4** 2–9 p 140
[7] Devyatova E A, Chernyagina O A and Abramova L M 2016 List of alien flora of Petropavlovsk-Kamchatsky *Vestnik of NEFU* 4 (54) pp 5–16
[8] Glazkova E A and Liksakova N S 2020 New and rare alien vascular plant species of the Kuril Islands *Botan. journal* **105** 12 pp 1226–34
[9] Smirnov A A 2012 Adventive plant species new to Sakhalin *Botan. journal* 10 **97** pp 1366–67
[10] Sabirova N D and Sabirov R N 2018 New adventive species in the flora of Sakhalin *Botan. journal* 7 **103** pp 930–933
[11] Sabirov R N, Sabirova N D and Lozhnikova O O 2019 The adventive species of vascular plants in the flora of the city of Yuzhno-Sakhalinsk and its suburbs *Journ. of Sakh. museum* 4 (29) pp 53–65

[12] Lozhnikova O O 2019 *Abstracts of the III National scientific Conf. Geodinamic. Processes and Natural Hazards* Adventive component of the flora of the city Dolinsk and its suburds: (Yuzhno-Sakhalinsk: IMGG FEB RAS) p 167
[13] Lozhnikova O O 2020 Adventive species of vascular plants in the flora of the Dolinsky district (Sakhalin region) *J. of Sakh. museum* **4** pp 52–63
[14] Tolmachev A I 1955 Geobotanical division of Sakhalin Island (Moscow-Leningrad: AN USSR) 80 p
[15] Barkalov V Yu and Taran A A 2004 List of species of vascular plants of Sakhalin island *Growing and fauna of Sakhalin island* (Vladivostok: Dalnauka) **1** pp 39–66
[16] Sabirova N D, Sabirov R N, Lozhnikova O O 2019 *IOP Conf. Series: Earth and Environmental Science* The invasive plant species in the flora of Sakhalin
[17] Schroeder F G 1969 Zur Klassifizierung der Anthropochoren *Vegetatio* Bd **16** ss 225–238
[18] Raunkiaer C 1937 *Plant Life Forms* (Oxford: Clarendon Press) 104 p
[19] Taran A A 2018 *Proc. 5th Int. scientifical-practical Conf. Urboecosystems: problems and perspectives of development* The flora of the city of Yuzhno-Sakhalinsk (Ishim, Tyumen: State University) pp 134–136
[20] Czernjaeva A M and Krapivina A M 1976 Experience of cultivation Heracleum sosnowskii on Sakhalin *Rastitelnye resursy* 3 **12** pp 433–438
[21] Taran A A 2016 New and rare vascular plant species in the flora of Sakhalin Island *Bull. of the Main Botan. Garden* 2 (202) pp 29–32

Trophic migrations of the brown bear in the forest landscapes of the Vostochny Nature Reserve on Sakhalin Island

I V Nikulina[1] **and R N Sabirov**[2]

[1]Sakhalin State University, Yuzhno-Sakhalinsk, Russia
[2]Institute of Marine Geology and Geophysics Far Eastern Branch of Russian Academy of Science, Yuzhno-Sakhalinsk, 693022, Russia

E-mail: irinkaeremenko@yandex.ru

Abstract. The article deals with trophic migrations of the brown bear in the characteristic forest landscapes of the Vostochny Nature Reserve. Active life of a bear in the reserve begins after hibernation and is directly related to foraging in different seasons of the year. In the spring-early summer period, bears concentrate in valley forest complexes and coastal-marine landscapes, where, along with plant food, they obtain various products of animal origin thrown out of the sea. During the summer period, from the beginning of the course of salmon and until the end of their spawning, the animals mainly live in the mouths of rivers, and as the fish move, they go upstream and are distributed among their numerous tributaries. During the autumn period, bears concentrate mainly in mountainous forest landscapes and high-altitude natural complexes of the reserve, where they eat ripe berries and pine nuts, their trophic migrations are significantly reduced.

1. Introduction

Many specially protected natural territories of the Sakhalin Region were created, as a rule, without a detailed survey of local landscapes and ecosystems, without establishing all the available biological diversity in them, therefore, without preparing an appropriate scientific justification. In addition, a significant part of them, due to the above and a number of other reasons, is designed to preserve already transformed and, therefore, not unique and not so reference ecosystems. Perhaps one of the few specially protected natural territories of the region, which differs from others in the preservation of forest ecosystems with all its components, is the Vostochny Nature Reserve. This reserve is located in the middle part of Sakhalin Island (Smirnykhovsky district) and covers the territory of the basins of two spawning rivers Pursh-Pursh and Vengeri, with a total area of 68 080 hectares. The reserve presents intact forest landscapes and other natural complexes with an original species composition, complex structure, great biological diversity and a high concentration of rare and endangered species [1, 2, 3, etc.].

The natural conditions of the territory under consideration, in comparison with other regions of the region, are characterized by a number of specific features. The highest mountain system of the island - the East Sakhalin Mountains - is located here, which is oriented mainly in the meridional direction and consisting of two main ridges, Nabil and Central. The cascade of mountain structures and their spurs alternating with intermountain depressions, as well as the proximity of the Sea of Okhotsk created a peculiar climatic regime in this area with numerous variants of meso- and microclimate, and also caused special soil-forming processes [4, 5, 6, etc.].

Due to these specific natural conditions, very original and diverse mountain taiga landscapes were formed on the territory of the reserve, which predetermined the composition, number and distribution of wildlife, as well as the structure of ecological and faunal complexes as a whole. As part of the theriofauna of the Vostochny Nature Reserve, the brown bear is a constantly inhabiting and widespread species, which inhabits all its natural landscapes and forms the highest population density here. Having a high mobility and flexible position in the trophic chain, the bear occupies a wide variety of biotopes of the reserve, therefore, acts not only as a link between individual elements of landscapes, but also is a kind of indicator of the state of the habitat. The purpose of this work is to determine the role of the most characteristic forest landscapes of the Vostochny Nature Reserve in trophic migrations of the brown bear, taking into account its seasonal activity.

2. Materials and Methods

Field complex observations of the brown bear and traces of its vital activity in the Vostochny Nature Reserve were carried out during 2010–2020. The research was based on visual registration of animals using the global satellite positioning system (GPS), as well as descriptions of the behavior and activity of the individuals encountered in the reserve. The main stationary studies (from May to October) were carried out mainly in the area of the mouth of the river. They included accounting for bears, monitoring their behavior, intraspecific relationships, the method of obtaining and the nature of consumption of food resources, as well as other aspects of animal life [7]. In the estuary zone of the Pursh-Pursh River, individuals were recorded using camera traps (Seelock company) in shooting mode – one frame every 5 minutes. The determination of the food ration was carried out on the basis of the excrement analysis, while taking into account the percentage of the volume of individual food components in each sample, as well as the description of feeding sites and traces of vital activity associated with the food production activities of the brown bear.

The study of the vegetation of the landscapes of the reserve was carried out according to standard methods widely used in geobotanical and forest-typological surveys [8, 9, 10, etc.]. Due to the vastness of the territory of the nature reserve, studies were carried out mainly by route, with detailed work on key sites, geobotanical profiles and test areas.

The data obtained were processed by the methods of geoinformation biogeographic mapping, using the software package ArcGIS Desktop 10, ScanEx Image Processor, Statistica Ultimate Academic 13 [11, 12 etc.]. To build a map of forest landscapes, the results of field research, satellite images of the Landsat series, Sentinel, digital terrain model (SRTM, absolute elevation marks) were used [13, 14, etc.].

3. Research results

Active life of a brown bear in the reserve begins after hibernation and is mainly associated with the search for appropriate food items. The latter are unevenly distributed in the natural landscapes of the reserve, change in different seasons of the year and as a result cause trophic migrations of the bear. In this regard, in the intraseasonal activity of a brown bear, we have identified several periods characterized with the changes in eating behavior and territorial movements: spring-early summer (average duration of 70–80 days, from the first decade of May to the second decade of July); summer (40–50 days, from the third decade of July to the third decade of August); autumn (50–60 days, from the first decade of September to the third decade of October). At the same time, manifestations of a brown bear vital activity were observed in the following forest landscapes of the reserve: coastal-marine, valley forest complexes, larch forests of intermountain depressions, low-mountain dark coniferous forests, stone-birch forests of the middle-altitude belt of mountains, subalpine belt dominated by cedar elfin, high-mountain landscapes with mountain tundra vegetation (figure 1).

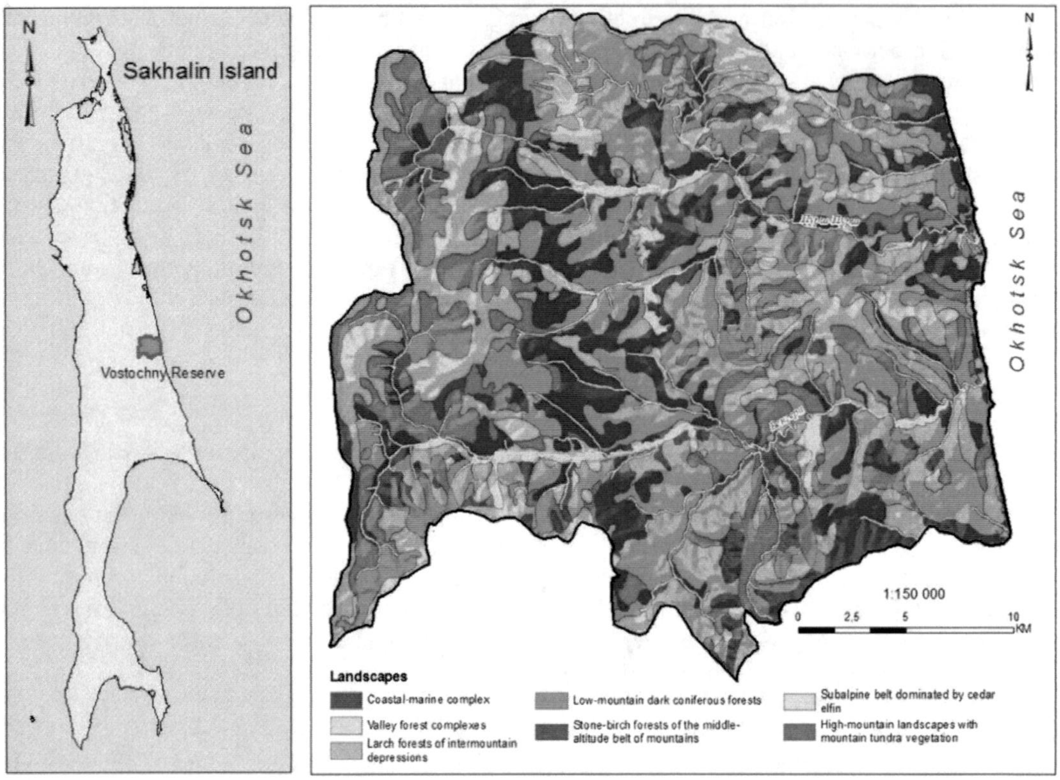

Figure 1. Location and structure of forest landscapes of the Vostochny Nature Reserve.

Of course, coastal and marine landscapes, which occupy 0.4 % of the area of the nature reserve, play an important role in the life of a brown bear. Despite the insignificant territorial distribution, limited by a narrow strip of the seashore and estuarine areas of rivers, small rivers and streams, these landscapes have many life opportunities suitable for bears to live in the post-winter period. It is in the spring-early summer period that animals actively eat thrown ashore: sea cabbage, crab caviar, shellfish, capelin suitable for spawning, etc. In addition, we have repeatedly observed very noticeable clusters of animals on whale carcasses, sometimes numbering up to 13 individuals at a time. Along with this, on the sea terraces, due to the earlier and intense snowfall, the first plant groupings appear, which allows the bear to use both underground and aboveground parts of plants for food. During this period, bears often visit floodplains and floodplain terraces, where they willingly eat early-growing plants, including their bulbs and rhizomes (figure 2 A).

The mass exits of bears to the coastal zone are also timed to the beginning of the course of salmon in spawning rivers (end of July – beginning of August). However, due to the current trend of reducing the number of pink salmon (the main mass species of spawning salmon in the rivers of the reserve), in the period from 2007 to 2016 there was an increase in the concentration of animals in the estuary zones from 3 to 10 individuals, and from 2017 to 2020 – from 12 to 25 individuals (figure 3). At the same time, the length of stay of animals increased by an average of 10–14 days.

The valley forest complex of the nature reserve is formed by communities dominated by Maksimovich poplar, chozenii, willow, hairy alder (*Populus maximowiczii, Chosenia arbutifolia, Salix udensis, S. cardiophylla, Alnus hirsuta*) and their mixed variants. These communities are found mainly on the second and third river terraces, flat elevated areas of the floodplain, on gentle coastal slopes. Valley complexes are characterized by good development and abundance of herbaceous plants, mainly hygrophytes and mesohygrophytes. Of these, the most commonly used in a brown bear's diet are white-tailed (*Petasites amplus, P. tatewakianus*), krestovnik (*Senecio cannabifolius*), laburnum

(*Filipendula camtschatica*), cacalia (*Cacalia auriculata, C. hastata*), dudnik (*Angelica ursina, A. genuflexa*), Weyrich's ram (*Aconogonon weyrichii*), kaluzhnitsa (*Caltha fistulosa*), hogweed (*Heraculum lanatum*), sedges (*Carex sp.*) and a number of other species. In addition, a bear willingly eats the fruits of bird cherry (*Padus avium*) here. During the summer period, from the beginning of the salmon course until the end of their spawning, a bear consumes fish together with plant feeds, so the animals often go to the rivers, and as the fish move, they go upstream and are distributed along their main tributaries (figure 2 B).

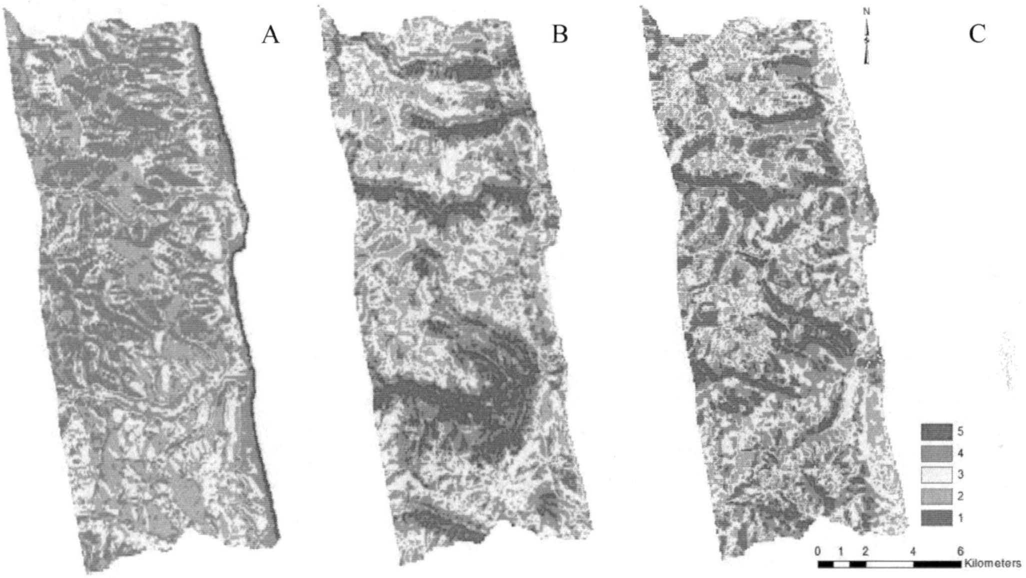

Figure 2. Fragments of maps of brown bear distribution on the territory of the Vostochny Nature Reserve in different seasons of the year: A – spring-early summer; B – summer; C – autumn. The color scale reflects the level of probability of the presence of a bear: 5 – high, 1 – low.

Larch forests in the Vostochny Nature Reserve are located mainly in the intermountain depression between the Nabil and Central Ridges and occupy about 30 % of the area. Larch (*Larix cajanderi*) grows here in both extremely dry and damp, stagnant moisture and swampy habitats, thereby forming the appropriate types of forest landscapes. There are about two dozen species in the shrubby tier of larch trees, individual representatives of which are often used in the bear's diet. Oval-leaved blueberries (*Vaccinium ovalifolium*) are quite abundantly represented in the corresponding types of forest, which, after ripening of berries, belong to the dominant feeds in the diet of a brown bear. In summer and autumn, the bear's diet also includes cloudberries (*Rubus chamaemorus*), mountain ash (*Sorbus sambucifolia*), raspberries (*Rubus sachalinensis*), pine nuts (*Pinus pumila*), cranberries (*Vaccinium vitis-idaea*), etc.

Low-mountain dark coniferous forests dominated by spruce (*Picea ajanensis*) and fir (*Abies sachalinensis*) cover about 20 % of the territory of the nature reserve. Of the feeding objects of a brown bear, blueberries, mountain ash, lingonberries, rose hips (*Rosa acicularis, R. amblyotis*), etc. are most often found here. Among other things, a bear uses woodlands as shelters, arranges beds and dens here. And he prefers fir as marking trees to mark his territory.

In the autumn period, after the completion of salmon spawning, the bear's diet changes and is largely formed already from plant resources. In search of food, animals move to high altitudes from the middle-altitude belt of mountains with stone-birch forests to the subalpine belt dominated by cedar elfin and high-altitude landscapes with mountain tundra vegetation (figure 2 C). Here bears graze mainly on berry bushes and pine nuts. In the late autumn–pre-autumn period, bear food activity is noticeably reduced. At this time, the animals are mostly confined to the mountain forest landscapes of the reserve, preparing for the winter season, equipping their dens and no longer make significant movements.

Figure 3. Bears at the mouth of the Hungeri during the spawning of pink salmon in the summer of 2019. Photo by I. Nikulina.

4. Conclusion

In the Vostochny Nature Reserve, the brown bear is a constantly inhabiting and widely distributed species, which inhabits all its characteristic landscapes and forms the highest population density here. However, the distribution and number of bears in the landscapes of the reserve varies significantly in different seasons of its life and depends mainly on the composition and structure of the food supply.

In the spring-early summer period, the bear's diet is dominated by herbaceous vegetation, the early-vegetating groups of which are concentrated in valley forest complexes and coastal-marine landscapes. In this regard, the diet of animals, of course, is dominated by herbaceous vegetation, the frequency of occurrence of which in excrement reaches 99 %. Along with herbs, bears willingly eat various animal products thrown ashore, even whale carcasses. Accordingly, the localization and concentration of bears during this period is observed in these two types of landscapes.

In summer, the intensity of trophic bear migrations is usually associated with the arrival of salmon to spawn. From the beginning of the salmon course until the end of their spawning, the animals consume fish as the main bait food, and as the fish move, they go upstream and are distributed along their main tributaries. At the same time, bears visit larch and dark coniferous forests along the way and eat ripe berries. The area of seasonal distribution of animals in summer is maximum.

In the autumn season, the localization of bears occurs in mountainous forest landscapes and high-altitude complexes, with regular exits to the riverbeds and rivers. Characteristic food components of animals during this period are berries and nuts of cedar elfin. In the late autumn–pre-autumn period, the trophic migrations of the bear noticeably weaken, they prepare for the winter season, equip their dens and no longer make significant movements. Of course, a brown bear, due to its eurytopicity, is characterized by high mobility in general and uses all the available landscapes of the Vostochny Nature Reserve for life.

References

[1] Sabirov R N, Sabirova N D and Voronov G A 2011 Nature Reserve «Vostochny» *Bull. of the Sakhalin Museum* 18 (Yuzhno-Sakhalinsk) pp 375–399

[2] Sabirov R N, Sabirova N D and Voronov G A 2017 Modern state of the biota of «Vostochny» Nature Reserve in Sakhalin Island *Vestnik of the FEB RAS* **1** 191 pp 108–115

[3] Sabirova N D and Sabirov R N 2018 Flora of vascular plants of the nature zakaznik «Vostochny» (Sakhalin Island) *V L Komarov Memorial Lecture* **66** (Vladivostok: FSC Biodiversity FEB RAS) pp 65–100

[4] Ivlev A M 1965 *Soils of Sakhalin* (Moscow: Nauka) p 116

[5] *Atlas of the Sakhalin region* 1967 (Moscow: GUGK) p 135

[6] Zemtsova A I 1968 *Climate of Sakhalin* (Leningrad: Hydrometeoizdat) p 197

[7] Seredkin I V, Lisitsin D V and Borisov M Yu 2012 The study of the brown bear on Sakhalin *Proc. Samara Scientific Center of the Russian Academy of Sciences* **14** 1 (8) pp 1925–28

[8] Sukachev V N and Sonn S V 1961 *Sciences with Methodological guidelines for the study of forest types* (Moscow: Publishing House of the USSR Academy of Sciences) p 144

[9] *Field geobotany* 1964 (Leningrad: Nauka) **3** p 532

[10] Methods of investigation of forest communities 2002 (Saint-Petersburg: SPbSU) p 240

[11] Nikulina I V, Minervin I G, Melky V A and Radchenko A V 2020 Geoinformation biogeographic mapping of the state of island ecosystems according to Earth remote sensing data *Vestnik SGUGiT* **25** 1 pp 194–210

[12] Guisan A and Zimmermann N E 2000 Predictive habitat distribution models in ecology *Ecological Modelling* **135** 2–3 pp 147–186

[13] Belyanina Ya P 2015 *Proc. of Dagestan state pedagogical university Natural and exact sciences* Landscapes of the south-eastern part of the Island of Sakhalin **32** 3 pp 87–92

[14] Verkhoturov A A and Melky V A 2020 *Proc. of the Int. Conf. InterKarto InterGIS Geoinformation support of sustainable development of territories* Mapping of plant communities of the subzone of dark coniferous forests of the south of Sakhalin on the basis of space surveys **26** (Moscow: MSU Publishing House) part 4 pp 60–72

Steppe communities within the coast of the Taman Peninsula

N A Pikalova[1,*], **S V Krylenko**[2] **and T A Volkova**[1,3]

[1]Kuban State University, Krasnodar, Russia
[2]Institute of Global Climate and Ecology named after Yu A Israel, Moscow, Russia
[3] P P Shirshov Institute of Oceanology RAS, Gelendzhik, Russia

E-mail: *pna678@mail.ru

Abstract. The article presents the results of field research in 2019–2020 of plant communities of the abrasive coast of the Taman Peninsula. The biomorphological structure of populations is dominated by annual species, which indicates a high dynamism associated with exogenous processes. Ecologically, the dominant species are xerophytes and xeromesophytes, as well as species with a wide eco-amplitude. The process of unification of the studied flora is low and amounted to 17 %. The generalized floristic list includes 231 plant species from 48 families. In the study area, 7 species of Red Book plants were found, of which 2 species were included in the Red Book of the Russian Federation.

1. Introduction

The climate of the region belongs to the Black Sea subregion of the Atlantic-Continental region and is determined by the impact of circulation processes in the southern zone of temperate latitudes. The Taman Peninsula belongs to the area of insufficient moisture, the moisture coefficient is 0.25–0.30. During the year 200–250 mm of precipitation falls, with a maximum in spring and autumn. Precipitation in the eastern part of the Azov Sea is about 500 mm per year. Air humidity during the year is significant and is 81–86 % in the cold period, 72–79 % in the warm one. In some rainy years, precipitation falls three to four times more than the average long-term norm, and in dry years, three to four times less than this norm. Precipitation is short-lived, mostly rainstorm. The total number of days with dry winds is 50–75 days in most of the territory, 35–40 days on the coast. Intense and very intense dry winds are absent [1]. According to the existing landscape zoning schemes, the study area belongs to the same type. These are flat-hilly landscapes with mud volcanism with forb-hurb-grass steppes and copses of xerophilous shrubs and trees. However, there is a significant facial diversity within this type of landscape in the boundaries of the study area due to the peculiarities of its location in the coastal zone of the sea, the activity of exogenous and endogenous geological processes.

In most of the territory landslide processes and the associated facies of disturbed landslide and landslide slopes with various types of vegetation on raw soils have been actively developed. Here, the distribution of vegetation and soils is mosaic in nature and completely depends on the current activity of landslides and the duration of their formation. Active landslide areas are indicated by a weak development of soil and vegetation cover. The soil layer on the displaced landslide blocks does not exceed the first centimeters. As a rule, there is a thinned grass cover in such areas. In general, the relief of the study area is a complex combination of landslide bodies of various ages and degrees of activity with a characteristic hilly-depression surface. Amplitudes of heights within the study area are up to 60–70 m (from sea level to the edges of landslide combs). The erosional dissection of the territory is weak [2].

Erosional forms are represented by shallow gullies, erosional incisions, furrows and washouts. Channel runoff occurs periodically during prolonged rains and snowmelt. The most dominated are erosional forms located between landslide bodies, since they develop for a longer time and are not disturbed with periodic displacement of landslide masses. Landslide bodies on most of the coast are strongly eroded by the sea and in the lower part form abrasion scarps several meters high. Abrasion processes are most widespread along the coastline. The main factors contributing to abrasion in this section of the coast are the outcropping of easily eroded rocks (clays, clays with interlayers of sands and marls, mud volcanic breccias), the activity of landslide processes, as well as growing anticlinal structures. As a result, the coast is abrasive almost along its entire length [3]. According to various data, the rate of abrasion in the study area can reach values from 1 m/year to 3-4 m/year, and sometimes even more [4].

The soil cover of the study area is distributed unevenly. In many areas affected by landslide processes, it is severely disturbed and is characterized by a primitive soil with a poorly developed vegetation cover. The zonal type of vegetation is the steppe. Common species in the area are bristle yarrow (*Achillea setacea*), austrian flax (*Linum austriacum*), klaus parsnip (*Pastinaca clausii*), prickly sage (*Phlomis pungens*), steppificated sage (*Salvia tesquicola*), tansy (*Tanacetum millefolium*), yarrow (*Teucrium chamaedrys*); barilla (*Salsola*), southern reed (*Phragmites australis*), sand wormwood (*Artemisia arenaria*) and marine artemisia (*Artemisia maritima*), sweet clover (*Melilotus officinalis*) and other steppe species grow in saline areas.

2. Materials and methods

The material for the study was the plant species collected in different ecosystems of the Taman Peninsula within the abrasion coast of the Black Sea. The studies were carried out in 2019-2020, and studies from previous years were also used. The study of the vegetation cover was carried out by the route method. The selected routes were visited repeatedly during the study period. This made it possible to identify the species composition, as well as to establish the confinement of species to certain habitats, plant communities in order to figure out or clarify their ecological features, which is most consistent with the research objectives. Descriptions of the vegetation cover of abrasion coasts were carried out according to standard approaches at survey sites up to 100 m^2. When establishing the systematic affiliation of plants, the following determiners were used: Kosenko I S (1970) [5], Zernov A S (2006) [6].

In accordance with the botanical and geographical zoning, the territory of our research (from Volna village to Cape Tuzla) (figure 1) is located in the Mediterranean region of the Crimean-Novorossiysk province of the Kerch-Taman district of the Taman region [7].

3. Results and its discussion

It was found that the main type of vegetation in the study area is represented by steppe communities (figure 2): wheatgrass-herbs, herbs-feather grass, cereal-herbs, and herbs-wheatgrass. Such communities are zonal vegetation types on the Taman Peninsula. There is the most powerful soil cover with steppe and shrub vegetation in the areas not involved in active landslide processes. About 70 % of the studied area is involved in agriculture. In the unplowed areas adjacent to the abrasion shore, wheatgrass-herb communities dominate. Floristic richness is about 45 species. In such an association, the dominant species is the crested wheatgrass (*Agropyron pectinatum* (M. Bieb.) P. Beauv.) and the following species are recorded in the herbs: crimean wormwood (*Artemisia taurica* Willd.), austrian artemisia (*Artemisia austriaca* Jacq.), eastern cornflower (*Centaurea orientalis* L.), galatella (*Galatella linosyris* s(L.) Rchb.f.), green oregano (*Origanum vulgare* ssp.*viride* (Boiss.) Hayek), russian bedstraw (*Galium ruthenicum* Willd.) yellow mignonette (*Reseda lutea* L.), etc. The total projective cover is up to 80 %.

Herbs and wheatgrass communities are in the second place in terms of occurrence. The first tier is dominated by steppe wheatgrass (*Elytrigia×tesquicola* (Prokudin) Klokov). In the second tier, the following species were recorded: purple-stem (*Phleum phleoides* (L.) H. Karst), bromegrass divaricate (*Bromus squarrosus* L.), ethiopian sage (*Salvia aethiopis* L.), jointed goat grass (*Aegilops cylindrica* Host), crimean wormwood (*Artemisia taurica* Willd.), Bieberstein tulip (*Tulipa biebersteiniana* Schult. et Schult. f.), splendid bellevalia (*Bellevalia speciosa* Woronow ex Grossh), iris humilis (*Iris pumila* L.), canadian conyza (*Conyza canadensis* (L.) Cronquist), blueweed (*Echium vulgare* L.), etc. A total of about 30 species, the total projective cover is up to 40 %.

Figure 1. Planning route of research.

Figure 2. Steppe communities of abrasion coasts.

Also, a significant area of growth is occupied by cereal-herbs communities in the vicinity of Volna village. In the first layer the following species are noted: slender harem (*Bupleurum tenuissimum* L.), soft brome (*Bromus mollis* L.), geniculate barley (*Hordeum geniculatum* All.) yellow everlasting (*Helichrysum arenarium* (L.) Moench), steppe euphorbia (*Euphorbia stepposa* Zoz ex Prokh.), crimean sage (*Phlomis taurica* Hartwiss ex Bunge), austrian artemisia (*Artemisia austriaca* Jacq.), Lerch sagebrush (*Artemisia lercheana* Weber ex Stechm.), etc. In the second layer, the followings are recorded: white-toment germander (*Teucrium polium* L.), mountain iron woundwort (*Sideritis montana* L.), Marshall thyme (*Thymus marschallianus* Willd.), astrakhan potentilla (*Potentilla astracanica* Jacq.).

About 40 species are found in such communities. The height of the first layer is 50 cm, the second is 20 cm. The total projective cover is 60–70 %.

Along the route of the study, abrasion coasts of the herb-feather grass community are found, consisting of the following species of the first layer: crimean sage (*Phlomis taurica* Hartwiss ex Bunge), ethiopian sage (*Salvia aethiopis* L.), steppe sage (*Salvia tesquicola* Klokov & Pobed.) esparto grass (*Stipa capillata* L.), sprawling cornflower (*Centaurea diffusa* Lam.), etc. In the second layer, the followings were noted: iris humilis (*Iris pumila* L.), yellow star of Bethlehem (*Gagea lutea* (L.) Ker Gawl., austrian wormwood (*Artemisia austriaca* Jacq.), alison hirsute (*Alyssum hirsutum* M. Bieb.) The total projective cover is 80–100 %, about 30 species in total.

The generalized floristic list, compiled from the results of a survey of the abrasion coasts of the Taman Peninsula, confined to the Black Sea coast, includes 231 plant species from 48 families. 11 families have a level of species richness higher above average, the leading families are Asteraceae – 41 species, Poaceae – 38 species, Lamiaceae – 16 species, Fabaceae – 15 species, Rosaceae – 12 species.

On the roadsides of unpaved roads that cut the study area, there are communities with a predominance of weed species: ragweed (*Ambrosia artemisiifolia*), canadian conyza (*Conyza canadensis*), couch grass (*Elytrigia repens*), sow thistle (*Sonchus arvensis*), shepherd's purse (*Capsella bursa-pastoris*), rugose rapistrum (*Rapistrum ugosum*), species that are rarely found in disturbed habitats: lepidium draba (*Cardaria draba*), winterweed (*Stellaria media*) tectorial anisantha (*Anisantha tectorum*), delphinium (*Delphinium consolida*), wild camomile (*Tripleurospermum inodorum*), stinkweed (*Thlaspi arvense*), spread cornflower (*Centaurea pseudosquarrosa* Mikheev ex Gabrieljan et Mikheev) and other species that tolerate trampling well. The process of unification of the studied flora was 17 %. Despite the fact that the study area is located near the "Port Taman", as well as a settlement, the unification process affected the flora being characterized to an insignificant extent. The territory is characterized by high species richness; it is a kind of ecotone zone, which is of great importance for the protection of plants listed in the Red Books of Russia and Krasnodar region.

One of the largest populations of the iris humilis (*Iris pumila* L.) (figure 3) grows in the surveyed area, which belongs to the objects of international agreements and conventions ratified by the Russian Federation [8, 9]. The dynamics of the range and number of the species is stable. The species is found almost throughout the Eastern Azov Sea region and in the north of the Black Sea coast of the Kuban. Its frequency and abundance is high.

Figure 3. Full-member population of *Iris pumila* L. in communities of abrasion coasts.

In total, 7 species of Red Book plants from 6 families have been registered: Black Sea mustard (*Cakile euxina* Pobed.), ethiopian sage (*Salvia aethiopis* L.), crimean sage (*Phlomis taurica* Hartwiss ex Bunge), yellow everlasting (*Helichrysum arenarium*) (L.) Moench, Bieberstein tulip (*Tulipa biebersteiniana* Schult. et Schult. f.), splendid bellevalia (*Bellevalia speciosa* Woronow ex Grossh.) and iris humilis (*Iris pumila* L.). According to the categories of the Red Book of the Krasnodar region (2017), the largest number of species are listed with the status "Exposed" or 3UV – 5 species, and with the status 2 – "Endangered" or 2 IP – 2 species. The Red Book of the Russian Federation includes two species of splendid bellevalia (*Bellevalia speciosa* Woronow ex Grossh.) and iris humilis (*Iris pumila* L.).

4. Conclusion

With a relatively small area of study, a high value of steppe plant communities was revealed, which contain a unique plant gene and cenofund that require protection measures. The biomorphological structure of populations is dominated by annual species, which indicates a high dynamism associated with exogenous processes. Ecologically, the predominant species are xerophytes and xeromesophytes, as well as species with a wide eco-amplitude. The process of unification of the studied flora is low and amounted to 17 %. The generalized floristic list includes 231 plant species from 48 families. The study area is characterized by the presence of protected plant species with fairly large populations.

Acknowledgments

The article was prepared with the financial support of the Russian Foundation for Basic Research, project 19-05-00716 "Exogenous processes as a factor in the spatial differentiation of plant communities of abrasive sea coasts".

References

[1] *Agroclimatic resources of the Krasnodar region* 1975 56 p
[2] Kosyan R D and Krylenko V V 2018 The main criteria for the complex classification of the Azov-Black Sea coasts of Russia *Oceanology* **58** 3 pp 501–511
[3] Volkova T A, Mishchenko A A, Antiptseva Yu O and Lipilin D A 2017 *Coastal geosystems in space and time: based on materials of Krasnodar region* 275 p
[4] Peshkov V M 2003 *Coastal zone of the sea* 350 p
[5] Kosenko I S 1970 *Keys to Plants of the Northwestern Caucasus and Ciscaucasia* 613 p
[6] Zernov A S 2006 *Flora of the Northwest Caucasus* 464 p
[7] *Atlas Krasnodar region and the Republic of Adygea* 1996 48 p
[8] *The Red Book of the Russian Federation* 2008 (*Plants* vol 2) (Moscow: Ministry of Natural Resources of the Russian Federation; RAS; RBO; MSU; Chairman: Head of the State Service for Environmental Protection of the MPR of the Russian Federation) 855 p
[9] *The Red Book of the Krasnodar Territory. Plants and fungi* 2017 ed S A Litvinskaya et al (Krasnodar: Administration of Krasnodar region) 2nd ed 848 p

IOP Conf. Series: Earth and Environmental Science **946** (2021) 012044 doi:10.1088/1755-1315/946/1/012044

The role of anthropogenic factors in forest transformation in Southern Sakhalin

R N Sabirov

Institute of Marine Geology and Geophysics Far Eastern Branch of Russian Academy of Science, Yuzhno-Sakhalinsk, 693022, Russia

E-mail: r.sabirov@imgg.ru

Abstract. The article analyzes the forests changes in the south part of Sakhalin Island from the original, natural state as a result of its habitation and economic development. The most essential disturbances of forest cover occurred in the first half of the last century when 9 pulp-and-paper mills were built by Japanese in southern island. All acceptable and more productive dark coniferous forests for providing these mills were cut down. Moreover, significant share of the forests was destroyed because of repeated and large-scale fires, creating agricultural lands and habitable territories, construction of roads, power lines, oil-and-gas pipelines, etc. The most considerable forests transformation was on the territory of Yuzhno-Sakhalinsk, and also in Korsakovsky, Dolinsky, Anivsky, Kholmsky and Nevelsky districts.

1. Introduction

The result of settlement and development of any region, along with creating the habitable territories with related infrastructure objects, construction of the industrial establishments and roads, another anthropogenic changes lead to unavoidable disturbance of the local landscapes, significant fragmentation and transformation, or even total destruction of the natural vegetation. Undoubtedly, the forests as the most acceptable and needed natural resources from the time of appearing a human being on any plot of the Earth are exposed to the anthropogenic impact to a great extent.

The forests of Southern Sakhalin as well as all over the island are predominant type of vegetation attributed to taiga zone [1, 2], and have practically covered all its territory before the active colonization and scaled using forest resources. It was due to, first of all, favorable climate conditions for coniferous forests and particular geographical location. As a result, in the southern part of Sakhalin the most productive forests were formed mainly from the ayan spruce (*Picea ajanensis*) and Sakhalin fir (*Abies sachalinensis*) with a slight admixture in the southwestern regions of broadleaf species: *Quercus mongolica*, *Phellodendron sachalinense*, *Acer mayrii*, *Kalopanax septemlobus*, etc. However, now the structure of the forest cover in the southern part of the island is differ from its natural state. In this regard, the purpose of this work is to assess the role of various anthropogenic factors in the transformation of the original forests of the southern part of Sakhalin as a result of many years of economic development and habitation of its territory.

2. Materials and Methods

To establish a whole range of anthropogenic factors and the degree of their influence on the change in the structure of the forest cover in the southern part of Sakhalin, numerous archival, historical and fund materials, forest management, statistical and other accounting data, plans of afforestations, topographic maps and atlases of various years were engaged in. In addition, changes in zonal vegetation were assessed based on the results of our own studies of the region forests, carried out over more than 30 years. At the same time, field research of forests was carried out according to standard methods in forestry and phytocenology [3–6 and others]. The available satellite images from the Landsat 8 spacecraft, taken in different periods, were also used to analyze the transformation of primary forests.

3. Results

The habitation and use of the natural resources of the southern part of Sakhalin took place most intensively in the first half of the last century, after the transfer in 1905 of its territory south of the 50th parallel to the jurisdiction of Japan. In a very short period of time, the Japanese built 9 pulp-and-paper mills in various regions, focused on the processing of valuable conifers, mainly spruce and fir. These mills were erected in the cities of Korsakov, Yuzhno-Sakhalinsk, Dolinsk, Kholmsk, Chekhov, Tomari, Uglegorsk, Makarov, Poronaysk, which led to almost complete coverage of the forest resource base. Place in operation all pulp-and-paper mills at full capacity led to a sharp increase in forest felling, the volume of which reached 5–7 million cubic meters by the mid-1930s of the last century. Moreover, most of the mills were erected in settlements with access to the sea, which ultimately ensured uninterrupted and very cheap export of finished products to Japan. These mills became one of the powerful city-forming factors and contributed to the active colonization of the south of Sakhalin. At the same time, other branches of the woodworking industry developed quite well: sawmilling, cooper, shipbuilding, plywood, where cut timber was processed on site, and then exported, along with cellulose and paper, in the form of finished products and other products of its processing.

Quite considerable volumes of forest resources, besides providing for the woodworking and pulp-and-paper industries, were also used for the manufacture of sleepers for the construction of railways. During the "Japanese" period, 120 stations, 24 tunnels, 618 bridges were built on Sakhalin, and the total length of railways was more than 700 km. Along with this, a significant amount of cut timber was used for the manufacture of mine racks and fastening beams in coal mines, the number of which reached 30. There 48 million tons of coal were mined [7, 8 and others].

In addition to being used at Sakhalin production facilities, timber was also prepared for export to Japan. In the first 15 years of colonization of the southern part of Sakhalin, the scale of export was not so significant and varied in the range of 4.2–84.4 thousand cubic meters per year. Then the volume of export of round timber to Japan began to increase sharply and in 1923 already amounted to 1.858 million cubic meters, in 1924 – 2.638, in 1926 – 3.002, in 1928 – 3.191 million cubic meters, etc. (figure 1). In 1921–1926, Japan exported 30 times more round timber from Sakhalin than from Korea and 80 times more than from Taiwan [8, 9, etc.]. With such volumes and intensive felling rates, by 1945 all available and most productive dark coniferous forests in the southern part of Sakhalin were under principal felling. Over the 40 years of the Japanese stay in the south of the island, more than 117 million cubic meters of timber were harvested [10].

Significant contribution to the transformation of the forests of the southern part of Sakhalin was made by agricultural development and creating the habitable territories. Until 1945, about a quarter of the local population was employed in agriculture, and it was almost completely self-sufficient in food. Sown areas in the south of Sakhalin at the apogee of agricultural development in 1935–1938 were 29.5 thousand hectares, the number of cattle – 8 thousand heads, pigs – up to 10 thousand heads, poultry – over 80 thousand heads, many indicators of which have not been achieved yet. During the "Japanese" period, in the valleys or at the mouth of almost every spawning river, settlements with corresponding infrastructure facilities were created and interconnected by a dense network of dirt roads. If in 1907 20.5 thousand people lived in the settlements of the south of the island, and in 1920 – 105.9, then by 1940 their number had already reached 415 thousand people [8, 11, 12, etc.], while at present there are slightly more than 470 thousand people all over the island.

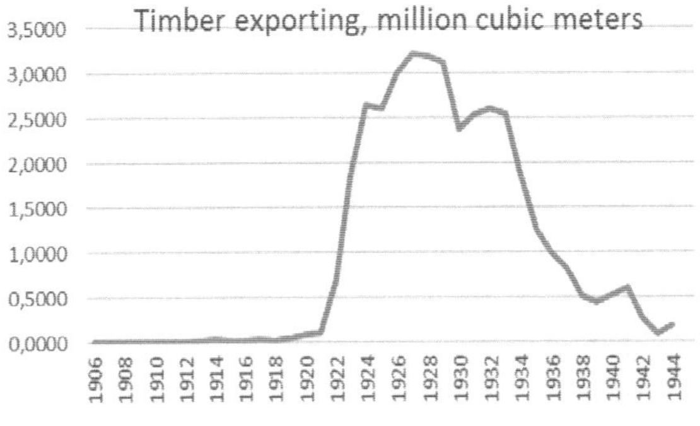

Figure 1. Dynamics of round timber export (*million cubic meters*) from Sakhalin to Japan in 1906 – 1944.

Quite active deforestation in Sakhalin also lasted during the Soviet period of time. In 1946–1991 the forestry enterprises all over the island harvested in average from 2 till 3.5 million cubic meters of timber per year. Of course, at this time, special logging enterprises were created on Sakhalin – integrated timber industry enterprises with powerful technical equipment, as well as associated forest villages. In the Soviet period, about 20% of the population worked in the forestry sector of the Sakhalin Region economy. At the same time, the old Japanese-built pulp-and-paper mills, using only coniferous wood, practically did not stop their activities. To ensure their uninterrupted supply of raw materials, the forests of the southern part of the island were under repeated principal felling. In addition, forests were cut down for the rapidly developing construction industry of Sakhalin. At the same time, the population of the region did not stop harvesting wood for household needs. As a result of long-term and large-scale felling, the available dark coniferous forests of the southern part of the island were almost completely cut down.

Along with felling, the fires, mainly of anthropogenic origin, had a significant impact on the transformation of forests [13, 14]. During the "Japanese" period, 1107 cases of fires were registered in the south of Sakhalin. They destroyed a forest of 802 thousand hectares with a total reserve of over 8 million cubic meters [10]. In 1949, about 354 thousand hectares of forests burned out in the south of the island, in 1950 – 207, and in 1954 – 435 thousand hectares, which later, together with the felling, turned into wastelands overgrown with Kuril bamboo and reed grass, practically unusable for natural reforestation.

Certainly, the transformation or even elimination of forest vegetation in the southern part of the island during the entire past and at the beginning of the 21st century, in addition to large-scale felling and fires, also took place in the process of building roads, industrial enterprises and structures, expanding settlements, carrying out agricultural land reclamation, creating hayfields. and cultivated fodder lands for animal husbandry, laying of power lines and oil and gas pipelines, development of quarries, coal mines, open-pit mines, etc. [7, 8, 11, etc.].

A set of main anthropogenic factors and an assessment of their impact on forest vegetation in various regions of Southern Sakhalin are shown in Table 1. The degree of impact of the most significant anthropogenic factors on the structure of forest cover was assessed on a three-point scale: 1 – weak, 2 – moderate, 3 – strong. With a weak integral influence, disturbance of forest vegetation does not exceed 25% of the area of the analyzed territory, with an average – by 26–50% of the area, and with a strong one – forest transformation is observed on an area of more than 50%.

As you can see from the above materials, the greatest impact on the initial forest cover was made by industrial felling, fires, as well as the construction of roads and settlements. The frequency of occurrence of the latter in the south of the island is much higher than in the northern regions. Of course, during the construction of roads and settlements, the vegetation and soil cover was completely eliminated.

Zonal forests turned out to be most transformed on the territory of Yuzhno-Sakhalinsk, as well as in Korsakovsky, Kholmsky, Nevelsky, Anivsky and Dolinsky municipalities, where the main production facilities are located, agriculture is well-developed, large ports and cities of the region are with high population density. At the same time, the original forest vegetation on the territory of the regional capital, in Yuzhno-Sakhalinsk, was radically transformed (Table 1). Indigenous coniferous forests here were completely destroyed, and in their place forest cultures were created, both from aboriginal and foreign forest-forming species (*Larix leptolepis, Pinus sylvestris, Picea abies*, etc.), which now represent suburban forest landscapes (figure 2). A similar situation developed on the territory of the Korsakovsky District, from where the Japanese began to develop the southern part of Sakhalin.

Table 1. Impact of anthropogenic factors on the forests of Southern Sakhalin.

Administrative Districts (municipalities)	Square of the District, km^2	Types of anthropogenic impact								
		Forest felling	Forest fires	Plough lands for agriculture	Creation of pastures, forage lands	Construction of settlements, roads	Recreational loads	Construction of industrial enterprises etc.	Construction of oil and gas pipelines	Cumulative anthropogenic impact
Smirnykhovsky	10457.4	2	1	1	1	2	1	1	1	10
Poronaysky	7284.3	3	3	1	1	2	1	3	1	15
Makarovsky	2148.4	3	3	1	1	2	1	3	1	15
Dolinsky	2441.6	3	2	3	3	3	2	3	2	21
Anivsky	2684.8	3	3	3	3	3	2	1	3	21
Korsakovsky	2623.6	3	2	3	3	3	2	3	3	22
Tomarinsky	3169.3	3	3	3	3	3	1	2	0	18
Uglegorsky	3965.6	3	3	3	3	3	1	3	0	19
Kholmsky	2279.0	3	3	3	3	3	2	3	0	20
Nevelsky	1445.4	3	3	3	3	3	2	3	0	20
Yuzhno-Sakhalinsk	905.0	3	3	3	3	3	3	3	3	24
Total effect of the factor		32	29	27	27	30	18	28	14	

A very moderate transformation of the forest cover in the Smirnykhovsky District is due to the fact that the northern part of its territory was Soviet until 1945 and, therefore, was less developed. In addition, a pulp-and-paper mill and other large industrial enterprises were not built in the region, and agriculture was not developed. Throughout the last century, the district was engaged in logging, for the implementation of which the corresponding settlements were created, their inhabitants mainly worked in the forest sector of the economy. Moreover, a significant part of the coniferous forests is concentrated on the hard-to-reach steep slopes of the East Sakhalin Mountains contributing to their preservation to a certain extent. It was the Smirnykhovsky District that we managed to create the Vostochny Nature Reserve with a total area of over 68 thousand hectares in order to preserve intact forest areas [15].

Figure 2. The forests around Yuzhno-Sakhalinsk have been transformed to the greatest extent and are mainly represented by artificial plantations and secondary communities.

4. Conclusion

Consequently, the transformation of zonal forests in the southern part of Sakhalin actively started in the last century and was associated with the colonization and development of natural resources. The most significant anthropogenic disturbances of forest cover occurred in the first half of the past century, when the Japanese built 9 pulp-and-paper mills in the south of the island, for which all available and most productive dark coniferous forests were under principal felling. The volume of forest felling in the south of Sakhalin by the middle of the 30s of the last century reached 5–7 million cubic meters per year. In addition to cellulose and paper, a significant amount of round timber was exported from Sakhalin to Japan. Subsequently, during the Soviet period, even on the entire Sakhalin Island, annual logging did not exceed 2–3.5 million cubic meters. Undoubtedly, large-scale and repeated logging during the past century has significantly wrecked the forest resources of the region and radically affected the formational composition and structure of the forest cover.

Moreover, huge areas of forest land have been destroyed by fires because of human activities. Only in the "Japanese" period, more than one thousand forest fires occurred in the southern part of Sakhalin and led to the accumulation in 1945 of about 1.5 million hectares of unforested cutting areas, glades, wastelands and burnt areas. Along with felling and fires, the transformation or even complete destroy of zonal forests took place in the process of creating agricultural lands and habitable territories, building industrial enterprises, railways and highways, developing coal mines, quarries and open-pit mines, laying oil and gas pipelines and other related infrastructure facilities.

Maximum anthropogenic transformation of forest landscapes occurred on the territory of Yuzhno-Sakhalinsk. With forming the urban agglomeration, the natural forests were completely destroyed and on the related to the city plots were created the artificial afforestations from both the indigenous and non-regional conifers. Essential anthropogenic transformation of zonal forests or even their total elimination was also in Korsakovsky, Kholmsky, Nevelsky, Anivsky and Dolinsky districts, where the main production facilities are located, agriculture is well-developed, ports and large cities of the region are with a very high population density.

IOP Conf. Series: Earth and Environmental Science **946** (2021) 012044 doi:10.1088/1755-1315/946/1/012044

References

[1] Tolmachev A I 1955 *Geobotanical zoning of the Sakhalin Iisland* (Moscow-Leningrad: Publishing House of the USSR Academy of Sciences) p 80
[2] Krestov P V, Barkalov V Yu and Taran A A 2004 *Materials of the International Sakhalin Project* Botanical and geographical zoning of Sakhalin Island Flora and fauna of Sakhalin Island (Vladivostok: Dalnauka) Part 1 pp 67–90
[3] Sukachev V N and Sonn S V 1961 *Sciences with Methodological guidelines for the study of forest types* (Moscow: Publishing House of the USSR Academy of Sciences) p 144
[4] *Field geobotany* 1964 (Leningrad: Nauka) **3** p 532
[5] Rabotnov T A 1978 *Phytocenology* (Moscow: Publishing House of Moscow State University) p 384
[6] *Methods of investigation of forest communities* 2002 (Saint-Petersburg: SPbSU) p 240
[7] Vysokov M S, Vasilevsky A A, Kostanov A I and Ishchenko M I 2008 *The history of Sakhalin and the Kuril Islands from ancient times to the beginning of the XXI century* (Yuzhno-Sakhalinsk: Sakhalin Publishing House) p 712
[8] Burykin A N 2010 The policy of the Japanese government on the formation of the permanent population of the south of Sakhalin (1905–1945) *Sociology of Power* **6** pp 159–165
[9] Lee Ben Dyu 1973 Predatory exploitation of the forest resources of Southern Sakhalin during the Japanese occupation *Materials on the history of the Far East* (Vladivostok: Far Eastern Research Center of the USSR Academy of Sciences) pp 186–188
[10] Vlasov S T 1959 *Forests of Sakhalin Island. Reference materials* (Yuzhno-Sakhalinsk: Sakhalin Book Publishing House) p 108
[11] Kozynyuk V M 1975 *Agriculture of the Sakhalin region* (Yuzhno-Sakhalinsk: Sakhalinsk. Department of the Far Eastern Bookstore. publishing House) p 192
[12] Bychkova A A, Gabrikov Ya E and Din Yu I 2017 *The Karafuto population Census of 1920* (Yuzhno-Sakhalinsk: JSC «Sakhalin Regional Printing House») p 490
[13] Ageenko A S and Klintsov A P 1969 Forests of Sakhalin Island and the Kuril Islands (Sakhalin region) *Forests of the Far East* (Moscow: Forest Industry) ed A S Ageenko pp 228–263
[14] Sabirov R N and Sabirova N D 2015 *All-Rus. Sci. Conf. with Int. Part. Geodynamic processes and natural disasters. The experience of Neftegorsk* Causes and prerequisites of forest fires on Sakhalin (Vladivostok: Dalnauka) **2** pp 391–394
[15] Sabirov R N, Sabirova N D and Voronov G A 2017 Modern state of the biota of «Vostochny» Nature Reserve in Sakhalin Island *Vestnik of the FEB RAS* **1** 191 pp 108–115

IOP Conf. Series: Earth and Environmental Science **946** (2021) 012045 doi:10.1088/1755-1315/946/1/012045

Biodiversity of the flora of vascular plants of the Makarovsky Nature Reserve on Sakhalin Island

N D Sabirova and R N Sabirov*

Institute of Marine Geology and Geophysics Far Eastern Branch of Russian Academy of Science, Yuzhno-Sakhalinsk, 693022, Russia

E-mail: *r.sabirov@imgg.ru

Annotation. The article is devoted to the research results concerning the flora of vascular plants of the Makarovsky Nature Reserve situated in the middle part of Sakhalin Island. Nowadays, 407 species of vascular plants have been identified on its territory, and 26.8 % of the total number of taxa known on the island. Among them, 13 species of rare plants included in the Red Data Book of the Sakhalin Region have been identified. The structure of the leading families and genera is indicated, the flora is analyzed by life forms, geographical elements and ecological-cenotic groups.

1. Introduction

At present, there are 57 specially protected natural reservations (SPNR) in the Sakhalin Region. Of these, 4 have the federal status, including 2 state nature reserves (Kurilskiy and Poronayskiy), 1 state nature reserve (Small Kuriles) and 1 health recreation area and the resort (Lake Izmenchivoe). The total area of SPNR with the federal status is 192 665 hectares, or 1.74 % of the territory of the Sakhalin Region. In addition to the federal ones, there are 53 SPNR with the regional status with the total area 685 986 hectares (7.81 %). They include 11 nature reserves, 1 natural park (Moneron Island) and 41 natural monuments. However, many SPNR in the region were created without the preparation of appropriate scientific substantiation and, therefore, without examining the ecosystems existing on their territory, without establishing the general biological diversity, as well as the structure and state of populations of rare species. Nowadays, only some of them have determined the full species structure of flora or fauna [1–3].

The Makarovsky Nature Reserve was created in 1992 as "forest hunting". Then, after several reorganizations in 2003, it was authorized in the status of an integrated biological nature reserve for protecting and reproducing rare and endangered species of plants and animals, preserving their biotopes, as well as the water bodies being spawning grounds for valuable salmon fish. Later, the boundaries of the reserve were changed from 2007 to 2009 under the results of an integrated environmental investigation. The most disrupted eastern sections have been excluded, and the area has been adjusted accordingly and now amounts to 34685 hectares (figure 1). At the same time, in order to identify the floristic and cenotic diversity, we studied the flora and vegetation of the reserve; some of the results are shown below.

2. Materials and methods

Standard methods were used for the collection and factual material processing on the determination of the biodiversity of vascular plants and the assessment of the current state of the vegetation of the reserve. In particular, field study was carried out mainly by route and detailed route methods and was aimed at maximum coverage of various types of landscapes and natural complexes of the reserve.

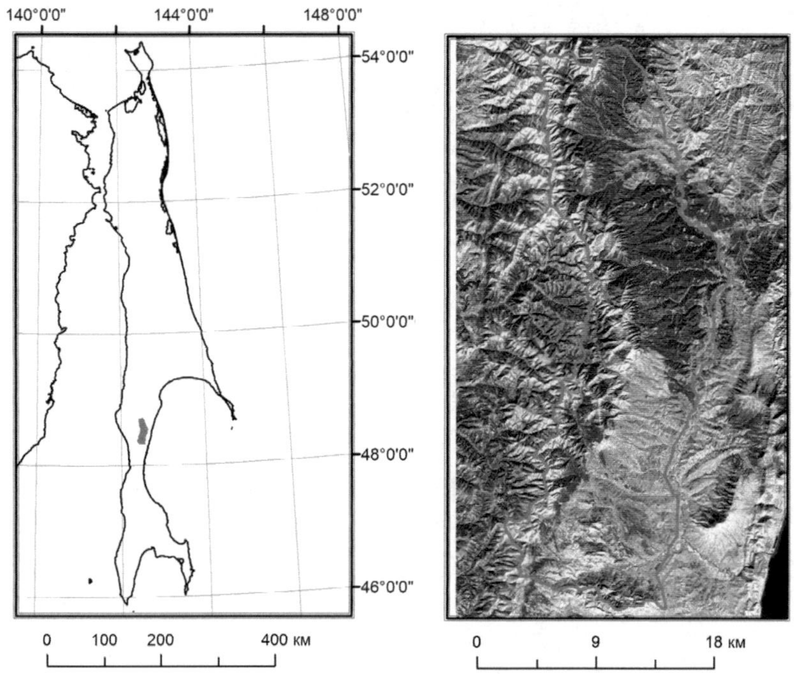

Figure 1. Location and recent boundaries of the Makarovsky Nature Reserve on Sakhalin Island.

Besides collecting herbarium, the index plots were under the establishment of test plots and geobotanical descriptions of plant communities, with the identification of the species structure of all layers. Performing these works, widely known methods in botany, phytocenology and forestry were used [4–7]. To identify the structure of the flora of the reserve and the subsequent taxonomic analysis, we followed the guidance papers by Tolmachev A.I. [8–10], Malyshev L.I. [11]. In the geographical analysis of the flora, the classification of plant areas worked out by Tolmachev A.I. [9, 10], Yurtsev B.A. [12], Yurtsev B.A., Kamelin R.V. [13] was used. The life forms of plants were established according to Serebryakov I.G. [14]. The materials of the Scientific Herbarium Institute of Marine Geology and Geophysics FEB RAS (SAK) were also additionally and critically revised.

3. Results of researches

The reserve is located in the southern part of the subzone of dark-pine forests with green mosses with a predominance of *Picea jezoensis* (Siebold et Zucc.) Carrière and within its limits in the West Sakhalin mountain geobotanical region [9], which generally determines the structure of its flora and the vegetation cover. The reserve is dominated by forest vegetation including fir-spruce (*Abies sachalinensis* F. Schmidt, *Picea jezoensis*), stone birch (*Betula ermanii* Cham.), valley willow (*Salix rorida* Laksch., *S. schwerinii* E.L. Wolf, S. *udensis* Trautv. et C.A. Mey.), poplar (*Populus maximowiczii* A. Henry), alder (*Alnus hirsuta* (Spach) Turcz. ex Rupr.) forests and their mixed variants, thickets of dwarf pine (*Pinus pumila* (Pall.) Regel). In addition, bamboos (*Sasa kurilensis* (Rupr.) Makino et Shibata), as well as fragments of alpine vegetation and secondary meadow communities, are very widely represented here.

Nowadays, there are 407 species of vascular plants from 279 genera and 90 families in the flora of the Makarovsky Nature Reserve covering 26.8 % of the modern taxonomic diversity of the flora of Sakhalin Island [15]. Flowering plants (64.6 %) are the significant share in the flora of the reserve, 40.3 % are dicotyledonous and 24.3 % are monocotyledonous plants. The flora of the natural reserve is characterized by a combination of various taxonomic groups, geographical elements and ecological-floristic complexes. The greatest idea of the systematic structure of the flora is given by the species and generic richness of the leading families, the spectrum is shown in the table 1.

Table 1. The spectrum of leading families and genera in the flora of the Makarovsky reserve.

№	Family	Quantity of species	Share of participation, %	Genera	Quantity of species	Share of participation, %
1	Asteraceae	43	10.6	Carex	20	4.9
2	Poaceae	39	9.6	*Poa*	8	2.0
3	Rosaceae	24	6.0	*Salix*	6	1.5
4	Ranunculaceae	23	5.6	*Galium*	6	1.5
5	Cyperaceae	20	4.9	*Viola*	6	1.5
6	Brassicaceae	14	3.4	*Artemisia*	5	1.2
7	Apiaceae	13	3.2	*Allium*	5	1.2
8	Ericaceae	13	3.2	*Agrostis*	4	0.9
9	Caryophyllaceae	11	2.7	*Equisetum*	4	0.9
10	Salicaceae	10	2.4	*Lycopodium*	4	0.9
	Total:	210	52.0		68	16.7

According to the set of leading families, the flora of the reserve is similar to the typical flora of the Boreal floristic area. The leading position of the families Asteraceae and Poaceae characterizes the floras of the Boreal region [12, 14]. The share of these families on the territory of the reserve is 20.2 % of the species diversity. At the same time, the high ranking position of the families Cyperaceae, Ericaceae, Ranunculaceae and Rosaceae emphasizes the peculiarity and inconsistency of the look of the reserve flora combining typical boreal features with East Asian ones. The generic spectrum as well as the family spectrum, indicates the heterogeneity of the flora of the study area and the significant participation of genera in it. Multi-species families are mostly multi-genus too. At the same time, the leading position is retained by Asteraceae (27 genera, or 9.7 % of their total number in the reserve), the second – by Poaceae family (7.9 %), the third – by Rosaceae (5.4 %).

If the family spectra reflect mainly the external, physiognomic features of the flora, then the generic spectra characterize its internal structure. Ten leading genera of the reserve include 68 species in total (16.7 % of the total taxonomic structure of the reserve). The generic spectrum also clearly demonstrates the originality and contradictory the flora look of the characterized territory, noted by Tolmachev A.I. [16] for all Sakhalin. The genus *Carex* is distinguished in this spectrum numbering 20 (4.9 %) species. The leading role of this genus is achieved by the high ecological plasticity of the enclosed species allowing them to occupy a wide variety of niches. In this regard, species of the genus *Carex* are represented in almost all plant formations widespread in the reserve. In addition to the genus *Carex*, the boreal features of the local flora are also determined by the high rank positions of the genera *Salix* and *Poa*. But along with this, the East Asian features of the reserve flora are exhibited by the presence in its structure the genera *Aconitum, Diphylleia, Spiranthes, Weigela* and others from the East Asian floristic region, as well as the presence here of 39 (9.6 %) single-species families (Aristolochiaceae, Cupressaceae, Empetraceae, Schisandraceae, Taxaceae, etc.). Thus, a systematic analysis has shown that the flora of the Makarovsky natural reserve possesses characteristic boreal features with the East Asian ones. The presence in it of such genera as *Aralia, Eleutherococcus, Hydrangea, Ilex, Schisandra* further enhances these specific features.

The East Asian geographical element clearly dominates in the reserve flora with 182 species (44.7 % of the total structure of the identified flora). East Asian species have a significant impact on

IOP Conf. Series: Earth and Environmental Science **946** (2021) 012045 doi:10.1088/1755-1315/946/1/012045

the formation of the forest (147 species; 36.1 %) of the ecological-cenotic complex. In addition to the East Asian geoelement, widespread circumpolar species (99; 24.3 %) make up a fairly significant proportion in the chorological structure of the flora of the studied area, with a noticeable participation of Asian-American (42; 10.3 %) and Eurasian representatives (32; 7.9 %). The leading ecological-cenotic complexes of the natural reserve are forest (262 species or 64.4 %), meadow-bog (42; 10.3 %) and coastal-seaside (40; 9.8 %). Species from the East Asian and circumpolar chorological groups play an important role in the formation of the above mentioned floristic complexes.

In the flora of the reserve, taking into account the ratio of life forms of plants, the overwhelming number of species belongs to perennial grasses – 313 (76.9 %), with the great role in the formation of meadow cenoses, as well as in the structure of the lower layers of forest vegetation. The participation of herbaceous monocarps (mainly one- and two-year-old species) is much less – 21 (5.1 %). They are widespread mainly in non-forest types of vegetation: in meadows, along pebbles, and are mostly represented by adventive species. At the same time, the flora of the reserve contains a relatively large number of woody plant species accounting for 15.0 % of the total number of species identified here and 13.6 % of the total number of genera (figure 2). The most common trees and shrubs of the reserve are: *Abies sachalinensis, Alnus hirsuta, Betula ermanii, Duschekia maximowiczii* (Callier) Pouzar, *Picea jezoensis, Pinus pumila, Rosa acicularis* Lindl., *Salix caprea* L., *S. rorida, S. udensis, Vaccinium ovalifolium* Smith, etc.

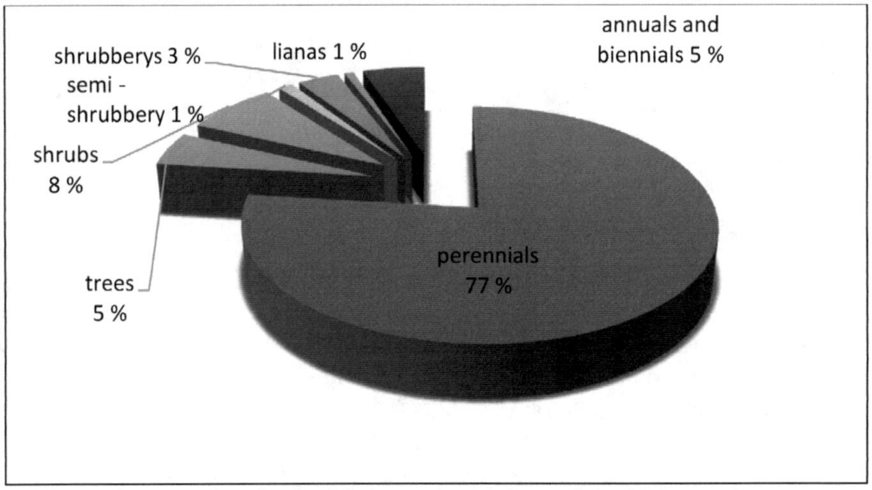

Figure 2. The ratio of plant life forms in the flora of the Makarovsky reserve

13 species of rare and endangered species of vascular plants, included in the Red Data Books of Russia and the Sakhalin Region [17, 18], grow in the Makarovsky Natural Reserve: *Aralia cordata* Thunb., *Cypripedium macranthon* Sw. (figure 3), *Cypripedium calceolus* L., *Diphylleia grayi* F. Schmidt, *Macropodium pterospermum* F. Schmidt, *Mecodium wrightii* Bosch, *Neottianthe cucullata* (L.) Schlechter, *Paeonia obovata* Maxim., *Platanthera camtschatica* (Cham. et Schlecht.) Makino, *Poa radula* Franch. et Sav., *Pulsatilla tatewakii* Kudo, *Rhodiola rosea* L., *Taxus cuspidata* Siebold et Zucc. (Figure 3). The share of rare plant species in the Makarovsky Natural Reserve is, in general, 7.1 % of their structure in the flora of the Sakhalin Region. 5 species of endemic plants were found on the territory of the reserve: *Aconitum neosachalinense* Lévl., *Corydalis multiflora* Mikhailova, *Pulsatilla tatewakii*, *Saussurea duiensis* F. Schmidt, *Trollius miyabei* Sipliv. Together with a rare category of plants, they increase the nature conservation and scientific importance of the Makarovsky Natural Reserve.

Figure 3. *Cypripedium macranthon* (on the left), *Taxus cuspidata.* Photo by R. Sabirov.

The adventive element of the reserve flora is represented by 34 plant species: *Capsella bursa-pastoris* (L.) Medik., *Galeopsis bifida* Boenn., *Gnaphalium uliginosum* L., *Melandrium album* (Mill.) Garcke, etc. They are found mainly along roadsides, in slashes, on pebbles along rivers and streams, as well as in secondary meadow communities.

4. Conclusion

The flora of the Makarovsky reserve despite its small area (0.44 % from Sakhalin Island square), is characterized by a fairly large biodiversity and includes 407 species of vascular plants. It's 26.8 % of the structure of the Sakhalin flora. The location of the reserve on the border of the boreal and East Asian floristic regions predetermined the taxonomic and chorological structure of the local flora under the physical and geographical conditions of the area. At the same time, the boreal features of the reserve flora are manifested through the high species richness of the genera *Carex, Poa, Salix*; the East Asian features are determined by the large number of representatives of the families Ericaceae, Ranunculaceae, Rosaceae, as well as the presence of 38 (9.3 %) single-species families (Actinidiaceae, Schisandraceae, Taxaceae, Vitaceae, etc).

The aerographic structure is characterized by the predominance of East Asian (182 species; 44.7 %) and circumpolar (99 species; 24.3 %) species, with a marked role of Asian-American (42 species; 10.3 %) and Eurasian (32 species; 7.9 %). The leading ecological-cenotic complex is undoubtedly forest (262 species or 64.4 %). This complex is dominated by species with East Asian and circumpolar spreading. In the meadow-bog complex, because of its intrazonal features, species with wide ranges prevail.

13 species of rare and endangered plants grow on the territory of the reserve. Their share reaches 16 % of their listed structure on Sakhalin Island. Such a concentration of Red Data Book species in a relatively small area is due not only to the peculiarities of florogenesis and the presence of a fairly wide range of ecotopes, but also to the long-term preservation of the indigenous forest communities of the reserve being as a kind of refugium for rare plant species.

References

[1] Sabirova N D and Sabirov R N 2007 Vascular plants of the Schmidt Peninsula (Northern Sakhalin) *V L Komarov Memorial Lectures* (Vladivostok: Dalnauka) **54** pp 82–114

[2] Sabirova N D and Sabirov R N 2018 Flora of vascular plants of the nature zakaznik "Vostochny" (Sakhalin island) *V L Komarov Memorial Lectures* (Vladivostok: FSC Biodiversity FEB RAS) **66** pp 65–100

[3] Barkalov V Yu, Koldaeva M N, Korznikov K A and Taran A A 2019 Vascular plants of nature monument "Gora Vaida" (Sakhalin, Russian Far East) *V L Komarov Memorial Lectures* (Vladivostok: FSC Biodiversity FEB RAS) **67** pp 14–53

[4] Sukachev V N and Sonn S V 1961 *Sciences with Methodological guidelines for the study of forest types* (Moscow: Publishing House of the USSR Academy of forest types) p 144

[5] *Field geobotany* 1964 (Leningrad: Nauka) 3 p 532

[6] Rabotnov T A 1978 *Phytocenology* (Moscow: Publishing House of Moscow State University) p 384

[7] *Methods of investigation of forest communities* 2002 (Saint-Petersburg: Nauka) p 240

[8] Tolmachev A I 1955 *Geobotanical division of Sakhalin Island* (Moscow-Leningrad: Izd-vo AN SSSR) p 80

[9] Tolmachev A I 1962 *Fundamentals of the doctrine of areas* (Leningrad: Leningrad University) p 100

[10] Tolmachev A I 1974 *Introduction to the geography of plants* (Leningrad: Leningrad University) p 243

[11] Malyshev L I 1975 Quantitative analysis of flora: spatial diversity level of species richness and representativeness of survey sites *Botan. journal* 11 **60** pp 1537–50

[12] Yurtsev B A 1974 *Problems of botanical geography of Northeast Asia* (Leningrad:Nauka) p 160

[13] Yurtsev B A and Kamelin R V 1991 *Basic concepts and terms of floristry* (Perm: Perm University) p 80

[14] Serebryakov I G 1962 *Ecological morphology of plants Life forms of angiosperms and conifers* (Moscow) p 378

[15] Barkalov V Yu and Taran A A 2004 *List of species of vascular plants of Sakhalin island Growing and fauna of Sakhalin island* (Vladivostok: Dalnauka) 1 pp 39–66

[16] Tolmachev A I 1959 *About the flora of Sakhalin Island* (Moscow-Leningrad: Izd-vo AN SSSR) p 102

[17] *Red Book of the Russian Federation: plants and fungi* 2008 (Moscow: Association of Scientific Publications of the CMC) p 855

[18] *Red Book of the Sakhalin region: plants and fungi* 2019 (Kemerovo: LLC Technoprint) p 352

Secondary phloem traits in mature stems of *Betula ermanii* in stressful environments of volcanic activity

A I Talskikh[*]**, A V Kopanina and I I Vlasova**

Institute of Marine Geology and Geophysics Far Eastern Branch of Russian Academy of Science, Yuzhno-Sakhalinsk, 693022, Russia

E-mail: [*]anastasiya_talsk@mail.ru

Abstract. This research summarizes the results of study of structural peculiarities of *Betula ermanii* growing around Baransky Volcano close to the thermal Golubye Ozera. We have performed statistical analysis of parameters of the secondary phloem of multiyear shoots and stems and found some features tied to extreme environmental conditions. We have found that structural reaction of the secondary phloem of multiyear shoots and stems of *B. ermanii* to the extreme conditions of post-volcanic activity of Baransky Volcano manifests itself as changes in the geometry of the conductive elements and dilatation of the radial parenchyma in the non-conductive phloem. We believe these changes to be adaptive in nature due to the need for increased conductivity in volcanic landscapes.

1. Introduction

The Kuril Islands have unique ecosystems shaped by various environmental factors. One of the key environmental factors changing the islands' natural systems is magma volcanism [1–3]. Volcanism is responsible for the dynamics of relief-forming processes; it ensures specific geochemical processes, influences ambient air transparency and composition, and it is also one of the key factors of soil formation which impacts vegetation. History of vegetation in these conditions is closely related to life strategies of the plants aimed at survival of certain species in specific environmental conditions. Such species have developed various adaptive mechanisms ensuring that their life processes adjust to stressful living conditions [4].

Plants with a wide reaction norm and high adaptivity to all kinds of habitats include *Betula ermanii* Cham. (fam. *Betulaceae* S. F. Grai). *B. ermanii* is a monoecious, foliage, anemophilous tree up to 30 m in height or large brush (at the vegetation limit up to three or five meters high). It forms an independent belt of stone birch forests in the mountains and foothills, brushwood at seashores and in the highlands; it can also be found in landscapes changed by current volcanic activity [5]. Area of species includes south of Siberia and a lot of area in the Russian Far East. The species grows in large quantities on Sakhalin Island and the Kuril Islands (Iturup, Kuhashir, Shikotan, Rasshua, Ushishir, Ketoi, Simushir, Urup) [6]; it is also found in China, the Korean Peninsula, and in Japan [5].

The purpose of this research is to find adaptive peculiarities in the secondary phloem of multiyear shoots (10 years, 23–24 years) and stem (56 years) of *B. ermanii* growing in conditions of post-volcanic activity of Baransky Volcano.

The information on the organization of phloem and sieve tubes as its conductive elements in trees is necessary to model long-distant transport of carbon in woody plants in various environmental conditions. That data is extremely useful for assessment of changes in carbon production of forests, namely, stone birch ones and those including stone birch, in conditions of changing climate. Right now, those issues are far from certainty. The peculiarities of phloem transport and mechanisms of its modeling in trees were described in a number of studies based on Munch's theory of phloem transport [7–9].

2. Materials and methods

Samples were collected during expeditions from 2015 to 2018. *B. ermanii* samples were collected in conditions typical for the species in fir–stone birch shrub–forbs forest on Krasnaya Mountain (Susunai Ridge, Sakhalin Island) on October 18, 2015 (figure 1). Samples for reference were collected in stone birch–larch–bamboo forest nearby hot springs around Golubye Ozera (Baransky Volcano, Iturup Island) on July 31, 2018 (figure 1). Golubye Ozera hot springs are two deep funnels filled with opalescent bluish water, fringed with bright dispersed sulfur rims. The water is of the sulfate-chloride type, ultra-acidic (pH = 1.2), up to 107.5°C hot. The emission gases are nitrogen (51 %); carbon dioxide (38 %), oxygen (9 %), methane and other hydrocarbons (0.05 %) [10].

Figure 1. *Betula ermanii* in various environmental conditions: on the left – fir–stone birch shrub–forbs forest on Krasnaya Mountain (Susunai Ridge, Sakhalin Island); on the right – stone birch–larch–bamboo forest nearby hot springs around Golubye Ozera (Baransky Volcano, Iturup Island).

Stem samples for anatomical analysis were selected and fixed following the standard techniques suggested in R.P. Barykina et al. [11] and modified in our laboratory [12, 13]. Transverse, radial and tangential 10–25 μm thick microsections of the stems were made with a sliding microtome Microm HM 430c (Thermo scientific) with a fast freezing unit Microm KS34 (Thermo Scientific). The microsections were stained regressively using Safranin and Nile blue. The bark samples were also macerated to reveal the structure of conducting elements of the phloem. Section images were processed with ZEN 2 lite software under light microscope Axio Scope.A1, CarlZeiss for measuring and photography. We analyzed 15 characters of secondary phloem in transverse and longitudinal sections for each model tree. The dataset for each character of each model tree comprised at least 32 measurements. The sample mean and its confidence interval (for 95 % probability) were estimated for each character. Bark tissue descriptions were made relying on the analytical approaches common in xylotomy, and according to current guidelines on bark anatomy of woody plants by the International Association of Wood Anatomists – IAWA [14].

3. Results and discussion

Secondary phloem is a tissue derivative of cambium. It transports products of photosynthesis. Functionally it can be divided into conductive and non-conductive phloem. It includes segments of sieve tubes with companion cells, the axial and radial parenchyma [15]. In the conducting phloem of *B. ermanii* the axial parenchyma is striped in structure; the sieve tubes are located in groups between the radial parenchyma. Sieve tubes contain complex sieve plates of step and net structure. Rays in the phloem are homo and heterocellular. Rays are single-row, double-row, triple-row, rarely even four rows. The non-conductive phloem, in addition to the elements listed before, contains multiple sclereid groups formed by sclerification of parenchyma cells [16].

During ontogenesis, secondary phloem in the bark of *B. ermanii* was found to increase in all studied habitats. Near thermal lakes the increase in the secondary phloem of *B. ermanii* is even more significant. It exceeds the norm by 34 % and can be as great as 1553.12±35.64 μm (while the norm is 1172.12±36.71 μm). At 24 years of age, conductive phloem width in samples from thermal lakes exceeds the norm (102.15±3.51 μm) by 15 %. In the stem, secondary phloem width in the stem part near thermal lakes is 62 % less than normal (3772.58±143.28 μm), due to reduction of non-conductive phloem, because in the stem part the width of conductive phloem near thermal lakes is similar to background conditions. Increased width of non-conductive phloem in multiyear shoots in thermal lakes conditions may be linked to active dilatation of cells of the axial and radial parenchyma. The share of secondary phloem in the total width of the bark near thermal lakes in a ten-year-old shoot is 58 %; in a 24-year-old shoot it's 71 %; in the stem part (56 years) it's 55 %. At the same time, the share of non-conductive phloem in the bark width in background conditions is 47 % in a ten-year-old shoot; 46 % in a 24-year-old shoot; 71 % in the stem part. Therefore, the share of non-conductive phloem in the bark in conditions of thermal lakes increases expectedly, exceeding the norm by 11 % in a ten-year-old shoot, by 25 % in a 24-year-old shoot, and in the stem part its share in the bark is 16 % less than normal.

In multiyear stems close to thermal lakes, the radial diameter of sieve tubes increases by 45 % at ten years of age, up to 26.40±1.93 μm; at 24 years of age the increase is 37 %. In a year-old shoot of *B. ermanii*, we also observe the same increase in the radial size of sieve tubes [17, 18]. In the stem part the radial diameter of sieve tubes remains normal. However, we see a reduction in their tangential diameter by 26 % versus the norm (45.42±1.94 μm). Length of sieve tube segments near thermal lakes changes in different ways: at ten years of age, we see a 19 % decrease in this parameter versus the norm (511.73±35.94 μm); then, at 24 years of age, the length of the sieve tube segments exceeds the norm (553.95±37.09 μm) by 22 %, and in the stem part this parameter is within the norm (745.09±45.92 μm).

In the secondary phloem near thermal lakes, we see a decrease in the total number of phloem rays per 1 mm of cross-section versus the norm: by 10 % at ten years of age; by 23 % at 24 years of age. We also see a decreased number of single-row rays per 1 mm near thermal lakes: by 12 % in a ten-year-old shoot; by 27 % in a 24-year-old shoot. In the stem part, this parameter is normal (4.39±0.40 pcs).

Specific volume of elements of secondary phloem per 1 mm of cross-section in the conductive phloem at specific ages near thermal lakes remains within norm, while in non-conductive phloem specific volume of radial parenchyma per 1 mm near thermal lakes exceeds the norm by 15 % in multiyear shoots and by 25 % in the stem part. Increase in the specific volume of radial parenchyma is due to active dilatation of that tissue in non-conductive phloem. As a result, specific volume of sieve tubes in the non-conductive phloem in multiyear shoots and in the stem part reduces by 9 %.

4. Conclusion

Statistical analysis of structural parameters of the secondary phloem of *B. ermanii* growing around the thermal lakes of Baransky Volcano (Iturup Island) shows, that is a clear reaction of structural elements of that tissue. It manifests itself as increased parenchymatization of secondary phloem in the non-conductive zone. This may be related to the need for an additional reservoir for water since it is insufficient due to volcanic salinization. Another structural change in phloem of *B. ermanii* is increased size of sieve tubes. We believe these changes to be adaptive in nature due to the need for increased conductivity in volcanic landscapes. Our results match those found in some research of the recent years studying the role of phloem in how plants react to drought [19]. The research shows that the conductive channels of the phloem must be surrounded by living cells if they are to maintain transportation capability of the phloem at low water potential in tissues when water is scarce. Those groups of living cells, the phloem parenchyma, can control water content and serve as a water reservoir quite independently of wood during drought. We agree with the researchers [19] thinking that the reaction of the phloem to drought can be a key to predicting plants' survival time when water is scarce or the plants' restorative capability after drought.

Acknowledgments
The study was conducted with support from the Russian Foundation for Basic Research (bridging grant no. 15-04-04774) and as a part of the state order to the Institute of Marine Geology and Geophysics, Far Eastern Branch of the Russian Academy of Sciences.

References

[1] Rybin A V, Chibisova M V, Smirnov S Z, Martynov Yu A and Degterev A V 2018 Petrochemical features of volcanic complexes of Medvezh'ya caldera (Iturup Island, Kuril Islands) *Geosystems of transition zones* **4** 1 pp 93–102
[2] Ezhkin A K 2019 Lichens of wood substrates in areas of solfataric activity on Southern Kuriles *Geosystems of transition zones* **3** 2 pp 256–263
[3] Romanyuk F A and Degterev A V 2020 Transformation of the coastline of Raikoke Island after the explosive eruption on June 21–25, 2019 (Central Kuril Islands) *Geosystems of transition zones* **4** 3 pp 351–358
[4] Man'ko Y and Sidel'nikov N 1989 *The influence of volcanism on vegetation* (Vladivostok: FEB RAS) p 163
[5] Kabanov N E 1972 *Stone birch forests in botanical-geographical forestry relations* (Moscow: science) p 136
[6] Barkalov V Yu 2009 *Flora of the Kuril Islands* (Vladivostok: Dalnauka) p 468
[7] Savage J A, Beecher S D, Clerx L, Gersony J T, Knoblauch J, Losada J M, Jensen K H, Knoblauch M and Holbrook N M 2017 Maintenance of carbohydrate transport in tall trees *Nature Plants* pp 965–972
[8] Morgan E F, Trumbore S and Hartmann H 2018 Detours on the phloem sugar highway: stem carbon storage and remobilization *Current Opinion in Plant Biology* **43** pp 89–95
[9] Jensen K H 2018 Phloem physics: mechanisms, constraints, and perspectives *Current Opinion in Plant Biology* **43** pp 96–100
[10] Zharkov R V 2014 *Thermal springs of the Kuril Islands* (Vladivostok: Dalnauka) p 378
[11] Barykina R, Veselova T, Devyatov A, Dzhalilova K, Iljina G and Chubatova N 2004 *Mannual on botanical microtechnique. Basic principles and methods* (Moscow: MSU Publ.) p 312
[12] Vlasova I I and Kopanina A V 2019 Specific microtechniques for bark tissues of woody plants *Plant anatomy: traditions and perspectives. Materials of the International Symposium dedicated to the 90 anniversary of Prof. Ludmila Ivanovna Lotova. Semtember 16–22, 2019* In two part (Moscow: MAKS Press) pp 274–278
[13] Vlasova I I and Kopanina A V 2021 Features of sample preparation of shrubs and heaps of heaps for structural analysis *Botanical readings: collection of articles. materials Intern. scientific. conf., Brest, 25 Feb.* (Brest: Brest State University named after A.S. Pushkin) pp 31–35
[14] Angyalossy V, Evert R, Marcati C, Oskolski A, Terrazas T, Kotina E, Lens F, Cristina M V, Angeles G, Machado S, Crivellaro A, Rao K, Junikka L, Nikolaeva N and Baas P 2016 IAWA List of Microscopic Bark Features *IAWA journal. International Association of Wood Anatomists* **37** pp 517–615
[15] Evert R F 2016 *Esau's plant anatomy. Meristems, cells and tissues of the plants body: their structure, function, and development* (Moscow: BINOM. Knowledge Laboratory) p 600
[16] Kopanina A V, Talskikh A I and Vlasova I I 2018 Structure of the bark Betula ermanii in the conditions of Sakhalin Island *Botanika in the modern world. Proceedings of the XIV*

Congress of the Russian Botanical Society and the conference "Botany in the modern world" (Makhachkala, June 18–23, 2018). Vol 3 Spore plants. Mycology. Structural botany. Physiology and biochemistry of plants. Plant embryolog. (Makhachkala: ALEF) pp 171–173

[17] Talskikh A I, Kopanina A V and Vlasova I I 2019 Specific structural features of the bark in annual stems of Betula ermanii Cham. exposed to hydrothermal gases and fluids of Baransky volcano, Iturup island *Plant anatomy: traditions and perspectives. Materials of the International Symposium dedicated to the 90th anniversary of Prof. Ludmila Ivanovna Lotova. September 16–22, 2019* In two parts (Moscow: MAKS Press) pp 241–245

[18] Talskikh A I, Kopanina A V and Vlasova I I 2021 Structure of the secondary phloem in annual stem of Betula ermanii under the conditions of gas-hydrothermal outcrops of Baransky volcano (Iturup Island) *Botanical readings: collection of articles. materials Intern. scientific. conf., Brest, 25 Feb.* (Brest: Brest State University named after A S Pushkin) pp 54–58

[19] Sevanto S 2018 Drought impacts on phloem transport *Current Opinion in Plant Biology* **43** 7 pp 6–81

IOP Conf. Series: Earth and Environmental Science **946** (2021) 012047 doi:10.1088/1755-1315/946/1/012047

Structural deviations in secondary phloem of young shoots of *Spiraea beauverdiana* near magma volcanoes

E O Vatserionova[1,*], **A V Kopanina**[1] **and I I Vlasova**[1]

[1]Institute of Marine Geology and Geophysics, Far East Branch, Russian Academy of Sciences, Yuzhno-Sakhalinsk, 693022, Russia

E-mail: *vatserionova.katya.85@mail.ru

Abstract
We performed a comparative analysis of the internal structure of the secondary phloem of one, two and three-year-old stems of *Spiraea beauverdiana* growing in extreme conditions of solfataric fields of Golovnin Volcano caldera and Mendeleyev Volcano and in normal conditions. The combination of environmental factors in conditions of solfataric activity, such as high temperatures in the soil and in the near-surface air, as well as saturation with gases toxic to plants, rare elements accumulating in the nearby substrate, and lack of soil moisture, interfere with normal phellogen and cambium activity. Deviations from the normal structure involve changes in the following parameters of the internal structure of a year-old stem of *S. beauverdiana*. Secondary phloem parameters in the studied habitats are normal, except for the length of the segments of sieve tubes, the height of single-row rays, the length of parenchymal girder; these are shorter in *S. beauverdiana* stems from volcanos. At two and three years of age in volcanic conditions we see reduction in the width of the secondary phloem (both conductive and non-conductive) and the diameter of the segments of sieve tubes. In samples from Golovnin Volcano we see reduction in tangential diameter, while in the samples from Mendeleyev Volcano it's the radial diameter. We also see reduction in the height of multiple-row rays. At that age we see changes in the structure of the radial parenchyma; namely, we find no double-row rays in samples from Golovnin Volcano caldera. One of the signs of impact of volcanic activity on the bark structure is development of non-specific anomalies in the internal structure of the *S. beauverdiana* bark, namely, in the outer bark, or in deeper levels, such as the secondary phloem. That causes sclerification and dilatation of parenchyma, and multiple layers in some tissues.

1. Introduction

Vegetation of the Kuril Islands is influenced by complex and very unique natural conditions. One of the main factors impacting the plants is current volcanic and post-volcanic activity [1–2], as well as the impact from the cold of the Sea of Okhotsk and the Pacific [3]. Post-volcanic activity creates special landscapes with sufficient deficiency of soil moisture, high concentrations of salts of heavy metals, sulfur compounds, and high concentrations of sulfur oxides, nitrogen and other substances in the near-surface air [4]. Woody plants growing in these conditions have a whole set of adaptation mechanisms allowing them to survive in such harsh areas.

The object of this research is *Spiraea beauverdiana* S.K. Schneid. (Rosaceae Juss.), a woody plant up to 60 cm high, with fringed or bare young reddish-brown shoots. The bark of a multi-year shoot is brown. Its flowers are white, pinkish sometimes [5]. Depending on the environment, this species has two life-forms: a deciduous brush (up to 1 m tall) and a deciduous low shrub. The distribution area of the species includes Siberia, the Russian Far East, China, and Japan. The northern boundary of the area is in the Arctic latitudes [6]. *S. beauverdiana* grows on rocks and rocky slopes, in the shrub tundra and herb meadows, in the mountain forests and sparce forests [5].

The purpose of this research is to make a comparative analysis of the internal structure of the secondary phloem of young shoots of *Spiraea beauverdiana* growing in solfataric fields of active volcanoes (Golovnin and Mendeleyev Volcanoes) and in normal conditions on Kunashir Island.

As a connecting link in distribution of carbon in plants, the phloem and its network of living channels (sieve tubes) are at the heart of the mechanisms distributing the plant's resources. Understanding of the functioning of the phloem of a woody plant, including cenose-forming species (trees, brush, and low shrub) has a great environmental and agricultural value. Using optical and electronic microscopy, we obtain data on the structure of segments of sieve tubes of plants taken from various areas, on their structural relations with satellite cells, the axial and radial parenchyma, which expands and clarifies our knowledge of the structure of the long-range phloem network in woody plants [7].

2. Materials and methods

We took samples of one, two and three-year-old shoots of *S. beauverdiana* from three plants growing in the northern section of the Central Eastern solfataric field of Golovnin Volcano caldera, in the North-Eastern solfataric field of Mendeleyev Volcano, and in the bushy bamboo typical for that species inside the caldera of Golovnin Volcano (Kunashir island). We select and preserve shoot samples for anatomical analysis in line with the standard guidelines [8–10]. Brief geobotanical descriptions were made for, and *Spiraea beauverdiana* samples were taken from, 10×10 m sampling areas.

In the northern section of the Central Eastern solfataric field of Golovnin Volcano caldera there are a number of boiling potholes and active solfataras. The solfataric gases can be as hot as 90–100°C. The thermal springs contain an increased concentration of microelements, such as Pb–Cd and Rb–As. Gas of the thermal springs contains mostly carbon dioxide and hydrogen sulfide [11]. In these conditions, *S. beauverdiana* can be found as part of thin vegetation. Loose groups of *Reynoutria sachalinensis* (Fr. Schmidt) Nakai and very close-knit bushes of *S. beauverdiana*, *Ledum hypoleucum* Kom., and *Empetrum sibiricum* V. Vassil. not taller than 10–15 cm grow 10–15 m away from the heart of the solfataras. In these conditions, *S. beauverdiana* is represented by its low shrub life form.

In the North-Eastern solfataric field of Mendeleyev Volcano there are many solfataras with temperatures up to 99.6°C, and several places where solfataric gas condensate seeps out in the stream creeks. The rocks here are altered and enriched with sulfides of metals as a result of hydrogen sulfide and sulfur dioxide coming to the surface. The whole surface of the field is covered with a crust of sulfur [11]. The vegetation around the field grows in open communities and standalone groups of plants. *S. beauverdiana* can be found in plant communities of mountain pine and bamboo *(Pinus pumila, Sasa, Hydrangea paniculata, Menziesia pentandra, Ledum hypoleucum)* or brush, sword grass and knotweed (*Ledum hypoleucum, Gaultheria miqueliana, Sasa, Miscanthus sinensis, Reynoutria sachalinensis*).

Samples of *S. beauverdiana* growing in typical conditions were collected inside the caldera, but far from the center of solfataric activity (about 1 km away). Here the plant is found among mountain pines and bamboo brush (*Ledum hypoleucum*, *Spiraea beauverdiana*, *Empetrum sibiricum*).

We made a statistic analysis of 13 quantitative parameters of the secondary phloem of *S. beauverdiana* for every habitat. Sample size for each parameter was at least 30 measurements. When analyzing parameters of tissue and cells for one, two and three-year-old stems, we calculated the sample average, maximum, and minimum values, variation coefficient, the standard square deviation, expectation for the sample average (for confidence figure of 95 %).

IOP Conf. Series: Earth and Environmental Science **946** (2021) 012047 doi:10.1088/1755-1315/946/1/012047

3. Results and discussion

Cross-sectionally, the stem of *S. beauverdiana* has four or five sides. The central cylinder is formed around the procambial ring. Bark tissue composition changes significantly over the first vegetation season, due to early generation of phellogen in the deep layers of the bark and subsequent degeneration of the primary tissues. At the end of the first vegetation season, bark tissues located outside of the periderm die and get deformed, but remain on the stem. The endoderm is likely formed after suberinization of the phellem. Therefore, at the end of the vegetation season the bark of the *S. beauverdiana* stem contains the following layers, from the outside to the center: deformed epidermis; deformed cortex, including collenchyma, ground parenchyma, and endodermis; outer bark; protophloem, protophloem fibers and sclereids; metaphloem and secondary phloem [12]. In a year-old stem the secondary phloem includes only the conductive phloem made up of sieve tubes and companion cells, the axial and radial parenchyma. Sieve tubes in the cross-section are diffuse or found in small groups of two or three. The axial parenchyma is diffuse [13].

Comparative studies of the structure of secondary phloem of young stems of *S. beauverdiana* in contrasting environmental conditions revealed a number of structural peculiarities. In a year-old stem from the solfataric field of Mendeleyev Volcano, the width of the conductive phloem is 18 % more (21.9±1.46 microns) than the norm (18.5±0.72 microns). And in the solfataric field of Golovnin Volcano caldera the width of the conductive phloem is close to typical values (figure 1a). In a two or three-year-old stem the secondary phloem includes conducting and non-conducting phloem. At that age, in volcanic conditions, the width of conductive and non-conductive secondary phloem decreases. Conductive phloem is reduced by 42 % at Golovnin Volcano (16.1±0.89 microns), and by 23 % at Mendeleyev Volcano (18.5±1.00 microns) (figure 1a). Non-conductive phloem reduced by 57 % at Golovnin Volcano (14.2±0.84 microns), and by 27 % at Mendeleyev Volcano (17.5±1.80 microns), while the norm is 22.3±1.76 microns (figure 1b).

In a year-old stem the radial and tangential diameters of the sieve tube segments are within normal (figure 1c, d). At two and three years of age, under stress, we see the radial diameter in samples from Mendeleyev Volcano reduced by 24 % (3.7±0.17 microns) (figure 1c) and tangential diameter in samples from Golovnin Volcano reduced by 16 % (5.1±0.35 microns) (figure 1d) as compared to the norm (4.6±0.31 microns and 5.8±0.29 microns, respectively).

In a year-old stem, the segments of sieve tubes in conditions of the solfataric field of Mendeleyev Volcano are 27 % shorter (135.3±15.11 microns) that the norm (172.2±15.31 microns). In two and three-year-old stems in extreme conditions that parameter remains within the norm (figure 1e).

The radius in the phloem of year-old stems from the solfataric field of Golovnin Volcano and in typical conditions are only single-row and homocellular, and double-row ones can be found in the solfataric field of Mendeleyev Volcano. At two and three years of age, the structure of the radial parenchyma changes, and there are no instances of double-row radius in samples from Golovnin Volcano caldera. The total number of phloem rays in one, two and three-year-old stems in stress conditions is normal. In a year-old stem of *S. beauverdiana* the height of single-row rays in conditions of the solfataric field of Mendeleyev Volcano is 39 % less (169.9±16.45 microns) than the norm (236.0±17.83 microns). In two and three-year-old stems this value is close to the typical conditions for both habitats (figure 1f). Multiple-row rays are found only in two or three-year-old stems. In two and three-year-old stems the multiple-row rays are shorter in both extreme habitats. In samples from Golovnin Volcano they are 25 % shorter (214.3±15.39 microns), and in samples from Mendeleyev Volcano they are 22 % shorter (220.4±21.61 microns) than the norm (268.7±20.35 microns). The width of multiple-row rays in two and three-year-old shoots in the solfataric field of Golovnin Volcano is 24 % less (18.2±1.56 microns) than the norm (22.6±1.51 microns). In samples from Mendeleyev Volcano the value is normal.

The parenchymal girder in a year-old stem in extreme conditions of Golovnin Volcano is 31 % shorter (82.7±5.15 microns), and in two and three-year-old stems 14 % shorter (113.1±5.11 microns) than the norm (109.0±6.55 microns and 129.6±4.46 microns, respectively). In one, two and three-year-old shoots in the solfataric field of Mendeleyev Volcano the value is close to that for typical conditions.

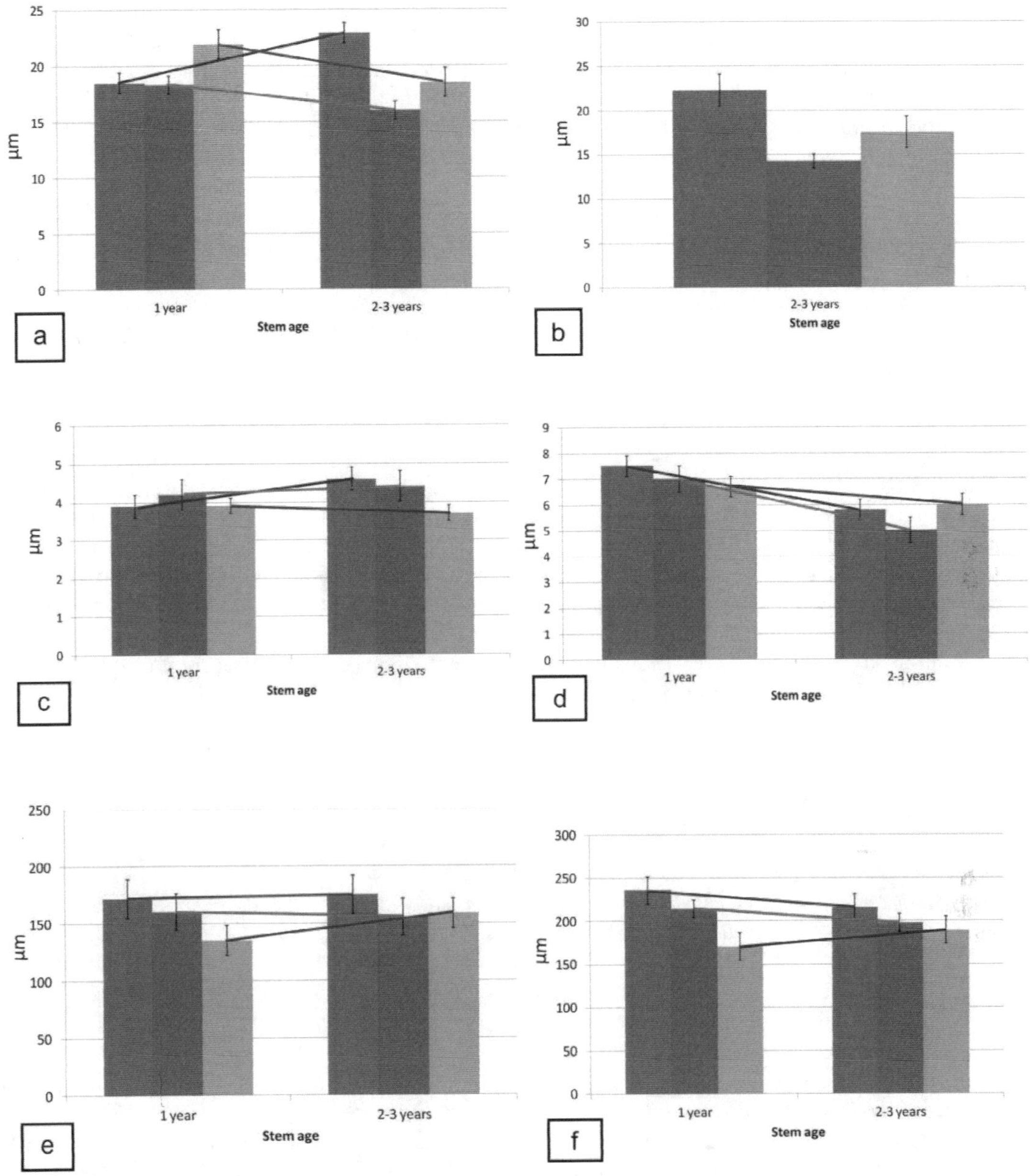

Figure 1. Anatomical features of the bark of 1 and 2–3 year old stems *Spiraea beauverdiana*. a – conductive phloem width; b – non-conductive phloem width; c – radial diameter of sieve tubes; d – tangent diameter of sieve tubes; e – sieve-tube element length; f – ray uniseriate height. ▬ Golovnin Volcano caldera (typical conditions); ▬ Central Eastern solfatara field of the Golovnin Volcano caldera; ▬ North-East solfatara field of the Mendeleyev Volcano.

One of the signs of impact of volcanic activity on the bark structure is development of non-specific anomalies in the internal structure of the *S. beauverdiana* bark, which develop most actively at a young age (three to five years) and reach maximum values in boughs by eight years [12]. We found anomalous areas in young stems of *S. beauverdiana* from solfataric fields of the caldera of Golovnin Volcano and Mendeleyev Volcano. Most commonly, we found that young stems have structural anomalies of the outer bark, such as multiple layers of phellem and phelloderm. Secondary phloem anomalies found in young stems of *S. beauverdiana* from solfataric fields of the researched volcanoes are mostly represented by small areas where axial and radial parenchyma includes groups of large sclereids. Tangential and radial sizes of such cells are two or three times bigger than normal tissue. Rays in such sections are heterocellular and dilated, and sieve tubes are deformed by the overgrown sections of parenchyma [14–15].

Our findings match the discovered trends of structural changes of secondary phloem of *Betula ermanii* Cham. in conditions of solfataric fields and fumarole fields of Baransky Volcano on Iturup island and *Toxicodendron orientale* Greene in conditions of Mendeleyev Volcano on Kunashir island [4, 16–18]. We saw similar results in year-old stems of *Betula ermanii* in conditions of Starozavodskoye solfataric field of Baransky Volcano, where the conductive phloem is wider than normal. The same trend was observed in year-old stems of *Toxicodendron orientale* in conditions of Verkhnedoctorsky hot springs at Mendeleyev Volcano [4]. This response is probably typical for many woody plant species growing in such extreme conditions. We think that further research is necessary for other species and other active volcanos.

4. Conclusion

The common trend in structural changes of the secondary phloem in conditions of solfataric fields of Golovnin and Mendeleyev Volcanos is adaptive reduction of conductive elements while maintaining their structural connection to parenchyma. We believe these peculiarities of conductive phloem to be adaptive in nature, as a response to the extreme conditions of gas and hydrothermal fields in volcanic landscapes. Perhaps lack of water due to high concentration of salts interferes with carbon transportation in the phloem. We think that development of anomalous areas of this tissue is the extreme form of structural and functional changes in the phloem in volcanic conditions.

Acknowledgements

The research was performed as part of a government task by the Institute of Marine Geology and Geophysics, Far East Branch, Russian Academy of Sciences.

References

[1] Ezhkin A K 2019 Lichens of wood substrates in areas of solfataric activity on Southern Kuriles *Geosystems of transition zones* **3** 2 pp 256–263

[2] Zharkov R V 2020 Thermal fields of the Ebeko volcano (Paramushir island, Kuril islands) and their recreational and tourist potential *Geosystems of Transition Zones* **4** 4 pp 514–25

[3] Manko Yu I and Sidelnikov A N 1989 *The Influence of Volcanism on Vegetation* (FEB RAS: Vladivostok) p 163

[4] Kopanina A V 2016 Structural ecological and anatomical studies of woody plants on Sakhalin and the Kuril Islands *Sci. J. Botanical Garden-Institute* **15** 2

[5] Yakubov V V, Nedoluzhko V A, Shantser I A, Tikhomirov V N and Rumyantsev S D 1996 *Family Pink – Rosaceae. Vascular Plants of the Soviet Far East* **8** (Saint Petersburg) pp 125–246.

[6] Bezdelev A B and Bezdeleva T A 2006 *Life Forms of Seed Plants of the Russian Far East* (Vladivostok) p 296

[7] Holbrook N M 2018 Physiology and metabolism Phloem a supracellular highway for the transport of sugars signals and pathogens *Current Opinion in Plant Biology* **43**:iii–vii

[8] Barykina R P, Veselova T D, Devyatov A G, Dzhalilova H Kh, Ilyina G M and Chubatova N V 2004 *Handbook of Botanical Microtechnics Basics and Methods* (Moscow) p 312

[9] Vlasova I I and Kopanina A V 2019 Specific microtechniques for bark tissues of woody plants *Plant Anatomy: Traditions and Perspectives. Materials of the Int. Symp. dedicated to the 90th anniversary of Prof. Ludmila Ivanovna Lotova* (September 16–22 Moscow) In two parts

Contributions in English ed by A C Timonin, D D Sokoloff (Moscow: MAKS Press) Part 1 pp 274–278

[10] Vlasova I I and Kopanina A V 2021 The peculiarity of sample preparation of shrubs and dwarf shrubs for structural analysis *Botanical Readings: Collection of Materials of sci. conf.* (February 25 Brest) Brest State University named after A S Pushkin ed by V I Boyko, N V Shkuratova et al (Brest: BrGU) pp 31–35

[11] Zharkov R V 2014 *Thermal Springs of the Southern Kuril Islands* (Dalnauka: Vladivostok) p 378

[12] Kopanina A V 2019 Structural features of the bark and wood of *Spiraea beauverdiana* (Rosaceae) in extreme conditions of the Arctic and post-volcanic activity in the Kuril Islands *Siberian Forest J.* **3** pp 52–63

[13] Vatserionova C O, Kopanina A V and Vlasova I I 2019 *Materials of the Int. Symp. dedicated to the 90 anniversary of Prof. Ludmila Ivanovna Lotova* Structural changes in the bark of Spiraea beauverdiana (Rosaceae) in Golovnin volcano caldera (Kunashir Island, South Kuriles) *In: Plant Anatomy: Traditions and Perspectives* (September 16–22 Moscow) In two parts Contributions in English ed by A C Timonin, D D Sokoloff (Moscow: MAKS Press) Part 1 pp 260–264.

[14] Vatserionova E O and Kopanina A V 2016 Structural features of young stems of *Spiraea beauverdiana* in the conditions of solfatar fields of the Golovnin caldera volcano, Kunashir Island *Sci. J. Botanical garden-institute* **15**

[15] Vatserionova E O, Kopanina A V and Vlasova I I 2021 The structure of the bark of the annual stem of *Spiraea beauverdiana* (Rosaceae) in the conditions of the solfatar fields of the Golovnin volcano caldera (Kunashir Island, South Kuril Islands) *Sci. J. Botanical Garden-Institute* **25**

[16] Talskikh A I, Kopanina A V and Vlasova I I 2018 Structural features of the annual stem *Betula ermanii* Cham. In the conditions of the Yuzhno-Sakhalin mud volcano (Sakhalin Island) *Materials of the IV (XII) Int. Botanical Conf. of Young Scientists in St. Petersburg* (Saint Petersburg) p 55

[17] Talskikh A I, Kopanina A V and Vlasova I I 2019a. Structural features of the bark in young stems of *Betula ermanii* Cham. in the conditions of Yuzhno-Sakhalinsky mud volcano (Sakhalin Island) *IOP Conf. Series Earthand Environmental Science* p 324 012033

[18] Talskikh A I, Kopanina A V and Vlasova I I 2019b *Materials of the Int. Symp. dedicated to the 90 anniversary of Prof. Ludmila Ivanovna Lotova* Specific structural features of the bark in annual stems of *Betula ermanii* Cham. exposed to hydrothermal gases and fluids of Baransky volcano, Iturup Island *In: Plant anatomy: traditions and perspectives* (September 16–22 Moscow) In two parts Contributions in English ed by A C Timonin, D D Sokoloff (Moscow: MAKS Press) Part 1 pp 241–245

Peculiarities of selecting woody plants for anatomy analysis in various environments

I I Vlasova* **and A V Kopanina**

Institute of Marine Geology and Geophysics Far Eastern Branch of Russian Academy of Science, Yuzhno-Sakhalinsk, 693022, Russia

E-mail: *iivlasova@gmail.com

Abstract. Methods for studying woody plants anatomy are now very diverse. We modified the guidelines for bark studies developed for wood analysis at all stages of preparing the woody plant samples for microscope study. For the first time, we used separation to separate hard and soft fractions of the bark tissue. Correct approach to selection of plant samples and further laboratory manipulations ensures validity of the results of our study. We select the size of the sampling area depending on the type of vegetation. In every habitat, we also include transects along the gradient of height above sea level or the impact from the source of natural stress. The results of our research will help study landscape changes during exogenous geological processes and phenomena using biological indication of geosystems.

1. Introduction

Structural environmental analysis of plant wood and bark is widely used for detecting extreme natural phenomena, such as fires, volcanic eruptions, or floods. In the last few decades study of wood from trees, brush, and low shrub is used to assess the current climate changes and past climate dynamics. Environmental plant anatomy benefits from technical and theoretical advancements in optical and electronic microscopy. There are lots of methods for preparing woody plants for microscopy. Methods for studying woody plants anatomy are now very diverse [1–10]. Most of those methods were developed for studying wood. Our study of woody plants around mud and magma volcanoes have proven to be suitable for studying bark. We apply microscopy, including biochemical analysis, to a specific object of woody plants study, namely, the bark. In our method, maceration of bark tissue is based on the method by Ge Wang et al [4]. This type of maceration was developed for analysis of wood fibers of mallow hemp, tracheids of the Chinese spruce (*Picea brachytyla*) and sclereids of bamboo.

Bark of woody plants is a very complex object for anatomical studies. The bark includes tissues, cell groups, and individual cells, all very different in terms of their physical and chemical properties. Some are very dense and sturdy, like sclerenchyma, phellem, and rhytidome, others are soft and water-saturated, like parenchyma and conductive elements of the phloem. Significant difference in mechanical and other properties of bark cells at a distance of 100–500 microns makes it very difficult to make a microslide or take a macerated sample. It also complicates further chemical manipulations required to obtain a microslide of the required quality. Therefore, we have developed special methods allowing us to single out a specific tissue complex in the bark and obtain a panoramic slide which would include various types of tissue. For instance, we separate conductive and non-conductive phloem both for maceration and for preparing permanent slides in the tree bark (figure 1). Bark from brush and low shrub, as well as bark from year-old shoots of the trees, does not require such separation, because cells and cell groups with different physical and chemical properties are never found in the same area of 50–100 microns. This provides for a good monolithic slide.

We developed some methods to adapt wood maceration guidelines of Wang et al [4]. For each type of woody plant, we found the optimal time of exposure to the macerating fluid, to separate the phloem into parts. We developed the guidelines for separating dense groups of cells and conductive elements and the parenchyma (figure 2). This helps us avoid errors when analyzing anatomical features of the bark [11–18].

We practice a comprehensive approach to studying the relation between geological environment and vegetation. Proper assessment of geological environment is critical for understanding structural changes in plants. That is why geochemical, geomorphological, and general geological characteristics of the habitat of the plants we study are of utmost importance in our approach [19–21]. Study of living and fossil organisms provides a picture of the mechanisms, age, and scope of geological processes [22–25].

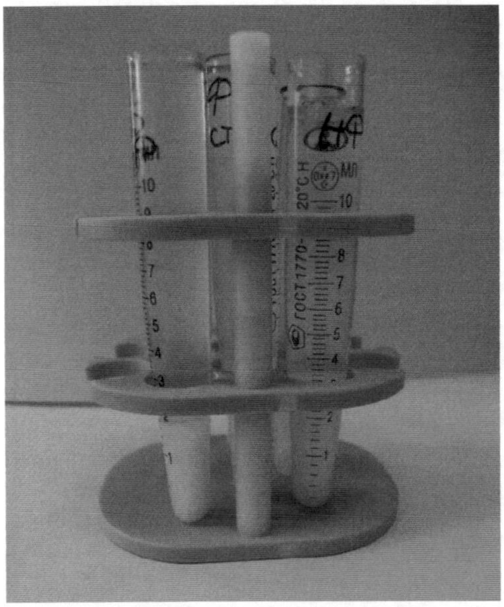

Figure 1. Separate conductive and non-conductive phloem.

Figure 2. Separating dense groups of cells and conductive elements and the parenchyma.

2. Materials and methods

Correct approach to selection of plant samples and further laboratory manipulations ensures validity of the results of our study. In the field, we take samples from three different specimens of tree, brush, and low shrub in the habitats we want to study. In extreme conditions, plants often grow in pessimum areas. In areas of volcanic activity manifested by gas and hydrothermal vents, the vegetation is sparce, therefore sampling area has to be expanded. The further from the source, the less the effect on the plants. First, we take samples from specimens closest to the source, where we can find sparce groups and synusia with areas of exposed substrate. Impact of the negative factor may vary depending on the wind direction, microrelief, and other landscape features.

We are governed by the methods and guidelines accepted in geobotanical studies when taking samples of woody plants for structural analysis [26–31]. Depending on the type of vegetation, we select the size of the sampling area, as small as 5×5 m for low shrubs and free-growing plant aggregations or as large as 20×20 m up to 100×100 m in the forest communities. Sampling areas are located inside the perimeter of gas and hydrothermal vents, thermal lakes, or stream banks. In every habitat we also include transects along the gradient of height above sea level or the impact from the source of natural stress. In each sampling area we select three sites where we sample woody plants, usually selected at random.

Collection of woody plants along the transect enables us to see how they adapt to various degrees of stress. Transect length and width, as well as the space between transects, are determined in situ. Samples of woody plants sprouts need to be dug out carefully, taken out, packaged, and transported. The time from the moment a plant is taken out of the substrate until the moment it is placed in the fixative solution should be as short as possible. A sample of brush or low shrub must be placed whole in a plastic bag along with the substrate remaining on its roots. Branches and fragments of larger shrub or splinters of tree trunks must be cut as long as practical (up to 20–30 centimeters), so that it is possible to cut out the middle part of a branch or trunk with live tissue before placing it in the fixative solution. This is especially important when studying bark.

It is important that the samples are taken at the end of the vegetation period. The best microslides of the highest quality come from samples taken in autumn, when the cambium becomes inactive until spring. On Sakhalin this period is usually the second half of September and all of October, in the southern Kurils it is October. Unfortunately, major comprehensive expeditions to remote areas such as the Kurils are rare in autumn. If samples are taken in summer, every stage of the process of making microslides, from identifying the plant age to placing the slide into the mounting medium, is more difficult.

It is important to choose the right stems ideal for studying their internal structure, especially young ones (1–10 years). We use classifications by I. Serebryakov [32] M. Mazurenko and A. Khokhryakov [33] to select and determine the types of shoots. For structural analysis, we use long branching shoots. Those are side shoots adding to the green matter of the assimilatory organs and seed productivity, because they contain the leaves and most of the flowers and blossom clusters. Long branching shoots develop from wintering buds. Forming shoots are not included in our samples. Such shoots are large due to wide germination of bark and wood because of intense growth, especially in their first few years. Undamaged shoots without visible stripping by insects are placed in the fixative. We take the first year-old stems when the buds open, and the last samples are taken at the end of the vegetation, after defoliation. Sampling frequency is about 7–10 days. This allows us to observe initiation, formation and development of bark tissue and elements, such as periderm, cortex and rays sclerification, initiation of the axial parenchyma layers in the conducting phloem, speed and intensity at which the conducting tissue grows and develops, the character of the phloem elements, generation of crystals. Sampling frequency depends on the species and its growth speed. The time between sample collection and fixation should not exceed 2–3 hours. After that the shoots can be stored in a fridge up to 8 hours, in a ziplocked plastic bag ensuring a moist chamber. Young shoots from the first batches of samples are covered with epidermis, so the best way to preserve those is by placing them in ethyl alcohol (50 %) right after sampling. For year-old stems and older stems covered with periderm, 70–95 % ethyl alcohol will do as well at the end of the vegetation season. Young stems covered outside with a thin layer of periderm should be fixed in 70 % alcohol. Samples covered with thick layer of periderm or crust and containing large sclerified areas in the non-conducting phloem can be placed in 95 % alcohol.

We are especially thorough and careful when taking woody plants samples in volcanic landscapes. Growth and development of plants as they adapt to adverse impact differ functionally and morphologically from plants of the same species growing in normal (background) natural conditions. In extreme conditions the plants tend to change from normal growth strategy to survival strategy. The plant body changes drastically, up to complete change of life form. For instance, impact of magmatic volcanoes or cold seashores causes the brush to take on a low, pillow-like shape closer to the substrate. Perennial branches of such plants become geophitization and burrow into the soil [17, 18, 34]. In such cases we take the whole specimen with the shoots buried into the soil and developing roots. The trees become crooked and turn into dwarves; they develop several skeletal axes and take on a bush form. The lower part of such plants is also covered under a layer of substrate [35], so first we need to clean the plants from the topsoil or substrate. Next, we cut off several large skeletal axes with a saw and preserve them as per our standard method.

Special care and lots of labor are required when taking samples for anatomical analysis from trees. We pick trees with the top shaped evenly (in the top, middle, or lower third of the trunk). Trees must be about the same age. The age of the tree is determined using a core taken with an increment borer. For us to assess age-related parameters and condition of the plant, the samples must be as comprehensive as possible. This means choosing an age increment ensuring that structural features of the bark and wood change in a successive order from the latest year-old shoot to the base of the stem or stipitate, with account for geophytization (burrowing of stems and stipitate into the substrate). We found this was a very common occurrence in woody plants growing in volcanic landscapes. Low shrubs are sometimes found in groves or parts thereof.

An important condition for preparing samples is ensuring that containers for fixation are airtight and leak-proof. The containers must be secure enough during transportation and must remain tight for several months.

3. Conclusion

So, we modified the guidelines for bark studies developed for wood analysis at all stages of preparing the woody plant samples for microscope study. We take a comprehensive approach to studying the relation between the geologic environment and the vegetation. We developed guidelines for taking woody plant samples for certain life forms. When selecting samples, we take into account the place of the stems in the system of shoots, its spatial location, and their location in the substrate (geophytization). We have a set of measures to preserve living bark tissue to the maximum extent possible before office studies. For the first time, we used separation to separate hard and soft fractions of the bark tissue.

Living organisms are involved in shaping the Earth surface, so the results of our research will help study landscape changes during exogenous geological processes and phenomena using biological indication of geosystems.

Acknowledgments

The study was carried out with support from RBFR (No. 15-04-04774) and within state order to the Institute of Marine Geology and Geophysics Far Eastern Branch of the Russian Academy of Sciences.

References

[1] Chamberlain C J 1935 *Methods in plant anatomy* (Chicago: University of Chicago Press) p 334

[2] Johansen D A 1940 *Plant microtechnique* (New York: McGraw-Hill) p 523

[3] Franklin G L 1945 *Nature J* **155** p 51

[4] Ge W, Sheldon Q S, Jinwu W, Yan Y, Shuangping C and Haitao C 2011 *Wood Fiber Sci.* **43** 4 pp 353–364

[5] Keating R C 2014 *Preparing plant tissues for light microscopy study: a compendium of simple techniques* (St Louis: Missouri Botanical Garden Press) 154 p

[6] Gärtner H and Schweingruber F H 2013 *Microscopic preparation techniques for plant stem analysis* (Birmensdorf: Verlag Dr. Kessel) 78 p

[7] Prozina M N 1960 *Botanical microtechnique* (Moscow: Vysshaya shkola) 206 p

[8] Kashina E N 1970 *Method for maceration of plant tissues* (Avt No 325938 USSR 07 06 1970)

[9] Kashina E N 1971 Maceration of tissues of leaves, roots and embryos of seeds of Heliantus annuus L *Botanicheskii zhurnal* (Saint Petersburg) **565** 10 p 1477

[10] Barykina R P, Veselova T D, Devyatov A G, Dzhalilova Kh Kh, Ilyina G M and Chubatova N V 2004 *Handbook of Botanical microtechniques Basics and Methods* (Moscow: Publishinghouse of Moscow State University) 312 p

[11] Kopanina A V, Vlasova I I 2019 Structural changes of bark of the woody liana Toxicodendron orientale Greene (Anacardiaceae) in the extreme environments of gas-hydrothermal volcanic activity *BP* **8** 2 pp 3–17
[12] Vatserionova C O, Kopanina A V and Vlasova I I 2019 *Proc. Int. Sym. dedicated to the 90th anniversary of Prof Ludmila Ivanovna Lotova* September 16–22, 2019 (Moscow: MAKS Press) ed A C Timonen, D D Sokoloff pp 274–278
[13] Vlasova I I 2016 *Byulleten' botanicheskogo sada instituta FEB RAS 2021* (*Electronic Materials* vol 15) pp 11–12
[14] Kopanina A V, Vlasova I I and Vatserionova E O 2017 *Vestnik FEB RAS* **1** pp 88–96
[15] Kopanina A V, Talskikh A I and Vlasova I I 2018 *Proc. 6th Int. Sym. B N Ugolev, dedicated to the 50th anniversary of the Regional Coordination Council on Modern Problems of Wood Science* September 10–16, 2018 (Krasnoyarsk/Novosibirsk: SB RAS) pp 111–115
[16] Talskikh A I, Kopanina A V and Vlasova I I 2018 Proc. *4th (12th) Int. Botanical Conf. Young Scientist*s April 22–28, 2018 (St Petersburg: Komarov Botanical Institute of the RAS) pp 54–55
[17] Talskikh A I, Kopanina A V and Vlasova I I 2019 *IOP Conf. Ser: Earth Environ. Sci.* 324012033
[18] Talskikh A I, Kopanina A V and Vlasova I I 2021 *Proc. Sci. Conf. Botanical readings* February 25, 2021 (Brest State University named after A S Pushkin) ed V I Boyko, N V Shkuratova, M V Levkovskaya pp 54–59
[19] Nikitenko O A and Ershov V V 2019 *Geosystems of transition zones* (Yuzhno-Sakhalinsk: IMGG FEB RAS) **3** 3 pp 325–332
[20] Nikitenko O A and Ershov V V 2020 *Geosystems of transition zones* (Yuzhno-Sakhalinsk: IMGG FEB RAS) **4** 3 pp 321–350
[21] Zharkov R V 2020 *Geosystems of transition zones* (Yuzhno-Sakhalinsk: IMGG FEB RAS) **4** 4 pp 514–525
[22] Rybin A V, Degterev A V, Dudchenko I P, Guryanov V B, Romanyuk F A and Klimantsov I M 2019 *Geosystems of transition zones* (Yuzhno-Sakhalinsk: IMGG FEB RAS) **1** 4 pp 21–29
[23] Razjigaeva N G, Ganzey L A, Grebennikova T A, Kaistrenko V M, Kharlamov A A, Arslanov K A and Maksimov F E 2019 *Geosystems of transition zones* (Yuzhno-Sakhalinsk: IMGG FEB RAS) **3** 2 pp 219–236
[24] Razjigaeva N G, Ganzey L A, Makarova T R, Kornyushenko T V, Kudryavtseva E P, Ganzey K L, Sudin V V and Kharlamov A A 2019 *Geosystems of transition zones* (Yuzhno-Sakhalinsk: IMGG FEB RAS) **4** 2 pp 230–249
[25] Ezhkin A K 2019 *Geosystems of transition zones* (Yuzhno-Sakhalinsk: IMGG FEB RAS) **3** 2 pp 256–263
[26] Ramenskiy L G 1937 *Accounting and description of vegetation (based on the projective method)* p 100
[27] Alekhin V V 1938 *Methodology for field study of vegetation and flora* (Moscow) p 201
[28] Lavrenko E M 1938 *Methodology of field geobotanical studies* (Moscow, Leningrad) pp 43–55
[29] Shennikov AP 1938 *Method of field geobotanical research* (Moscow, Leningrad) p 5–26
[30] Rabotnov T A 1983 *Phytocenology* (Moscow) p 292
[31] Barkman J J 1989 *Vegetatio* **85** pp 89–104
[32] Serebryakov I G 1949 *Bulletin of Moscow Society of Naturalists Biological Series* **54** pp 47–63
[33] Mazurenko M T and Khokhryakov A P 1977 *Structure and morphogenesis of shrubs* (Moscow: Nauka) p 160
[34] Kopanina A V 2019 *Sibirskij lesnoj zhurnal* (Krasnoyarsk Scientific Center Novosibirsk) **3** pp 52–63
[35] Talskikh A I, Kopanina A V and Vlasova I I 2021 *Rastitelnye resursy* (St Petersburg) **57** 2 pp 124–144

AUTHOR INDEX